Compact Textbooks in Mathematics

Birkhäuser

Compact Textbooks in Mathematics

This textbook series presents concise introductions to current topics in mathematics and mainly addresses advanced undergraduates and master students. The concept is to offer small books covering subject matter equivalent to 2- or 3-hour lectures or seminars which are also suitable for self-study. The books provide students and teachers with new perspectives and novel approaches. They may feature examples and exercises to illustrate key concepts and applications of the theoretical contents. The series also includes textbooks specifically speaking to the needs of students from other disciplines such as physics, computer science, engineering, life sciences, finance.

- **compact:** small books presenting the relevant knowledge
- **learning made easy:** examples and exercises illustrate the application of the contents
- **useful for lecturers:** each title can serve as basis and guideline for a semester course/lecture/seminar of 2–3 hours per week.

More information about this series at http://www.springer.com/series/11225

Liviu C. Florescu

Lebesgue Integral

Birkhäuser

Liviu C. Florescu
Faculty of Mathematics
"Al. I. Cuza" University
Iaşi, Romania

ISSN 2296-4568 ISSN 2296-455X (electronic)
Compact Textbooks in Mathematics
ISBN 978-3-030-60162-1 ISBN 978-3-030-60163-8 (eBook)
https://doi.org/10.1007/978-3-030-60163-8

Mathematics Subject Classification: 26-01, 26Axx

This book is published under the imprint Birkhäuser, www.birkhauser-science.com, by the registered company Springer Nature Switzerland AG.
The registered company address is: Gewerbestrasse 11, 6330 Cham, Switzerland

Introduction

Until the end of the nineteenth century, mathematical analysis focused on the study of continuous functions and on the Riemann integral as the main measuring instrument. The need to model increasingly complex phenomena justified the efforts of mathematicians to widen the class of integrable functions.

Inspired by the work of E. Borel and C. Jordan, H. Lebesgue built in around 1900 a measure theory, which is used later to define a more general integral than the Riemann integral, an integral which bears his name.

If $f : [a, b] \to [m, M]$ is a bounded function, then G. Darboux introduced the sums $s_\delta = \sum_{k=1}^{n} m_k \cdot (x_k - x_{k-1})$ and $S_\delta = \sum_{k=1}^{n} M_k \cdot (x_k - x_{k-1})$, where $\delta = \{x_0, x_1, \ldots, x_n\}$ is a partition of the interval $[a, b]$, $m_k = \inf_{x\in[x_{k-1},x_k]} f(x)$, and $M_k = \sup_{x\in[x_{k-1},x_k]} f(x)$, for any $k = 1, \cdots n$.

The function f can be integrated in the Riemann sense if the distance between the two sums can be as small as we want.

Henry Lebesgue had the idea to reverse things: let $\Delta = \{y_0, y_1, \ldots, y_n\}$ be a partition of the interval $[m, M]$ and, for any $k = 1, \cdots n$, let $E_k = \{x \in [a, b] : y_{k-1} \leq f(x) < y_k\}$. If we can give a meaning to the "length" ("measure") of the sets E_k, $\lambda(E_k)$, then we can consider the sums $\sigma_\Delta = \sum_{k=1}^{n} y_{k-1} \cdot \lambda(E_k)$ and $\Sigma_\Delta = \sum_{k=1}^{n} y_k \cdot \lambda(E_k)$. The function f will be "integrable" if the difference $\Sigma_\Delta - \sigma_\Delta$ can be small enough for fairly fine partitions Δ.

The development of this simple idea leads to a significant extension of the Riemann integral.

In the figure below, we have represented the Darboux sums attached to a function f and to the partition $\delta = \{x_0 = 0, x_1, x_2, x_3, x_4, x_5\}$ and next the Lebesgue sum relating to the partition $\Delta = \{y_0 = 0, y_1, y_2, y_3, y_4\}$ for the same function (Fig. 1).

In the left, S_δ is the area of the polygon upper bounded by the thickened line and s_δ is the area of the polygon upper bounded by the dashed line. In the right, the area of the polygon upper bounded by the thickened line is equal to the Lebesgue sum Σ_Δ.

To give us a first idea of the difference in approach between the two constructions, let us evaluate the sums of Darboux and the sums of Lebesgue for the Dirichlet function: $f : [0, 1] \to [0, 1]$,

$$f(x) = \begin{cases} 1\ , & x \in \mathbb{Q} \cap [0, 1], \\ 0\ , & x \in (\mathbb{R} \setminus \mathbb{Q}) \cap [0, 1], \end{cases}$$

and for the partitions $\delta = \{x_0, x_1, \cdots, x_n\}$ and $\Delta = \{y_0, y_1, \cdots, y_p\}$ of $[0, 1]$. It is obvious that $S_\delta = 1, s_\delta = 0$; therefore, f is not Riemann

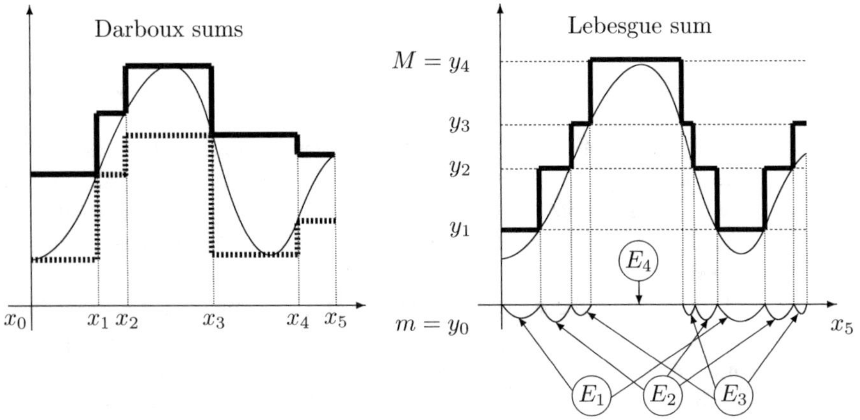

Fig. 1 Riemann and Lebesgue sums

integrable. However, $E_1 = (\mathbb{R} \setminus \mathbb{Q}) \cap [0, 1]$ and $E_k = \emptyset$, for any $k = 2, \cdots, p$. By anticipating certain results of measure theory, we can evaluate $\lambda(E_1) = 1$ and $\lambda(E_k) = 0$, for $k = 2, \cdots, p$ and then $\Sigma_\Delta = y_1$ and $\sigma_\Delta = y_0 = 0$. Thus, $\Sigma_\Delta - \sigma_\Delta = y_1$ is less than the mesh of the partition Δ. So, f is integrable in the sense of Lebesgue.

The theory of measure and integration is mainly the work of H. Lebesgue. Many other mathematicians, among whom we cite E. Borel, G. Vitali, F. Riesz, D. Egoroff, N. Lusin, J. Radon, M. Fréchet, H. Hahn, O. Nikodym, and C. Carathéodory, contributed to the development of this theory.

The introduction into this theory is as necessary (because of its multiple applications) as difficult for the uninitiated. Most of the measure theory treaties involve a large accumulation of knowledge and some important theoretical difficulties. Hence, the idea of providing a less knowledgeable reader with a material allowing him or her to easily access the definition and main properties of the Lebesgue integral.

The results presented in this book refer to the measure and integral on $\mathbb{R}$ and $\mathbb{R}^2$. In order to take a broader look at the measure theory, we have introduced, at the end of each chapter, a section entitled "Abstract Setting," where the properties of the measure on abstract spaces are presented (without demonstrations).

As we have noted, for the introduction of the Lebesgue integral, it is necessary to measure certain subsets of $\mathbb{R}$ that are more complicated than the intervals. First of all, we must establish the conditions under which a definition of "measurability" can be considered acceptable. Can we measure all the subsets of $\mathbb{R}$? The answer is no! What are the properties of the sets to which we can assign a measure? These questions are answered in the first chapter of the book. The measure of open sets is introduced as a natural consequence of the structure of open sets in $\mathbb{R}$. The outer measure is then defined, and finally the class of the Lebesgue measurable sets $\mathscr{L}$. $\mathscr{L}$ is strictly included in the family of all subsets of $\mathbb{R}$ but has the same cardinal number as the latter.

In ▶ Chap. 2, we deal with the so-called measurable functions. For a measurable function f, it is possible to assign a measure to sets of form $E = \{x : c \leq f(x) <$

$d\}$. An important class of measurable functions is the class of functions continuous almost everywhere, a class that contains continuous functions and monotonic functions. A corollary of the Lebesgue criterion (criterion presented in ▶ Sect. 7.1) guarantees that the Riemann integrable functions are also measurable in the Lebesgue sense. In this chapter, we will study different types of convergence for sequences of measurable functions and also the structure of these functions.

▶ Chapter 3 is devoted to the construction and properties of the Lebesgue integral on $\mathbb{R}$. ▶ Section 3.3 highlights the complete seminormed structure of the Lebesgue integrable function space. ▶ Section 3.4 compares the Riemann integral with the Lebesgue integral. To understand the generality and flexibility offered by the Lebesgue tool, it is suitable for the reader to know the basics of Riemann theory. However, in order to facilitate the reading of the book, ▶ Sect. 7.1 presents the basic elements of the theory of Riemann integral. In ▶ Sect. 3.5, we study some properties of the Lebesgue integral: the change of variable, the derivation under the integral, and the Jensen inequality.

In the fourth chapter, the L^p spaces are defined for $1 \leq p \leq +\infty$. In the particular case $p = 2$, some elements from the theory of Fourier series are presented.

▶ Chapter 5 presents Lebesgue's measure and integral on $\mathbb{R}^2$. The construction can easily be adapted in the general case of the $\mathbb{R}^n$ spaces. The representation of open sets in $\mathbb{R}^2$ (as a countable union of two-dimensional nonoverlapping closed rectangles) leads to a natural definition of their measure. As in the one-dimensional case, the outer measure is introduced; with it, the measurable sets in $\mathbb{R}^2$ are defined. One calculates the measure of a measurable set by integrating its one-dimensional sections; in particular, this allows an easy demonstration of Cavalieri's principle. The double integral is defined, and Fubini's theorem is proved; this allows its calculation by means of its iterated integrals.

Finally, in the last chapter, "Signed Measures," we will use the concepts and results of the sections entitled "Abstract Setting" to study measures with real values on abstract spaces. We prove the decomposition theorems of Hahn and Jordan, as well as the representation theorem of Radon-Nikodym. At the end of the chapter, we present the integration of derivatives and the Leibniz-Newton formula. An important role in this of latter is played by the class of absolutely continuous functions. These functions are differentiable almost everywhere; they are exactly those for which the Leibniz-Newton formula takes place. Cantor's staircase function or "Devil staircase" is a function that is uniformly continuous but not absolutely continuous on $[0, 1]$; consequently, it will not satisfy the Leibniz-Newton formula. Differentiable functions on an interval whose derivative is Lebesgue integrable provide an important example of absolutely continuous functions. Pompeiu's function, presented in ▶ Sect. 7.2, is an example of such a function. Using the absolutely continuous functions, an integration by parts formula is given at the end of the section.

The book ends with three appendices. In the first one, basic elements of the Riemann integral are presented. This appendix also contains the Lebesgue criterion concerning the integrability of Riemann, criterion indicating to us to what extent the Riemann integrable functions are discontinued. The second appendix presents an interesting example given by Pompeiu of a function whose derivative is not Riemann integrable in any subinterval of its domain of definition. The third appendix deals with the famous Lebesgue theorem of derivation of monotonic functions.

Each chapter includes a number of exercises; in addition to highlighting certain results, they highlight certain new and interesting aspects of the theory.

The bibliography presented is summarized to the books and works to which direct reference is made or from which some concrete methods or demonstrations have been taken. The purpose of such a work is to provide the reader with a material as complete as possible and to prevent him or her from going through a great diversity of styles and notations, a task that remains the responsibility of the author.

The book is intended as a course at the graduate school level or at the senior undergraduate level. The readers targeted by this volume do not necessarily have to have in-depth knowledge of classical analysis. However, some knowledge of elementary analysis is required; these include real number system, sequences and series of numbers and functions, limits, continuity, differentiability, and functions of several variables. Reasonable knowledge of point set topology on $\mathbb{R}$ and $\mathbb{R}^2$ is also required. The book addresses a mature audience who needs knowledge for a better understanding of certain concepts and results of probability theory, statistics, economic equilibrium theory, game theory, etc., concepts where the Lebesgue integral makes its presence felt.

This volume represents an improved and added version of the book "Teoria Măsurii" (in Romanian), which was published in 2012 at The Publishing House of the "Al.I. Cuza" University of Iaşi. The interest that this book aroused (cf. https://www.researchgate.net/publication/233980873_Teoria_masurii/stats) was the reason why I decided to present it to a wider audience.

Iaşi, Romania
May 02, 2020

Liviu C. Florescu

Contents

List of Figures

1 Lebesgue Measure on $\mathbb{R}$

Let $\mathbb{R}$ be the set of real numbers. Throughout this book, the extended set of real numbers is $\bar{\mathbb{R}} = \mathbb{R} \cup \{-\infty, +\infty\}$ and the extended subset of real positive numbers is $\bar{\mathbb{R}}_+ = \mathbb{R} \cup \{+\infty\}$. An **interval** in $\bar{\mathbb{R}}$ is a subset $J \subseteq \bar{\mathbb{R}}$ with the property that, for all $x, y \in J$ and for all z with $\min\{x, y\} < z < \max\{x, y\}$, it follows that $z \in J$. If $a, b \in \bar{\mathbb{R}}, a < b$, then $]a, b[= \{x \in \bar{\mathbb{R}} : a < x < b\}$ is the **open interval** and $[a, b] = \{x \in \bar{\mathbb{R}} : a \leq x \leq b\}$ is the **closed interval** of extremities a and b. Obviously $\mathbb{R} =] - \infty, +\infty[$ and $\bar{\mathbb{R}} = [-\infty, +\infty]$.

Let J be an interval and $a = \inf J$, $b = \sup J \in \bar{\mathbb{R}}$; then $]a, b[\subseteq J \subseteq [a, b]$. Therefore, J is one of the following intervals of extremities a and b: $]a, b[,]a, b], [a, b[, [a, b]$, e.g., $\bar{\mathbb{R}}_+ = [0, +\infty]$. If $a, b \in \mathbb{R}$, then J is a **bounded interval**, and if $a = -\infty$ or $b = +\infty$, the J is an **unbounded interval**.

Let $\mathcal{J} = \{J \subseteq \mathbb{R} : J \text{ is an interval}\}$ be the family of all the intervals of $\mathbb{R}$. Thus $\mathcal{J}$ consists of all bounded intervals and by those unbounded intervals of type $]-\infty, b[,]-\infty, b],]a, +\infty[, [a, +\infty[,]-\infty, +\infty[, a, b \in \mathbb{R}$. For any interval $J \in \mathcal{J}$ of extremities a and b, let $|J| = |b - a|$ be the **length** of the interval J ($|J| = +\infty$ if J is unbounded). We will agree that $\emptyset =]a, a[\in \mathcal{J}$, $\{a\} = [a, a] \in \mathcal{J}$, and then $|\emptyset| = |\{a\}| = 0$.

Let $J \in \mathcal{J}$ and $x \in \mathbb{R}$; then $x + J = \{x + y : y \in J\} \in \mathcal{J}$ and $|x + J| = |J|$. If $x \in \mathbb{R} \setminus \{0\}$, then $x \cdot J = \{x \cdot y : y \in J\} \in \mathcal{J}$ and $|x \cdot J| = |x| \cdot |J|$.

The questions we want to answer in this chapter are:

(1) Is there a set function λ defined on the family of all the subsets of $\mathbb{R}$, $\mathcal{P}(\mathbb{R})$, which satisfies the following properties?
   - (a) $\lambda\left(\bigcup_{n=1}^{\infty} A_n\right) = \sum_{n=1}^{\infty} \lambda(A_n)$, for every sequence $(A_n) \subseteq \mathcal{P}(\mathbb{R})$, $A_n \cap A_m = \emptyset$ and for all $n \neq m$,
   - (b) $\lambda(J) = |J|$, for every $J \in \mathcal{J}$,
   - (c) $\lambda(x + A) = \lambda(A)$, for every $A \subseteq \mathbb{R}$ and every $x \in \mathbb{R}$?

We want to say from the start that the answer to the above question is negative; in other words, we cannot assign a "measure" to any subset of $\mathbb{R}$.

L. C. Florescu, *Lebesgue Integral*, Compact Textbooks in Mathematics,
https://doi.org/10.1007/978-3-030-60163-8_1

A second question therefore arises:

(2) What is the largest class $\mathcal{A} \subseteq \mathcal{P}(\mathbb{R})$ whose sets can be "measured?" That is, what is the largest class $\mathcal{A}$ to which we can extend the interval length function so that its extension satisfies the three properties above on $\mathcal{A}$?

Naturally, a first extension will be made to the class of countable unions of mutually disjoint open intervals; these sets are the open sets of $\mathbb{R}$. This extension will be studied in the first paragraph.

In the second paragraph, we will introduce the outer Lebesgue measure λ^*; it is an extension of the interval length to all the subsets of $\mathbb{R}$. We will show (see 1.2.7) that this function is not σ-additive (so do not check the above property a)).

In the third paragraph, we will restrict λ^* to the class of Lebesgue measurable sets, which gives an answer to question (2). In this section, we will study the main properties of measurable sets and of the Lebesgue measure.

The fourth paragraph is dedicated to the definition and study of measure in an abstract framework. We will mention certain properties of the Lebesgue measure which are preserved in the general case.

Finally, in the last paragraph, exercises highlight additional properties of the Lebesgue measure.

1.1 Measure of Open Sets

Let $\mathcal{I} \subseteq \mathcal{J}$ be the family of all open intervals of $\mathbb{R}$ (bounded or not), that is to say the intervals of form $]a, b[$, $]-\infty, b[$, $]a, +\infty[$ and $]-\infty, +\infty[= \mathbb{R}$. Throughout the following, $\mathbb{N}$ is the set of natural numbers, and $\mathbb{N}^* = \mathbb{N}\setminus\{0\}$ is the set of natural numbers without zero.

Lemma 1.1.1 *Let* $\{I_p : p \in \mathbb{N}\} \subseteq \mathcal{I}$.

(1) If $I_0 \subseteq \bigcup\limits_{p=1}^{\infty} I_p$, *then* $|I_0| \leq \sum_{p=1}^{\infty} |I_p|$.

(2) If $\bigcup\limits_{p=1}^{\infty} I_p \subseteq I_0$ *and* $I_p \cap I_q = \emptyset$, *for all* $p \neq q$, *then* $\sum_{p=1}^{\infty} |I_p| \leq |I_0|$.

Proof

(1) If there is I_p unbounded, then $|I_p| = +\infty$, and the inequality is obvious.
Let us assume that $I_p =]a_p, b_p[$ is bounded, for all $p \geq 1$.
   (a) If the interval $I_0 =]a, b[$, $a < b$ is bounded, then, for all $\varepsilon > 0$, $[a+\varepsilon, b-\varepsilon] \subseteq \bigcup_{p=1}^{\infty}]a_p, b_p[$, and therefore there is $p_0 \in \mathbb{N}^*$ such that

$$[a+\varepsilon, b-\varepsilon] \subseteq \bigcup_{k=1}^{p_0} I_k. \qquad (*)$$

Indeed, if we claim that $[a+\varepsilon, b-\varepsilon]$ has no finite subcover, then, for all $p \in \mathbb{N}^*$, there exists $x_p \in [a+\varepsilon, b-\varepsilon] \setminus \bigcup_{k=1}^{p}]a_k, b_k[$. The sequence $(x_p)_p$ is bounded. Then let $(x_{k_p})_p$ be a subsequence convergent to $x \in [a+\varepsilon, b-\varepsilon]$ (Bolzano-Weierstrass theorem). Let $p_1 \in \mathbb{N}^*$ such that $x \in I_{p_1}$; as I_{p_1} is a neighborhood of x, there is $p_2 \in \mathbb{N}^*$, $p_2 > p_1$ such that $x_{k_p} \in I_{p_1}$, for all $p \geq p_2$. But $x_{k_{p_2}} \in [a+\varepsilon, b-\varepsilon] \setminus \bigcup_{k=1}^{k_{p_2}} I_k$ which, because of $p_1 < p_2 \leq k_{p_2}$, contradicts $x_{k_{p_2}} \in I_{p_1}$ (see also Lemma 7.1.6).

The relation $(*)$ allows us to renumber the finite family of intervals $\{I_k : k = 1, \cdots p_0\}$ so that $a_1 < a+\varepsilon < b-\varepsilon < b_{p_0}$. Then $b-a-2\varepsilon < b_{p_0} - a_1 \leq \sum_{k=1}^{p_0} |I_k| \leq \sum_{k=1}^{\infty} |I_k|$. Because ε is arbitrarily positive,

$$b - a = |I_0| \leq \sum_{p=1}^{\infty} |I_p|.$$

(b) If $I_0 =]a, +\infty[$, then, for any $n \in \mathbb{N}$, $]a, n[\subseteq \bigcup_{p=1}^{\infty} I_p$. Using the previous point, $n - a \leq \sum_{k=1}^{\infty} |I_k|$ from where $\sum_{k=1}^{\infty} |I_k| = +\infty = |I_0|$.

The same reasoning exists in other possible cases for the unbounded interval I_0.

(2) If I_0 is unbounded, then $|I_0| = +\infty \geq \sum_{p=1}^{\infty} |I_p|$.

Let us assume that $I_0 =]a, b[$ is bounded; then the intervals $I_p =]a_p, b_p[$ are all bounded.

For every $n \in \mathbb{N}^*$, $\bigcup_{p=1}^{n} I_p \subseteq I_0$. Since $I_1, \ldots, I_n$ are pairwise disjoint, we can renumber them so that

$$a \leq a_1 < b_1 \leq a_2 < b_2 \leq \cdots \leq a_n < b_n \leq b.$$

Then $|I_0| = b - a \geq (b_1 - a_1) + (b_2 - a_2) + \cdots + (b_n - a_n) = \sum_{p=1}^{n} |I_p|$, for any $n \in \mathbb{N}^*$, from where $\sum_{p=1}^{\infty} |I_p| \leq |I_0|$. ■

Definition 1.1.2

A subset $A \subseteq \mathbb{R}$ is said to be **open** if each time that it contains a point x, it contains also an open interval which itself contains x (for all $x \in A$, there is $I \in \mathcal{I}$ such as $x \in I \subseteq A$). The family τ_u of all open subsets of $\mathbb{R}$ is a **topology** on $\mathbb{R}$, that is to say it satisfies the following properties:

(T_1) $D \cap G \in \tau_u$, for every $D, G \in \tau_u$;

(T_2) $\cup_{\gamma \in \Gamma} D_\gamma \in \tau_u$, for every $\{D_\gamma : \gamma \in \Gamma\} \subseteq \tau_u$;

(T_3) $\emptyset, \mathbb{R} \in \tau_u$.

τ_u is called the **usual topology** of $\mathbb{R}$.

We note that the open intervals are open sets, and therefore, according to (T_2), the countable unions (even uncountable ones) of open intervals are open sets. We can show that each open set is a countable union of open intervals.

Theorem 1.1.3 (structure of open sets in $\mathbb{R}$)
Any open set $D \subseteq \mathbb{R}$ is a countable union of pairwise disjoint open intervals. This representation of D is unique, up to the order of the family intervals.

Proof
If $D = \emptyset$, then it is a countable union of empty open intervals of form $]a, a[$.

If D is not empty, for every $x \in D$, there exist $a_0, b_0 \in \mathbb{R}$ such that $x \in]a_0, b_0[\subseteq D$. We denote

$$A_x = \{a \in \mathbb{R} : \text{ there exists } b \in \mathbb{R} \text{ such that } x \in]a, b[\subseteq D\},$$

$$B_x = \{b \in \mathbb{R} : \text{ there exists } a \in \mathbb{R} \text{ such that } x \in]a, b[\subseteq D\}.$$

We note that $a_0 \in A_x$, $b_0 \in B_x$, and then $A_x \neq \emptyset \neq B_x$. Let $a_x = \inf A_x \in [-\infty, +\infty[$, $b_x = \sup B_x \in]-\infty, +\infty]$, and let $I_x =]a_x, b_x[$. Let's show that $x \in I_x \subseteq D$. Indeed, $a_x \leq a_0 < x < b_0 \leq b_x$, and then $x \in I_x$. For every $y \in I_x$, $a_x < y < b_x$, and then there is $a \in A_x$ and $b \in B_x$ such that $a_x \leq a < y < b \leq b_x$. Given the definitions of the sets A_x and B_x, there exist $a_1, b_1 \in \mathbb{R}$ such that $x \in]a, b_1[\subseteq D$ and $x \in]a_1, b[\subseteq D$. $]a, b_1[$ and $]a_1, b[$ are non-disjoint intervals, and then $]a, b_1[\cup]a_1, b[=]a_2, b_2[$, where $a_2 = \min\{a, a_1\}$, $b_2 = \max\{b, b_1\}$. It's obvious that $]a_2, b_2[\subseteq D$ and $a_2 \leq a < y < b \leq b_2$ from where $y \in D$. It follows that $I_x \subseteq D$. Then I_x is the largest open interval that contains the point x and is included in D. This maximility character of I_x allows us to show that, for every $x, y \in D$, $I_x = I_y$, or $I_x \cap I_y = \emptyset$. Indeed, if we assume that $I_x \cap I_y \neq \emptyset$, then $I = I_x \cup I_y$ is an interval and $x \in I \subseteq D$, $y \in I \subseteq D$. Using the maximality of intervals I_x and I_y, we obtain $I \subseteq I_x$ and $I \subseteq I_y$, which leads us to $I_y \subseteq I_x$ and respectively to $I_x \subseteq I_y$; therefore $I_x = I_y$.

The family of these maximal intervals $\mathcal{I}_D = \{I_x : x \in D\}$ is countable. Indeed, for every $x \in D$, let a rational number $q_x \in I_x$ and let the mapping $\varphi : \mathcal{I}_D \to \mathbb{Q}$, defined by $\varphi(I_x) = q_x$ (here $\mathbb{Q}$ denotes the set of rational numbers). It should be noted that if $I_x = I_y$, then I_x appears once as an element of the family $\mathcal{I}_D$, and then the same rational point q_x is chosen in $I_x = I_y$; therefore φ is well-defined. If $I_x \neq I_y$, then $I_x \cap I_y = \emptyset$, and then $\varphi(I_x) = q_x \neq q_y = \varphi(I_y)$. Therefore φ is injective and then $\mathcal{I}_D$ is countable. Let $\{I_n : n \in \mathbb{N}\}$ be an enumeration of family intervals $\mathcal{I}_D$, where $I_n \cap I_m = \emptyset$, for all $m, n \in \mathbb{N}$, $n \neq m$. Then $D = \cup_{x \in D} I_x = \cup\{I : I \in \mathcal{I}_D\} = \cup_{n=1}^{\infty} I_n$; hence D is a countable union of pairwise disjoint open intervals.

Let $\mathcal{I}' = \{I'_n : n \in \mathbb{N}\}$ be another representation of D with open and disjoint intervals: $D = \cup_{n\in\mathbb{N}} I'_n$. Then, for any $n \in \mathbb{N}$ and every $x \in I'_n$, $x \in I'_n \subseteq D$. By the maximal character of the interval I_x, $I'_n \subseteq I_x$. But $I_x \in \mathcal{I}_D$, and then there exists $m_n \in \mathbb{N}$ such that $I'_n \subseteq I_x = I_{m_n}$. Let $I'_n =]a'_n, b'_n[$; if we assume that $I'_n \neq I_{m_n}$, then $a'_n \in I_{m_n} \subseteq D$ or $b'_n \in I_{m_n} \subseteq D$. Let now $p \in \mathbb{N}$, $p \neq n$ such that $a'_n \in I'_p$ or $b'_n \in I'_p$, but this is nonsense because I'_p and I'_n are disjoint. Therefore $I'_n = I_{m_n}$, and then $\mathcal{I}' \subseteq \mathcal{I}_D$. On the other hand, for any $n \in \mathbb{N}$ and for every $x \in I_n \subseteq D$, there is $I'_{m_n} \in \mathcal{I}'$ such that $x \in I'_{m_n} \subseteq D$; it also follows that $I'_{m_n} \subseteq I_n$, and with a reasoning similar to the one above, it follows that $I_n = I'_{m_n} \in \mathcal{I}'$.
Therefore $\mathcal{I}_D = \mathcal{I}'$, which ensures the uniqueness of decomposition of D. ■

Definition 1.1.4

With the notations of the above theorem, we will say that $D = \bigcup_{n=1}^{\infty} I_n$ is the **representation** of the set D or that $\{I_n : n \in \mathbb{N}\}$ are the intervals of the representation of D.
The mapping $\lambda : \tau_u \to \bar{\mathbb{R}}_+$, defined by $\lambda(D) = \sum_{n=1}^{\infty} |I_n|$, where $\{I_n : n \in \mathbb{N}\}$ are the intervals of the representation of D, is said to be the **measure** of open sets.
We note that, because of the uniqueness of representation, the above definition is consistent (in a series with positive terms, we can sum the terms in an indifferent order and always get the same sum).

In the following theorem, we mention some important properties of the measure of open sets.

Theorem 1.1.5
The measure of open sets has the following properties:
(1) $\lambda(I) = |I|$, *for every* $I \in \mathcal{I}$,
(2) $\lambda(\emptyset) = 0$, $\lambda(\mathbb{R}) = +\infty$,
(3) $x + D \in \tau_u$ *and* $\lambda(x + D) = \lambda(D)$, *for every* $x \in \mathbb{R}$ *and* $D \in \tau_u$,
(4) $xD \in \tau_u$ *and* $\lambda(xD) = |x|\lambda(D)$, *for every* $x \in \mathbb{R} \setminus \{0\}$ *and* $D \in \tau_u$,
(5) $\lambda(D) \leq \lambda(G)$, *for every* $D, G \in \tau_u$, *with* $D \subseteq G$,
(6) $\lambda(\cup_{n=1}^{\infty} D_n) = \sum_{n=1}^{\infty} \lambda(D_n)$, $\forall (D_n) \subseteq \tau_u$, $D_n \cap D_m = \emptyset \ \forall n \neq m$,
(7) $\lambda(\cup_{n=1}^{\infty} D_n) \leq \sum_{n=1}^{\infty} \lambda(D_n)$, *for every* $(D_n) \subseteq \tau_u$.

Proof
Properties (1) and (2) are obvious.
To show (3), it suffices to note that if $D = \bigcup_{n=1}^{\infty} I_n$ is the representation of open set D, then $x + D = \bigcup_{n=1}^{\infty}(x + I_n) \in \tau_u$ and that $\{x + I_n : n \in \mathbb{N}^*\}$ are the intervals of the representation of $x + D$.

Then $\lambda(x+D) = \sum_{n=1}^{\infty} |x + I_n| = \sum_{n=1}^{\infty} |I_n| = \lambda(D)$.

(4) For every $x \in \mathbb{R} \setminus \{0\}$ and every $D \in \tau_u$, $x \cdot D = \{xy : y \in D\}$. If $D = \bigcup_{n=1}^{\infty} I_n$ is the representation of set D, then $x \cdot D = \bigcup_{n=1}^{\infty}(x \cdot I_n) \in \tau_u$ is the representation of $x \cdot D$, and so $|\lambda(x \cdot D)| = \sum_{n=1}^{\infty} |x \cdot I_n| = |x| \cdot \sum_{n=1}^{\infty} |I_n| = |x| \cdot \lambda(D)$.

(5) Let $D = \bigcup_{n=1}^{\infty} I_n \subseteq G = \bigcup_{m=1}^{\infty} J_m$ be the representations of the two open sets.

For every $m \in \mathbb{N}^*$, let $N_m = \{n \in \mathbb{N}^* : I_n \subseteq J_m\} \subseteq \mathbb{N}^*$. We note that some of the N_m can be empty (it could be that some J_m does not contain any interval I_n); let $M = \{m \in \mathbb{N}^* : N_m \neq \emptyset\} \subseteq \mathbb{N}^*$.

Then $\{N_m : m \in M\}$ is a partition of $\mathbb{N}^*$, which means that:

(a) $\mathbb{N}^* = \bigcup\limits_{m \in M} N_m$ and

(b) $N_m \cap N_p = \emptyset$, for all $m, p \in M, m \neq p$.

Indeed, for any $n \in \mathbb{N}^*$, $I_n \subseteq D \subseteq G = \bigcup\limits_{m=1}^{\infty} J_m$, and then there exists $m \in \mathbb{N}^*$ such that $I_n \cap J_m \neq \emptyset$. By the maximal character of J_m, we have $I_n \subseteq J_m$ and then $n \in N_m$ and $m \in M$.

If we assume that it exists $n \in N_m \cap N_p$, then $I_n \subseteq J_m$ and $I_n \subseteq J_p$, which contradicts that J_m and J_p are disjoint for $m \neq p$.

For every $m \in M$, $\bigcup\limits_{n \in N_m} I_n \subseteq J_m$. Since the intervals I_n are pairwise disjoint, according to (2) of Lemma 1.1.1, we obtain $\sum\limits_{n \in N_m} |I_n| \leq |J_m|$, for any $m \in M$. Then

$$\lambda(D) = \sum_{n=1}^{\infty} |I_n| = \sum_{m \in M} \sum_{n \in N_m} |I_n| \leq \sum_{m \in M} |J_m| \leq \sum_{m \in \mathbb{N}^*} |J_m| = \lambda(G).$$

(6) For every $n \in \mathbb{N}^*$, let $D_n = \bigcup_{k=1}^{\infty} I_k^n$ be the representation of D_n; then $D = \bigcup_{n=1}^{\infty} D_n = \bigcup_{n=1}^{\infty} \bigcup_{k=1}^{\infty} I_k^n$ is the representation of D, and therefore

$$\lambda(D) = \sum_{n=1}^{\infty} \sum_{k=1}^{\infty} |I_k^n| = \sum_{n=1}^{\infty} \lambda(D_n).$$

(7) Let $D = \bigcup_{n=1}^{\infty} D_n \in \tau_u$. Let $D = \bigcup_{p=1}^{\infty} I_p$ be the representation of D, and for any $n \in \mathbb{N}$, let $D_n = \bigcup_{k=1}^{\infty} I_k^n$ be the representation of D_n.

Then, for any $p \in \mathbb{N}$,

$$I_p = I_p \cap D = \bigcup_{n=1}^{\infty} (I_p \cap D_n) = \bigcup_{n=1}^{\infty} \bigcup_{k=1}^{\infty} (I_p \cap I_k^n),$$

where $I_p \cap I_k^n \in \mathcal{I}$, for every $p, k, n \in \mathbb{N}$.

According to (1) of Lemma 1.1.1, $|I_p| \le \sum_{n=1}^{\infty}\sum_{k=1}^{\infty}|I_p \cap I_k^n|$, and then

$$\lambda(D) = \sum_{p=1}^{\infty}|I_p| \le \sum_{p=1}^{\infty}\sum_{n=1}^{\infty}\sum_{k=1}^{\infty}|I_p \cap I_k^n| = \sum_{n=1}^{\infty}\sum_{k=1}^{\infty}\sum_{p=1}^{\infty}|I_p \cap I_k^n| =$$

$$= \sum_{n=1}^{\infty}\sum_{k=1}^{\infty}\lambda(D \cap I_k^n) = \sum_{n=1}^{\infty}\sum_{k=1}^{\infty}|I_k^n| = \sum_{n=1}^{\infty}\lambda(D_n).$$

In the above relations, we have used that for all $n, k \in \mathbb{N}$, $D \cap I_k^n = \bigcup_{p=1}^{\infty}(I_p \cap I_k^n)$ is the representation of open set $D \cap I_k^n$. ■

Definition 1.1.6

Property (3) of the previous theorem is called the property of **translation invariance** of the Lebesgue measure λ. Property (5) tells us that λ is **monotonic**. Property (6) is said to be the property of σ**-additivity** and (7) the property of σ**-subadditivity** of the measure λ.

In the spaces $\mathbb{R}^n, n \ge 2$, a result like the Theorem 1.1.3 does not work. For example, in $\mathbb{R}^2$, we cannot represent open sets as a countable union of a pairwise disjoint family of open two-dimensional intervals (rectangles). However, it will work a result of the representation of an open set as a countable union of the almost disjoint closed rectangles (the interiors of the rectangles are pairwise disjoint)—see Theorem 5.1.12. We have a similar result in the case of $\mathbb{R}$.

Definition 1.1.7

An interval $J \in \mathcal{J}$ is said to be **closed** if:

(a) J is bounded and of the form $J = [a, b]$, with $a, b \in \mathbb{R}$ or

(b) J is unbounded and of the form $]-\infty, b]$ or $[a, +\infty[$, with $a, b \in \mathbb{R}$.

A point x is **interior** for the interval $J = [a, b]$ if $a < x < b$ (if $J =]-\infty, b]$, then $x < b$, and if $J = [a, +\infty[$, then $a < x$).

Theorem 1.1.8

Any non-empty open set of $\mathbb{R}$ is a countable union of nonoverlapping closed intervals (without common interior points).

If $D = \bigcup_{n=1}^{\infty} J_n$, where $\{J_n : n \in \mathbb{N}\}$ is a countable family of almost disjoint closed intervals, then $\lambda(D) = \sum_{n=1}^{\infty}|J_n|$.

Proof

Let $D = \bigcup_{n=1}^{\infty} I_n$, where, for any $n \in \mathbb{N}^*$, $I_n =]a_n, b_n[$ is an open interval (bounded or unbounded) and $I_n \cap I_m = \emptyset$, for all $n \neq m$ (the representation of D).

It is enough to represent each open interval $]a_n, b_n[$ as a countable union of closed nonoverlapping intervals. Let $\alpha_p^n \downarrow a_n$ and $\beta_p^n \uparrow b_n$ such that $a_n < \alpha_p^n < \beta_q^n < b_n$, for all $n, p, q \in \mathbb{N}$. Then

$$]a_n, b_n[= \bigcup_{p=0}^{\infty} [\alpha_{p+1}^n, \alpha_p^n] \cup [\alpha_0^n, \beta_0^n] \cup \bigcup_{p=0}^{\infty} [\beta_p^n, \beta_{p+1}^n].$$

Now let $D = \cup_{n=1}^{\infty} J_n$, where $\{J_n : n \in \mathbb{N}\}$ is a family of almost disjoint closed intervals.

(a) If we assume that the intervals J_n are all bounded, then, for all $n \in \mathbb{N}^*$, $J_n = [a_n, b_n]$, where $a_n, b_n \in \mathbb{R}$. For every $\varepsilon > 0$ and every $n \in \mathbb{N}^*$, there exist the open intervals I_n, K_n such that $I_n \subseteq J_n \subseteq K_n$ and

$$|J_n| \leq |I_n| + \frac{\varepsilon}{2^n}, |K_n| \leq |J_n| + \frac{\varepsilon}{2^n}$$

(we choose $I_n = \left]a_n + \frac{\varepsilon}{2^{n+1}}, b_n - \frac{\varepsilon}{2^{n+1}}\right[$ and $K_n = \left]a_n - \frac{\varepsilon}{2^{n+1}}, b_n + \frac{\varepsilon}{2^{n+1}}\right[$).

Let $I = \bigcup_{n=1}^{\infty} I_n$ and $K = \bigcup_{n=1}^{\infty} K_n$; then $I \subseteq D \subseteq K$ and, since the intervals I_n are pairwise disjoint,

$$\lambda(D) \geq \lambda(I) = \sum_{n=1}^{\infty} |I_n| \geq \sum_{n=1}^{\infty} |J_n| - \varepsilon. \tag{1.1}$$

On the other hand, using the monotonicity and the σ-subadditivity of the measure λ, we get

$$\lambda(D) \leq \lambda(K) \leq \sum_{n=1}^{\infty} |K_n| \leq \sum_{n=1}^{\infty} |J_n| + \varepsilon. \tag{1.2}$$

Since ε is arbitrarily positive, according to (1) and (2), it follows that $\lambda(D) = \sum_{n=1}^{\infty} |J_n|$.

(b) If, among the J_n, there is an interval unbounded, $J_{n_0} = [a_{n_0}, +\infty[$, then, for all $\varepsilon > 0$, let $I_{n_0} =]a_{n_0} + \varepsilon, +\infty[\subseteq J_{n_0} \subseteq D$; then $+\infty = |I_{n_0}| = \lambda(D)$. Therefore $\lambda(D) = +\infty = |J_{n_0}| = \sum_{n=1}^{\infty} |J_n|$. Similarly, if $J_{n_0} =] - \infty, b_{n_0}]$ ∎

Remark 1.1.9 We note that an open set D can have several representations like countable union of almost disjoint closed intervals; for each of these representations, the sum of the lengths of the intervals is the same—the measure of D.

1.2 Lebesgue Outer Measure

We will now extend the measure of open sets to an application defined on $\mathcal{P}(\mathbb{R})$ (the family of all subsets of $\mathbb{R}$).

Definition 1.2.1

The mapping $\lambda^* : \mathcal{P}(\mathbb{R}) \to \bar{\mathbb{R}}_+$, defined by

$$\lambda^*(A) = \inf\{\lambda(D) : D \in \tau_u,\ A \subseteq D\}, \ \text{ for all } A \subseteq \mathbb{R},$$

is called Lebesgue **outer measure** on $\mathbb{R}$.
By definition, we immediately observe that

$$\lambda^*(A) = \inf\left\{\sum_{n=0}^{\infty}(b_n - a_n) : A \subseteq \bigcup_{n=0}^{\infty}]a_n, b_n[\right\}, \ \text{ for every } A \subseteq \mathbb{R}.$$

Remark 1.2.2 We note that for every open set D, $\lambda^*(D) = \lambda(D)$. Indeed, from the definition of outer measure, $\lambda^*(D) \le \lambda(D)$ (among the open sets containing D appears the set itself). On the other hand, for every open set G with $D \subseteq G$, $\lambda(D) \le \lambda(G)$ (see Property (5) of Theorem 1.1.5), and so $\lambda(D) \le \lambda^*(D)$.

It follows that the Lebesgue outer measure is an extension of measure of open sets to the family of all subsets of $\mathbb{R}$. It is not a measure because, as we will see in 1.2.7, it is not σ-additive.

The Lebesgue outer measure has the following properties:

Theorem 1.2.3
(1) $\lambda^*(\emptyset) = 0$,
(2) $\lambda^*(A) \le \lambda^*(B)$, *for every* $A, B \in \mathcal{P}(\mathbb{R})$, $A \subseteq B$,
(3) $\lambda^*(\bigcup_{n=1}^{\infty} A_n) \le \sum_{n=1}^{\infty} \lambda^*(A_n)$, *for every* $(A_n)_n \subseteq \mathcal{P}(\mathbb{R})$.

Proof
(1) Due to the previous remark, $\lambda^*(\emptyset) = \lambda(\emptyset) = 0$, because $\emptyset \in \tau_u$ and $\lambda(\emptyset) = 0$.
(2) Since $A \subseteq B$, $\{\lambda(D) : D \in \tau_u,\ B \subseteq D\} \subseteq \{\lambda(G) : G \in \tau_u,\ A \subseteq G\}$; hence, passing to infimum, $\lambda^*(A) \le \lambda^*(B)$.
(3) If there is $n \in \mathbb{N}^*$ such that $\lambda^*(A_n) = +\infty$, then the inequality is obvious.

Now assume that for any $n \in \mathbb{N}^*$, $\lambda^*(A_n) < +\infty$. For every $\varepsilon > 0$ and every $n \in \mathbb{N}^*$, there exists $D_n \in \tau_u$ such that $A_n \subseteq D_n$ and $\lambda(D_n) < \lambda^*(A_n) + \frac{\varepsilon}{2^n}$. Then $D = \bigcup_{n=1}^{\infty} D_n \in \tau_u$ and $\bigcup_{n=1}^{\infty} A_n \subseteq D$. Therefore $\lambda^*(\bigcup_{n=1}^{\infty} A_n) \le \lambda(D) \le \sum_{n=1}^{\infty} \lambda(D_n) \le \sum_{n=1}^{\infty} \lambda^*(A_n) + \varepsilon$. Because ε is positive and arbitrarily small, we get the σ-subadditivity of λ^*. ∎

Remarks 1.2.4

(i) $\lambda^*(\{x\}) = 0$, for every $x \in \mathbb{R}$.

Indeed, for every $x \in \mathbb{R}$ and for any $n \in \mathbb{N}^*$, $\{x\} \subseteq]x - \frac{1}{n}, x + \frac{1}{n}[$, from where $\lambda^*(\{x\}) \leq \frac{2}{n}$, for any $n \in \mathbb{N}^*$, and then $\lambda^*(\{x\}) = 0$.

(ii) $\lambda^*(A \cup B) \leq \lambda^*(A) + \lambda^*(B)$, for every $A, B \subseteq \mathbb{R}$.

Let $\{A_n : n \in \mathbb{N}^*\}$, so that $A_1 = A$, $A_2 = B$ and, for any $n \geq 3$, $A_n = \emptyset$; then due to (1) and (3) of the previous theorem,

$$\lambda^*(A \cup B) = \lambda^*\left(\bigcup_{n=1}^{\infty} A_n\right) \leq \sum_{n=1}^{\infty} \lambda^*(A_n) = \lambda^*(A) + \lambda^*(B).$$

This property is called finite subadditivity; it can be extended by complete induction to a finite number of subsets of $\mathbb{R}$.

(iii) For any interval $J \in \mathcal{J}$, $\lambda^*(J) = |J|$. Indeed, if J is an open interval, then the property results from Remark 1.2.2. If J is not open, then it differs from an open interval by two points, at most. The property results from (i).

The outer measure has a property similar to that of Theorem 1.1.8.

Proposition 1.2.5 *Let* $A = \bigcup_{n=1}^{\infty} J_n$, *where* J_n *are nonoverlapping closed intervals (the interiors of the intervals are pairwise disjoint); then*

$$\lambda^*(A) = \sum_{n=1}^{\infty} |J_n|.$$

Proof

If there is $n_0 \in \mathbb{N}^*$ so that the interval J_{n_0} is unbounded, then $\lambda^*(A) \geq \lambda^*(J_{n_0}) = |J_{n_0}| = +\infty$, and therefore equality is checked.

We now assume that $|J_n| < +\infty$, for all $n \in \mathbb{N}^*$. For every $\varepsilon > 0$ and every $n \in \mathbb{N}^*$, there exists an open interval $I_n \subseteq J_n$ such that $|J_n| < |I_n| + \frac{\varepsilon}{2^n}$. Let $D = \bigcup_{n=1}^{\infty} I_n \in \tau_u$. Then $D \subseteq A$, and, since the intervals I_n are pairwise disjoint, $\lambda^*(A) \geq \lambda^*(D) = \lambda(D) = \sum_{n=1}^{\infty} |I_n| \geq \sum_{n=1}^{\infty} |J_n| - \varepsilon$, from where $\lambda^*(A) \geq \sum_{n=1}^{\infty} |J_n|$. The σ-subadditivity of λ^* assures us the conversely inequality. ■

The following theorem shows that the outer measure is invariant by translations.

Theorem 1.2.6

$\lambda^*(x + A) = \lambda^*(A)$, *for every* $x \in \mathbb{R}$ *and every* $A \subseteq \mathbb{R}$.

$\lambda^*(x \cdot A) = |x| \cdot \lambda^*(A)$, *for every* $x \in \mathbb{R} \setminus \{0\}$ *and every* $A \subseteq \mathbb{R}$.

Proof
For every $D \in \tau_u$ with $A \subseteq D$, $x + A \subseteq x + D$; from (3) of Theorem 1.1.5, it results that $\lambda^*(x + A) \leq \lambda(D)$ and then $\lambda^*(x + A) \leq \lambda^*(A)$. Since this last inequality takes place for every $x \in \mathbb{R}$ and every $A \subseteq \mathbb{R}$, $\lambda^*(A) = \lambda^*(-x + (x + A)) \leq \lambda^*(x + A)$.

For every $D \in \tau_u$ with $A \subseteq D$, $x \cdot D \in \tau_u$ and $x \cdot A \subseteq x \cdot D$. According to (4) of Theorem 1.1.5, $\lambda^*(x \cdot A) \leq \lambda(x \cdot D) = |x| \cdot \lambda(D)$, and then

$$\lambda^*(x \cdot A) \leq |x| \cdot \lambda^*(A). \tag{*}$$

The inequality $(*)$ occurs for every $x \neq 0$ and every $A \subseteq \mathbb{R}$ and then

$$\lambda^*(A) = \lambda^*\left(\frac{1}{x} \cdot (x \cdot A)\right) \leq \frac{1}{|x|} \cdot \lambda^*(x \cdot A)$$

where we get the inverse inequality of $(*)$. ■

From the above, λ^* checks properties (b) and (c) presented at the beginning of this chapter; this function is an invariant σ-subadditive extension of the interval length.

The following example shows that λ^* is not σ-additive (does not check property a) formulated in the introduction to this chapter).

Vitali Set 1.2.7 On the set $I = [0, 1]$, we define a relation by $x \varrho y$ if and only if $x - y \in \mathbb{Q}$. We can easily see that ϱ is reflexive, symmetric, and transitive, therefore an equivalence relation on I. For every $x \in I$, the equivalence class of representative x is $[x] = \{y \in I : x \varrho y\} = \{y \in I : y - x \in \mathbb{Q}\} = \{y \in I : y \in x + \mathbb{Q}\} = I \cap (x + \mathbb{Q})$. It follows that $[x]$ is a countable set, for every $x \in I$. Two distinct equivalence classes are disjoint, and the union of these classes is I. Because each equivalence class is not empty, the axiom of choice assures us that there exists a set $V \subseteq I$ containing a single element of each equivalence class. So for any equivalence class $[x]$, $\mathrm{card}(V \cap [x]) = 1$, where $\mathrm{card}(A)$ denotes the cardinal number of A. Let us denote by $\mathbb{Q}_1 = \mathbb{Q} \cap [-1, 1]$ the set of rational numbers in the interval $[-1, 1]$; then

$$[0, 1] \subseteq \bigcup_{r \in \mathbb{Q}_1} (r + V) \subseteq [-1, 2]. \tag{*}$$

Indeed, for every $x \in [0, 1]$, there exists $x_1 \in V \cap [x]$; then $x \varrho x_1$, and so $x - x_1 = r \in \mathbb{Q}$. It follows that $x = r + x_1 \in r + V$ and $r = x - x_1 \in \mathbb{Q}_1$. The second inclusion is obvious if we note that for all $r \in \mathbb{Q}_1$, $r + V \subseteq r + [0, 1] \subseteq [-1, 2]$.

Using successive the results (iii) of Remarks 1.2.4 and (2) and (3) of Theorems 1.2.3 and 1.2.6, we will get from the first inclusion of $(*)$ that

$$1 = \lambda^*([0, 1]) \leq \lambda^*\left(\bigcup_{r \in \mathbb{Q}_1} (r + V)\right) \leq \sum_{r \in \mathbb{Q}_1} \lambda^*(r + V) = \sum_{r \in \mathbb{Q}_1} \lambda^*(V),$$

from where $\lambda^*(V) > 0$.

The sets $\{r + V : r \in \mathbb{Q}_1\}$ are pairwise disjoint. Indeed, if we assume that, for $r, s \in \mathbb{Q}_1, r \neq s$ there exists $x \in (r+V) \cap (s+V)$, then $x-r, x-s \in V$, but $(x-r)\varrho(x-s)$. Then $x-r$ and $x-s$ are two different elements of V belonging to the same equivalence class; this is a contradiction because V contains only one element in each equivalence class.

If we assume that λ^* is σ-additive,

$$\lambda^*\left(\bigcup_{r\in\mathbb{Q}_1}(r+V)\right) = \sum_{r\in\mathbb{Q}_1}\lambda^*(r+V) = \sum_{r\in\mathbb{Q}_1}\lambda^*(V) = +\infty. \qquad (**)$$

According to the second inclusion of $(*)$, $\lambda^*\left(\bigcup_{r\in\mathbb{Q}_1}(r+V)\right) \leq \lambda^*([-1, 2]) = 3$, which contradicts $(**)$. Therefore λ^* is not σ-additive.

It follows that the extension made in Definition 1.2.1 is too wide, λ^* not meeting the requirements specified at the beginning of this chapter.

In the following proposition, we give other calculation formulas for the outer measure of a set.

Proposition 1.2.8 *For every set $A \subseteq \mathbb{R}$,*

$$\begin{aligned}\lambda^*(A) &= \inf\{\textstyle\sum_{n=1}^{\infty}(b_n - a_n) : A \subseteq \cup_{n=1}^{\infty}]a_n, b_n]\} \\ &= \inf\{\textstyle\sum_{n=1}^{\infty}(b_n - a_n) : A \subseteq \cup_{n=1}^{\infty}[a_n, b_n]\}.\end{aligned}$$

Proof

Let $\lambda_1^*(A) = \inf\{\sum_{n=1}^{\infty}(b_n - a_n) : A \subseteq \cup_{n=1}^{\infty}]a_n, b_n]\}$. For every cover with open intervals $]a_n, b_n[, n \in \mathbb{N}^*$ of A, we have $A \subseteq \bigcup_{n=1}^{\infty}]a_n, b_n]$, and therefore $\lambda_1^*(A) \leq \sum_{n=1}^{\infty}(b_n - a_n)$, from where $\lambda_1^*(A) \leq \lambda^*(A)$.

If $\lambda_1^*(A) = +\infty$, the equality is demonstrated.

We assume that $\lambda_1^*(A) < +\infty$. For every $\varepsilon > 0$, there is a sequence of semi-closed intervals $(]a_n, b_n])_{n\geq 1}$ such that $A \subseteq \bigcup_{n=1}^{\infty}]a_n, b_n]$ and $\lambda_1^*(A) + \varepsilon > \sum_{n=1}^{\infty}(b_n - a_n)$. Then $A \subseteq \bigcup_{n=1}^{\infty}\left]a_n, b_n + \frac{\varepsilon}{2^n}\right[$, and therefore $\lambda^*(A) \leq \sum_{n=1}^{\infty}(b_n - a_n) + \varepsilon < \lambda_1^*(A) + 2\varepsilon$. Since ε is arbitrarily positive, we obtain the inverse inequality $\lambda^*(A) \leq \lambda_1^*(A)$.

The proof of the second formula is similar. ■

Although the outer measure is not, in general, σ-additive, on certain sequences of sets, it verifies the property of σ-additivity.

For two non-empty sets $B, C \subseteq \mathbb{R}$, let us denote with $d(B, C) =$ $= \inf\{|x - y| : x \in B, y \in C\}$ the **distance** between B and C. Obviously, if B and C have one point in common, then $d(B, C) = 0$. The converse is not true. Indeed, if $B = \{\frac{1}{n} : n \in \mathbb{N}^*\}$ and $C = \{0\}$, then $B \cap C = \emptyset$, and $d(B, C) = 0$. In general, we can show that $d(B, C) = 0$ if and only if there exist two sequences $(x_n)_n \subseteq B$, $(y_n)_n \subseteq C$ such that $x_n - y_n \to 0$.

If $B = \emptyset$ or $C = \emptyset$, we agree that $d(B, C) = +\infty$.

Theorem 1.2.9
(1) $\lambda^*(B \cup C) = \lambda^*(B) + \lambda^*(C)$, *for every* B, C, *with* $d(B, C) > 0$.
(2) $\lambda^*(\cup_{n=1}^{p} A_n) = \sum_{n=1}^{p} \lambda^*(A_n)$, *for every* $A_1, \ldots, A_n \subseteq \mathbb{R}$ *with* $d(A_n, A_m) > 0$, *for all* $n, m \in \{1, \cdots, p\}$, $n \neq m$.
(3) $\lambda^*(\cup_{n=1}^{\infty} A_n) = \sum_{n=1}^{\infty} \lambda^*(A_n)$, *for every sequence* $(A_n)_{n\geq 1}$ *of subsets of* $\mathbb{R}$ *with* $d(A_n, A_m) > 0$, *for all* $n \neq m$.

Proof
(1) From the property of finite subadditivity of λ^* (see (ii) of Remark 1.2.4), $\lambda^*(B \cup C) \leq \lambda^*(B) + \lambda^*(C)$. If $\lambda^*(B \cup C) = +\infty$, equality is evident.

Now let us assume that $\lambda^*(B \cup C) < +\infty$; using Proposition 1.2.8, for every $\varepsilon > 0$, there exists $\{]a_n, b_n] : n \in \mathbb{N}\}$ such that $B \cup C \subseteq \bigcup_{n=0}^{\infty}]a_n, b_n]$ and $\lambda^*(B \cup C) + \varepsilon > \sum_{n=0}^{\infty}(b_n - a_n)$. Without limiting the generality, it can be assumed that, for all $n \in \mathbb{N}$, $b_n - a_n < \delta \equiv d(B, C)$ (otherwise the intervals $]a_n, b_n]$ will be divided into a sufficient number of subintervals of the same kind, the lengths of which make it possible to verify the above requirement).

Then, for all $n \in \mathbb{N}$, the interval $]a_n, b_n]$ intersects only one of sets B and C. Let

$N_1 = \{n \in \mathbb{N} :]a_n, b_n] \cap B \neq \emptyset\}$ and

$N_2 = \{n \in \mathbb{N} :]a_n, b_n] \cap C \neq \emptyset\}$.

Note that $N_1 \cap N_2 = \emptyset$ and

$$B \subseteq \bigcup_{n\in N_1}]a_n, b_n], \; C \subseteq \bigcup_{n\in N_2}]a_n, b_n].$$

Therefore $\lambda^*(B) \leq \sum_{n\in N_1}(b_n - a_n)$, and $\lambda^*(C) \leq \sum_{n\in N_2}(b_n - a_n)$, and then

$$\lambda^*(B) + \lambda^*(C) \leq \sum_{n\in N_1}(b_n - a_n) + \sum_{n\in N_2}(b_n - a_n) = \sum_{n\in N_1\cup N_2}(b_n - a_n) \leq$$

$$\leq \sum_{n\in\mathbb{N}}(b_n - a_n) < \lambda^*(B \cup C) + \varepsilon.$$

Since ε is arbitrary positive, we obtain the inverse inequality and therefore the required equality.

(2) The reasoning is done by induction; for p = 2, the demonstration was made at the previous point. Suppose the property is checked for $p - 1$, and let $\{A_1, \cdots, A_p\}$ be a family of p subsets of $\mathbb{R}$ for which the distance between any two is strictly positive.

Let us denote $B = \bigcup_{n=1}^{p} A_n$ and $C = \bigcup_{n=1}^{p-1} A_n$. Then

$$d(C, A_p) = \min\{d(A_n, A_p) : n = 1, \cdots, p-1\} > 0.$$

Using (1) and the recurrence hypothesis, $\lambda^*(B) = \lambda^*(C \cup A_p) = \lambda^*(C)+$ $+\lambda^*(A_p) = \sum_{n=1}^{p-1} \lambda^*(A_n) + \lambda^*(A_p) = \sum_{n=1}^{p} \lambda^*(A_n)$.

(3) For every $p \in \mathbb{N}^*$, $\cup_{n=1}^{\infty} A_n \supseteq \bigcup_{n=1}^{p} A_n$, and then $\lambda^*(\cup_{n=1}^{\infty} A_n) \geq \lambda^*(\bigcup_{n=1}^{p} A_n) = \sum_{n=1}^{p} \lambda^*(A_n)$. Therefore $\lambda^*(\cup_{n=1}^{\infty} A_n) \geq \sum_{n=1}^{\infty} \lambda^*(A_n)$. The inverse inequality comes from the property of σ-subadditivity of λ^*. ■

At the end of this paragraph, we will present a very important concept in the theory of measure and integration—that of a null set or negligible set.

Definition 1.2.10

A subset $A \subseteq \mathbb{R}$ is a **Lebesgue null set** or a **Lebesgue negligible set** if $\lambda^*(A) = 0$. According to the definition, A is a null set if and only if, for every $\varepsilon > 0$, there exists a sequence of open intervals $(I_n)_{n \in \mathbb{N}} \subseteq \mathcal{I}$ such that $A \subseteq \bigcup_{n=0}^{\infty} I_n$ and $\sum_{n=0}^{\infty} |I_n| < \varepsilon$. Note that in the case of Jordan null sets, the cover by a family of open intervals is finite, so that any Jordan null set is a Lebesgue null set.
Because we will not be working with Jordan null sets, in what follows, we will use the term **null set** for a Lebesgue null set.

Examples 1.2.11

(i) For every $x \in \mathbb{R}$, $\{x\}$ is a null set (see (i) of Remark 1.2.4).

(ii) Any subset of a null set is a null set (see (2) of Theorem 1.2.3).

(iii) Any countable union of null sets is a null set. Indeed, let $A = \bigcup_{n=1}^{\infty} A_n$ where, for any $n \in \mathbb{N}^*$, $\lambda^*(A_n) = 0$; then $\lambda^*(A) \leq \sum_{n=1}^{\infty} \lambda^*(A_n) = 0$ and so A is a null set.

(iv) From (i) and (iii), every countable set is a null set. Particularly, $\mathbb{N}$, $\mathbb{Z}$ and $\mathbb{Q}$ are null sets (here $\mathbb{Z}$ denotes the set of the integers).

But there are also uncountable null sets; such is the triadic Cantor set that we will build in 1.3.16. Another example is presented in Exercise (4) of ▶ Sect. 1.5.

1.3 Lebesgue Measurable Sets

In this section, we will specify the subsets of $\mathbb{R}$ to whom a measure can be assigned and which properties have that measure.

For every $A \subseteq \mathbb{R}$, we defined $\lambda^*(A) = \inf\{\lambda(D) : D \in \tau_u, A \subseteq D\}$.

If we assume that $\lambda^*(A) < +\infty$, then, for every $\varepsilon > 0$, there exists $D \in \tau_u$ with $A \subseteq D$ such that $\lambda(D) < \lambda^*(A) + \varepsilon$ or $\lambda(D) - \lambda^*(A) < \varepsilon$.

On the other hand, $D = A \cup (D \setminus A)$, from where $\lambda(D) = \lambda^*(D) \leq \lambda^*(A) + \lambda^*(D \setminus A)$ and then $\lambda(D) - \lambda^*(A) \leq \lambda^*(D \setminus A)$.

Definition 1.3.1

A set $A \subseteq \mathbb{R}$ is **measurable** (in the sense of Lebesgue) if, for all $\varepsilon > 0$, there exists $D \in \tau_u$ such that $A \subseteq D$ and $\lambda^*(D \setminus A) < \varepsilon$.

We will denote with $\mathcal{L}(\mathbb{R})$ or $\mathcal{L}$ the class of measurable sets of $\mathbb{R}$. The restriction $\lambda = \lambda^*|_{\mathcal{L}}$ is called the **Lebesgue measure** on $\mathbb{R}$. For every $A \in \mathcal{L}$, we will denote with $\mathcal{L}(A) = \{B \subseteq A : B \in \mathcal{L}\}$ the family of all measurable subsets of A.

Remark 1.3.2 $\tau_u \subseteq \mathcal{L}$; indeed, if $G \in \tau_u$, then, for every $\varepsilon > 0$, there exists $D = G \supseteq G$ such that $\lambda^*(D \setminus G) = \lambda^*(\emptyset) = 0 < \varepsilon$.

From this follows that λ is an extension of the measure of open sets and then the notation performed is therefore not confusing.

Theorem 1.3.3

(1) *Every null set is a measurable set.*

(2) $\bigcup_{n=1}^{\infty} A_n \in \mathcal{L}$, *for every* $(A_n)_n \subseteq \mathcal{L}$.

Proof

(1) Let $A \subseteq \mathbb{R}$ be a null set ($\lambda^*(A) = 0$); for every $\varepsilon > 0$, there exists $D \in \tau_u$ such that $A \subseteq D$ and $\lambda(D) < \varepsilon$. Then $\lambda^*(D \setminus A) \leq \lambda^*(D) = \lambda(D) < \varepsilon$, and therefore $A \in \mathcal{L}$.

(2) Let $A = \bigcup_{n=1}^{\infty} A_n$, where $\{A_n : n \in \mathbb{N}^*\} \subseteq \mathcal{L}$; for every $\varepsilon > 0$ and every $n \in \mathbb{N}^*$, there exists $D_n \in \tau_u$ such that $A_n \subseteq D_n$ and $\lambda^*(D_n \setminus A_n) < \frac{\varepsilon}{2^n}$. Let $D = \bigcup_{n=1}^{\infty} D_n \in \tau_u$; then $A \subseteq D$ and $D \setminus A = \bigcup_{n=1}^{\infty}(D_n \setminus A) \subseteq \bigcup_{n=1}^{\infty}(D_n \setminus A_n)$, from where $\lambda^*(D \setminus A) \leq \sum_{n=1}^{\infty} \lambda^*(D_n \setminus A_n) \leq \varepsilon$. ■

Remarks 1.3.4

(i) $\emptyset \in \mathcal{L}$.

(ii) For every $A \in \mathcal{L}$ with $\lambda(A) = 0$ and every $B \subseteq A$, $\lambda^*(B) = 0$; therefore $B \in \mathcal{L}$. We will say that λ is **complete** on $\mathcal{L}$ or that $\mathcal{L}$ is **complete** with respect to λ.

(iii) $A \cup B \in \mathcal{L}$, for every $A, B \in \mathcal{L}$ $(A \cup B = A \cup B \cup \emptyset \cup \cdots \cup \emptyset \cup \cdots)$.

(iv) Each interval is a measurable set. Indeed, open intervals are open sets and therefore measurable; each other interval differs up to two points from an open interval.

We will show that the closed sets are also measurable.

We recall that $A \subseteq \mathbb{R}$ is a **closed** set if its complement is an open set ($\mathbb{R} \setminus A \in \tau_u$) or equivalent if, for any convergent sequence of A, the limit belongs to set A. In particular, any closed interval (see Definition 1.1.7) is a closed set. By passing to the complementary in the properties of open sets (see Definition 1.1.2), we obtain that the union of a finite family of closed sets is a closed set and that any intersection (finite or infinite) of closed sets is a closed set.

A set $A \subseteq \mathbb{R}$ is a **compact** set if it is bounded and closed, or equivalent, if any sequence of A has a convergent subsequence to a point of A. Any bounded closed interval (Definition 1.1.7) is a compact set.

First, we will present a lemma.

Lemma 1.3.5 *If F is a closed set and K is a compact set such that $F \cap K = \emptyset$, then* $d(F, K) = \inf\{|x - y| : x \in F, y \in K\} > 0$.

Proof

If we assume that $d(F, K) = \inf\{|x - y| : x \in F, y \in K\} = 0$, then there exist two sequences, $(x_n)_n \subseteq F$ and $(y_n)_n \subseteq K$, such that $x_n - y_n \to 0$. K being compact, $(y_n)_n$ has a subsequence $(y_{k_n})_n$ converging to $y \in K$. It follows that $x_{k_n} \to y$ and, since F is closed, $y \in F$. So $y \in F \cap K$, which contradicts the hypothesis that F and K are disjoint. ■

Theorem 1.3.6
Any closed set is a measurable set.

Proof

(a) First we assume that F is a bounded closed set; therefore F is a compact, and $\lambda^*(F) < +\infty$ (see Exercise (5) in ▶ Sect. 1.5).

According to the outer measure definition, for any $\varepsilon > 0$, there exists $D \in \tau_u$ such that $F \subseteq D$ and $\lambda(D) < \lambda^*(F) + \varepsilon$. Then $D \setminus F \in \tau_u$, and according to Theorem 1.1.8, $D \setminus F = \bigcup_{n=1}^{\infty} J_n$, where $\{J_n : n \geq 1\}$ is a family of nonoverlapping closed intervals; moreover, $\lambda(D \setminus F) = \sum_{n=1}^{\infty} |J_n|$.

For all $m \in \mathbb{N}^*$, $\bigcup_{n=1}^{m} J_n = J$ is a closed subset of $D \setminus F$ and then $F \cap J = \emptyset$. From the previous lemma, $d(F, J) > 0$, and then Theorem 1.2.9 assures us that $\lambda^*(F \cup J) = \lambda^*(F) + \lambda^*(J)$.

It follows that $\lambda(D) \geq \lambda^*(F \cup J) = \lambda^*(F) + \lambda^*(J)$.

Any two intervals of the family $\{J_n : n = 1, \cdots, m\}$ have at most one point in common (in which case their union is a closed interval) or are disjoint (the case where the distance between them is strictly positive); applying again Theorem 1.2.9, we get $\lambda^*(J) = \sum_{n=1}^{m} |J_n|$. Then $\sum_{n=1}^{m} |J_n| \leq \lambda(D) - \lambda^*(F) < \varepsilon$, for any $m \in \mathbb{N}^*$. Then $\lambda(D \setminus F) = \sum_{n=1}^{\infty} |J_n| < \varepsilon$, from where $F \in \mathcal{L}$.

(b) If F is an unbounded closed set, then $F = \bigcup_{n=1}^{\infty}(F \cap [-n, n])$ is a countable union of bounded closed sets, hence of measurable sets; from (2) of Theorem 1.3.3, $F \in \mathcal{L}$. ■

For every $A \subseteq \mathbb{R}$, we denote by $A^c = \mathbb{R} \setminus A$ the complement of A.

Theorem 1.3.7
The complement of a measurable set is a measurable set.

Proof
Let $A \in \mathscr{L}$; for any $n \in \mathbb{N}^*$, there exists $D_n \in \tau_u$ such that $A \subseteq D_n$ and $\lambda^*(D_n \setminus A) < \frac{1}{n}$. Then $A \subseteq \bigcap_{n=1}^{\infty} D_n$, and so $B = \bigcup_{n=1}^{\infty} D_n^c \subseteq A^c$. We can then write

$$A^c = B \cup (A^c \setminus B). \tag{*}$$

The sets $D_n^c = \mathbb{R} \setminus D_n$ are closed, for all $n \in \mathbb{N}^*$; according to Theorem 1.3.6, $D_n^c \in \mathscr{L}$; therefore $B = \bigcup_{n=1}^{\infty} D_n^c \in \mathscr{L}$ (see point (2) of Theorem 1.3.3).

On the other hand, $A^c \setminus B = A^c \cap B^c = \bigcap_{n=1}^{\infty} D_n \setminus A$, and then $A^c \setminus B \subseteq D_n \setminus A$, for any $n \in \mathbb{N}^*$. It follows that $\lambda^*(A^c \setminus B) < \frac{1}{n}$, for any $n \in \mathbb{N}^*$ from where $\lambda^*(A^c \setminus B) = 0$. Since $A^c \setminus B$ is a null set, it is measurable, and then, according to $(*)$, A^c is a union of the two measurable sets, so it is measurable. ■

Corollary 1.3.8
(1) $A \cap B \in \mathscr{L}$, *for every* $A, B \in \mathscr{L}$.
(2) $A \setminus B \in \mathscr{L}$, *for every* $A, B \in \mathscr{L}$.
(3) $\bigcap_{n=1}^{\infty} A_n \in \mathscr{L}$, *for every* $(A_n)_n \subseteq \mathscr{L}$.

Proof
The proof is an immediate consequence of the previous theorem and of the relations: $(A \cap B)^c = A^c \cup B^c$, $A \setminus B = A \cap B^c$, and $(\bigcap_{n=1}^{\infty} A_n)^c = \bigcup_{n=1}^{\infty} A_n^c$. ■

The following corollary gives a characterization of the measurable sets using closed sets.

Corollary 1.3.9 *$A \in \mathscr{L}$ if and only if, for every $\varepsilon > 0$, there exists a closed set $F \subseteq A$, such that $\lambda^*(A \setminus F) < \varepsilon$.*

Proof
$A \in \mathscr{L}$ if and only if $A^c \in \mathscr{L}$ and so, equivalent, if, for every $\varepsilon > 0$, there exists $D \in \tau_u$ such that $A^c \subseteq D$ and $\lambda^*(D \setminus A^c) < \varepsilon$ which is equivalent to the existence of the closed set $F = D^c \subseteq A$ with $\lambda^*(A \setminus F) = \lambda^*(A \cap F^c) = \lambda^*(A \cap D) = \lambda^*(D \setminus A^c) < \varepsilon$. ■

Theorem 1.3.10
*The set function $\lambda : \mathscr{L} \to [0, +\infty]$, $\lambda(A) = \lambda^*(A)$, for every $A \in \mathscr{L}$, is σ-**additive**, which means that, for every pairwise disjoint sequence $(A_n)_n \subseteq \mathscr{L}$,*

$$\lambda\left(\bigcup_{n=1}^{\infty} A_n\right) = \sum_{n=1}^{\infty} \lambda(A_n).$$

Proof

Let $(A_n)_n \subseteq \mathcal{L}$ be a pairwise disjoint sequence, and let $A = \bigcup_{n=1}^{\infty} A_n$.

(a) First we assume that all the sets A_n are bounded.
According to the previous corollary, for all $n \in \mathbb{N}^*$ and all $\varepsilon > 0$, there exists a closed set $F_n \subseteq A_n$ such that $\lambda^*(A_n \setminus F_n) < \frac{\varepsilon}{2^n}$. Then $\lambda^*(A_n) \le \lambda^*(A_n \setminus F_n) + \lambda^*(F_n) < \lambda^*(F_n) + \frac{\varepsilon}{2^n}$.
For every $n \neq m$, A_n is disjoint from A_m, and so $F_n \cap F_m = \emptyset$; the sets F_n being bounded and closed (therefore compact), we can apply the result of Lemma 1.3.5 to obtain $d(F_n, F_m) > 0$, for all $n \neq m$. According to Theorem 1.2.9, $\lambda^*\left(\bigcup_{n=1}^{\infty} F_n\right) = \sum_{n=1}^{\infty} \lambda^*(F_n)$. We deduce that

$$\lambda^*(A) = \lambda^*\left(\bigcup_{n=1}^{\infty} A_n\right) \ge \lambda^*\left(\bigcup_{n=1}^{\infty} F_n\right) = \sum_{n=1}^{\infty} \lambda^*(F_n) > \sum_{n=1}^{\infty} \lambda^*(A_n) - \varepsilon.$$

Since ε is arbitrarily positive, $\lambda^*(A) \ge \sum_{n=1}^{\infty} \lambda^*(A_n)$. The inverse inequality is always verified (see (3) of Theorem 1.2.3), and then we obtain the required equality.

(b) Suppose now that A_n are not all bounded. For every $p \in \mathbb{N}$, we denote $I_p = [-p, p]$; then $\bigcup_{p=0}^{\infty} I_p = \mathbb{R}$, and $I_p \subseteq I_{p+1}$. For every $p \in \mathbb{N}$, let $J_{p+1} = I_{p+1} \setminus I_p$, $J_0 = \{0\}$; J_p are measurables (unions of two intervals), pairwise disjoint, and $\bigcup_{p=0}^{\infty} J_p = \mathbb{R}$. For every $p \in \mathbb{N}$ and every $n \in \mathbb{N}^*$, let $A_n^p = A_n \cap J_p$; then $A = \bigcup_{n,p} A_n^p$ and the sets A_n^p are bounded and pairwise disjoint. Using case (a), we obtain

$$\lambda^*(A) = \sum_{n=1}^{\infty}\sum_{p=0}^{\infty} \lambda^*(A_n^p) = \sum_{n=1}^{\infty} \lambda^*\left(\bigcup_{p=0}^{\infty} A_n^p\right) = \sum_{n=1}^{\infty} \lambda^*(A_n).$$ ■

In the following theorem, we list some consequences of σ-additivity of Lebesgue measure on $\mathcal{L}$.

We recall that, for a sequence $(x_n)_n \subseteq \mathbb{R}$, $\liminf_n x_n = \sup_{n\in\mathbb{N}} \inf_{k\ge n} x_k$, and $\limsup_n x_n = \inf_{n\in\mathbb{N}} \sup_{k\ge n} x_k$. For every sequence of sets $(A_n)_n \subseteq \mathcal{P}(\mathbb{R})$, $\liminf_n A_n = \bigcup_{n=0}^{\infty} \bigcap_{k=n}^{\infty} A_k$, and $\limsup_n A_n = \bigcap_{n=0}^{\infty} \bigcup_{k=n}^{\infty} A_k$.

Theorem 1.3.11

The measure $\lambda : \mathcal{L} \to \bar{\mathbb{R}}_+$ has the following properties:

(1) $\lambda\left(\bigcup_{k=1}^{n} A_k\right) = \sum_{k=1}^{n} \mu(A_k)$, *for every* $(A_k)_{k=1}^{n} \subseteq \mathcal{L}$, *and* $A_k \cap A_l = \emptyset$, *for any* $k \neq l$.

(2) $\lambda(A) \le \lambda(B)$, *for every* $A, B \in \mathcal{L}$, $A \subseteq B$.

(3) $\lambda(B \setminus A) = \lambda(B) - \lambda(A)$, *for every* $A, B \in \mathcal{L}$, $A \subseteq B$, $\lambda(A) < +\infty$.

(4) $\lambda(A \cup B) + \lambda(A \cap B) = \lambda(A) + \lambda(B)$, *for every* $A, B \in \mathcal{L}$.

(Continued)

Theorem 1.3.11 (continued)

(5) $\lambda\left(\bigcup_{n=1}^{\infty} A_n\right) \le \sum_{n=1}^{\infty} \lambda(A_n)$, *for every* $(A_n) \subseteq \mathscr{L}$.
(6) $\lambda\left(\bigcup_{n=1}^{\infty} A_n\right) = \lim_n \lambda(A_n)$, $\forall (A_n) \subseteq \mathscr{L}$, $A_n \subseteq A_{n+1}$, *for any* $n \in \mathbb{N}^*$.
(7) $\lambda\left(\bigcap_{n=1}^{\infty} A_n\right) = \lim_n \lambda(A_n)$, *if* $(A_n) \subseteq \mathscr{L}$, $A_{n+1} \subseteq A_n$, *for any* $n \in \mathbb{N}^*$ *and* $\lambda(A_1) < +\infty$.
(8) $\lambda(\liminf_n A_n) \le \liminf_n \lambda(A_n)$, *for every* $(A_n) \subseteq \mathscr{L}$.
(9) $\limsup_n \lambda(A_n) \le \lambda(\limsup_n A_n)$, $\forall (A_n) \subseteq \mathscr{L}$, $\lambda\left(\bigcup_{n=1}^{\infty} A_n\right) < +\infty$.

Proof

(1) $\lambda(\bigcup_{k=1}^{n} A_k) = \lambda(\bigcup_{k=1}^{\infty} A_k)$, where, for all $k > n$, $A_k = \emptyset$. By applying the property of σ-additivity of the measure, we obtain $\lambda(\bigcup_{k=1}^{n} A_k) = \sum_{k=1}^{\infty} \lambda(A_k) = \sum_{k=1}^{n} \lambda(A_k)$ (for any $k > n$, $\lambda(A_k) = 0$).
(2) The monotonicity is the consequence of point (2) of Theorem 1.2.3 and of the fact that $\lambda = \lambda^*|_{\mathcal{L}}$.
(3) The property results from $\lambda(A) < +\infty$ and from $\lambda(B) = \lambda(A \cup (B \setminus A)) = \lambda(A) + \lambda(B \setminus A)$.
(4) If $\lambda(A \cap B) = +\infty$, then the relation is obvious. Suppose that $\lambda(A \cap B) < +\infty$, and apply the finite additivity of measure λ in the relation:

$$A \cup B = [A \setminus (A \cap B)] \cup (A \cap B) \cup [B \setminus (A \cap B)].$$

From property (3), we obtain

$$\lambda(A \cup B) = [\lambda(A) - \lambda(A \cap B)] + \lambda(A \cap B) + [\lambda(B) - \lambda(A \cap B)]$$

which immediately leads us to the desired relation.
(5) The property results from (3) of Theorem 1.2.3.
(6) Let $(A_n)_n \subseteq \mathscr{L}$ be an increasing sequence of sets, and let $A = \bigcup_{n=1}^{\infty} A_n$. If there is $n_0 \in \mathbb{N}^*$ such that $\lambda(A_{n_0}) = +\infty$, then, for any $n \ge n_0$, $\lambda(A_n) = +\infty$, and hence $\lambda(A) = +\infty = \lim_n \lambda(A_n)$.
Suppose that $\lambda(A_n) < +\infty$, for any $n \in \mathbb{N}^*$; then the associated disjoint sequence $(B_n)_n$ is defined by $B_1 = A_1$, $B_n = A_n \setminus A_{n-1}$, for any $n \ge 2$. We can easily show that $A = \bigcup_{n=1}^{\infty} B_n$ and that (B_n) are pairwise disjoint. Then, using (3),

$$\lambda(A) = \lambda\left(\bigcup_{n=1}^{\infty} B_n\right) = \sum_{n=1}^{\infty} \lambda(B_n) = \lim_n \sum_{k=1}^{n} \lambda(B_k) =$$

$$= \lim_n [\lambda(A_1) + \lambda(A_2 \setminus A_1) + \ldots + \lambda(A_n \setminus A_{n-1})] =$$

$$= \lim_n [\lambda(A_1) + \lambda(A_2) - \lambda(A_1) + \cdots + \lambda(A_n) - \lambda(A_{n-1})] = \lim_n \lambda(A_n).$$

(7) Let $(A_n) \subseteq \mathcal{L}$ be a decreasing sequence of sets with $\lambda(A_1) < +\infty$, and let $A = \bigcap_{n=1}^{\infty} A_n$; then the sequence $(B_n)_n$ defined by $B_n = A_1 \setminus A_n$, for any $n \in \mathbb{N}^*$, is an increasing one ($B_n \subseteq B_{n+1}$, for any $n \in \mathbb{N}^*$), and $\bigcup_{n=1}^{\infty} B_n = A_1 \setminus (\bigcap_{n=1}^{\infty} A_n)$. By applying property (6), we obtain $\lambda(\bigcup_{n=1}^{\infty} B_n) = \lim_n \lambda(B_n)$. Since $\lambda(A_1) < +\infty, \lambda(A_n) < +\infty$, for any $n \in \mathbb{N}^*$, and then we can use (3). It follows that $\lambda(A_1) - \lambda(\bigcap_{n=1}^{\infty} A_n) = \lim_n[\lambda(A_1) - \lambda(A_n)]$, from where $\lambda(\bigcap_{n=1}^{\infty} A_n) = \lim_n \lambda(A_n)$.

(8) Let (B_n), defined by $B_n = \bigcap_{k=n}^{\infty} A_k$, for any $n \geq 1$. It's easy to note that $B_n \subseteq B_{n+1}$, for any $n \in \mathbb{N}^*$ and $\bigcup_{n=1}^{\infty} B_n = \liminf_n A_n$. From property (6), it follows that $\lambda(\liminf_n A_n) = \lambda(\bigcup_{n=1}^{\infty} B_n) = \lim_n \lambda(B_n)$. But, for any $n \in \mathbb{N}^*$, $B_n \subseteq A_n$, from where $\lambda(B_n) \leq \lambda(A_n)$, for any $n \in \mathbb{N}^*$, and so, passing to the lower limit $\lim_n \lambda(B_n) = \liminf_n \lambda(B_n) \leq \liminf_n \lambda(A_n)$, which leads to announced inequality.

(9) Let (A_n) be a sequence with the properties required in the statement, and let $A = \bigcup_{n=1}^{\infty} A_n$. We define the sequence (B_n) by $B_n = A \setminus A_n$. Then $\liminf_n B_n = A \setminus \limsup_n A_n$; for any $n \in \mathbb{N}^*$, $\bigcap_{k=n}^{\infty} B_k = A \setminus \bigcup_{k=n}^{\infty} A_k$, and so, from (8), it follows that $\lambda(\liminf_n B_n) \leq \liminf_n \lambda(B_n)$, or $\lambda(A \setminus \limsup_n A_n) \leq \liminf_n \lambda(A \setminus A_n)$. Since $\lambda(A) < +\infty$, we can use (3)

$$\lambda(A) - \lambda(\limsup_n A_n) \leq \liminf_n[\lambda(A) - \lambda(A_n)] = \lambda(A) - \limsup_n \lambda(A_n). \qquad \blacksquare$$

Definition 1.3.12

Property (1) is said to be the property of **finite additivity** the Lebesgue measure λ; property (2) is the property of **monotonicity**, (5) is the **σ-subadditivity**, and (6) and (7) are the properties of **continuity** of the measure λ from **below**, respectively from **above**.

An immediate consequence of Theorem 1.3.11 is the property of continuity of Lebesgue outer measure from below.

Proposition 1.3.13 *Let $\lambda^* : \mathcal{P}(\mathbb{R}) \to \bar{\mathbb{R}}_+$ be the Lebesgue outer measure, and let $(A_n)_{n\in\mathbb{N}^*} \subseteq \mathcal{P}(\mathbb{R})$ (not necessarily measurable) such that $A_n \subseteq A_{n+1}$, for any $n \in \mathbb{N}^*$. Then*

$$\lambda^*\left(\bigcup_{n=1}^{\infty} A_n\right) = \lim_n \lambda^*(A_n).$$

Proof

Let $A = \bigcup_{n=1}^{\infty} A_n \subseteq \mathbb{R}$. Using the monotonicity property of the outer measure (see (2) of Theorem 1.2.3), there exists $\lim_n \lambda^*(A_n) \leq \lambda^*(A)$. If there were $n \in \mathbb{N}^*$ such that $\lambda^*(A_n) = +\infty$, then $\lim_n \lambda^*(A_n) = +\infty = \lambda^*(A)$.

Let us assume that $\lambda^*(A_n) < +\infty$, for any $n \in \mathbb{N}^*$. For every $n \in \mathbb{N}^*$, there exists $D_n \in \tau_u \subseteq \mathcal{L}$ such that $A_n \subseteq D_n$ and $\lambda(D_n) < \lambda^*(A_n) + \frac{1}{n}$. For every $n \in \mathbb{N}^*$, let

$G_n = \bigcap_{k=n}^{\infty} D_k \in \mathscr{L}$. Since, for any $n \in \mathbb{N}^*$ and any $k \geq n$, $A_n \subseteq A_k \subseteq D_k$, we obtain $A_n \subseteq G_n \subseteq D_n$, and therefore $A = \bigcup_{n=1}^{\infty} A_n \subseteq \bigcup_{n=1}^{\infty} G_n = \liminf_n D_n$. From (8) of Theorem 1.3.11,

$$\lambda^*(A) \leq \lambda^*(\liminf_n D_n) = \lambda(\liminf_n D_n) \leq \liminf_n \lambda(D_n) \leq$$

$$\leq \lim_n \lambda^*(A_n) \leq \lambda^*(A).$$ ■

Remark 1.3.14 The outer measure λ^* is not continuous from above (does not verify a property similar to property (7) in Theorem 1.3.11). Indeed, let $V \subseteq [0,1]$ be Vitali set (see 1.2.7), let $\mathbb{Q}_1 = \mathbb{Q} \cap [-1,1] = \{q_1, \cdots, q_n, \cdots\}$, and let $V_n = q_n + V \subseteq [-1,2]$; then, for any $n \in \mathbb{N}^*$, $\lambda^*(V_n) = \lambda^*(V) > 0$. The sequence $(A_n)_n$, defined by $A_n = \bigcup_{k=n}^{\infty} V_k$, is decreasing ($A_n \supseteq A_{n+1}$), and $\lambda^*(A_1) \leq 3$. Furthermore, $\bigcap_{n=1}^{\infty} A_n = \emptyset$. Indeed, if we assume that there is $x \in \bigcap_{n=1}^{\infty} A_n$, then $x \in A_1$, and so there is $k \geq 1$ such that $x \in V_k$; but $x \in A_{k+1}$, and then there is $l \geq k+1$ such that $x \in V_l$. So $V_k \cap V_l \neq \emptyset$ which contradicts the fact that sets V_n are pairwise disjoint (see 1.2.7).

Therefore $\lambda^*(\bigcap_{n=1}^{\infty} A_n) = 0 < \lambda^*(V) = \lim_n \lambda^*(V_n) \leq \lim_n \lambda^*(A_n)$.

In the following theorem, we underline the invariance of the Lebsegue measure by translations as well as the behavior with respect to homotheties.

Theorem 1.3.15

(1) $x + A \in \mathscr{L}$ *and* $\lambda(x + A) = \lambda(A)$, *for every* $A \in \mathscr{L}$ *and every* $x \in \mathbb{R}$.

(2) $x \cdot A \in \mathscr{L}$ *and* $\lambda(x \cdot A) = |x| \cdot \lambda(A)$, *for every* $A \in \mathscr{L}$ *and every* $x \in \mathbb{R} \setminus \{0\}$.

Proof

(1) Since A is measurable, for every $\varepsilon > 0$, there exists $D \in \tau_u$ such that $A \subseteq D$ and $\lambda^*(D \setminus A) < \varepsilon$. Then, according to (3) of Theorem 1.1.5, $x + D \in \tau_u$, $x + A \subseteq x + D$, and $\lambda^*((x + D) \setminus (x + A)) = \lambda^*(x + (D \setminus A)) = \lambda^*(D \setminus A) < \varepsilon$ (see Theorem 1.2.6); we deduce that $x + A \in \mathscr{L}$. Equality is also the consequence of Theorem 1.2.6.

(2) Let $x \in \mathbb{R} \setminus \{0\}$. For every $\varepsilon > 0$, there exists $D \in \tau_u$ with $A \subseteq D$ and $\lambda^*(D \setminus A) < \frac{\varepsilon}{|x|}$. From (4) of Theorem 1.1.5, $x \cdot D \in \tau_u$, and $x \cdot A \subseteq x \cdot D$. Then, according to Theorem 1.2.6, $\lambda^*((x \cdot D) \setminus (x \cdot A)) = \lambda^*(x \cdot (D \setminus A)) = |x| \cdot \lambda^*(D \setminus A) < \varepsilon$. Therefore, $x \cdot A \in \mathscr{L}$. ■

We have mentioned (see example (iii) of 1.2.11) that any countable set is a null set. There are also examples of uncountable sets which are null sets. One such example is the Cantor ternary set.

Cantor Ternary Set 1.3.16 Cantor set is a subset of $I = [0, 1]$ obtained by an iterative construction. We remove the central third of I, $J_1 = \left]\frac{1}{3}, \frac{2}{3}\right[$; then we repeat the operation

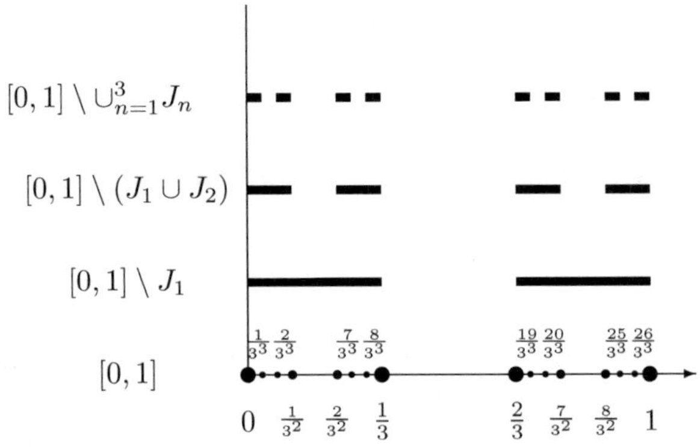

Fig. 1.1 Cantor's set construction

on the two remaining intervals $\left[0, \frac{1}{3}\right] \cup \left]\frac{2}{3}, 1\right]$ from where we remove $J_2 = \left]\frac{1}{3^2}, \frac{2}{3^2}\right[\cup \left]\frac{7}{3^2}, \frac{8}{3^2}\right[$.

At the third stage of the set $\left[0, \frac{1}{3^2}\right] \cup \left]\frac{2}{3^2}, \frac{1}{3}\right] \cup \left]\frac{2}{3}, \frac{7}{3^2}\right] \cup \left]\frac{8}{3^2}, 1\right]$, we remove the set $J_3 = \left]\frac{1}{3^3}, \frac{2}{3^3}\right[\cup \left]\frac{7}{3^3}, \frac{8}{3^3}\right[\cup \left]\frac{19}{3^3}, \frac{20}{3^3}\right[\cup \left]\frac{25}{3^3}, \frac{26}{3^3}\right[$ and so on.

In the above figure, we have marked with a bold line what remains of the interval I after we have removed J_1, J_2 and J_3 (Fig. 1.1).

Let $J = \bigcup_{n=1}^{\infty} J_n$ and $D = I \setminus J$. It is obvious that $\lambda(J_1) = |J_1| = \frac{1}{3}$, $\lambda(J_2) = \frac{2}{3^2}$, $\lambda(J_3) = \frac{2^2}{3^3}$. We can easily see that $\lambda(J_n) = \frac{2^{n-1}}{3^n}$. Since the sets J_n are pairwise disjoint, $\lambda(J) = \sum_{n=1}^{\infty} \lambda(J_n) = \frac{1}{2}\sum_{n=1}^{\infty}\left(\frac{2}{3}\right)^n = 1$, and then $\lambda(D) = 0$ (see (3) of Theorem 1.3.11). To establish the cardinal of set D, we will use the ternary number system; all $x \in I$ is written $x = \sum_{n=1}^{\infty} \frac{x_n}{3^n} = 0_3, x_1 \cdots x_n \cdots$, where $x_n \in \{0, 1, 2\}$. For the uniqueness of the writing, we agree that between two expressions of the same number, choose the one containing an infinite number of non-zero digits ($\frac{1}{3} = 0_3, 1$ will be written $0_3, 022\cdots$, $\frac{2}{3} = 0_3, 2$ will be written $0_3, 122\cdots$, etc.). We can see that $x \in J_1$ if and only if $x = 0_3, 1x_2x_3\cdots$ and $x \in J_2$ if and only if $x = 0_3, x_1 1 x_3 \cdots$, where $x_1 \in \{0, 2\}$. In general, $x \in J_n$ if and only if $x = 0_3, x_1x_2\cdots x_{n-1}1x_{n+1}\cdots$, where $x_1, x_2, \cdots x_{n-1} \in \{0, 2\}$. Then $x \in J = \bigcup_{n=1}^{\infty} J_n$ if and only if $x = 0_3, x_1x_2x_3\cdots$, and there is $k \in \mathbb{N}^*$ such that $x_k = 1$. Therefore $x \in D = I \setminus J$ if and only if $x = 0_3, x_1x_2x_3\cdots x_n\cdots$ and $x_n \in \{0, 2\}$ for all $n \in \mathbb{N}^*$.

Let $f : D \to I$ be the function defined by $f(x) = \sum_{n=1}^{\infty} \frac{x_n}{2^{n+1}} = 0_2, z_1z_2\cdots z_n\cdots$ (numbers written in base 2), where $x = 0_3, x_1x_2\cdots x_n\cdots$ and, for any $n \in \mathbb{N}^*$, $z_n = \frac{x_n}{2}$. Then f is a bijection:

f is strictly increasing: Let $x = \sum_{n=1}^{\infty} \frac{x_n}{3^n}$, $y = \sum_{n=1}^{\infty} \frac{y_n}{3^n} \in D$ with $x < y$; there exists $n_0 \in \mathbb{N}^*$ such that $x_n = y_n$, for any $n < n_0$ and $x_{n_0} < y_{n_0}$. Since $x, y \in D$, $x_{n_0} = 0$, and $y_{n_0} = 2$. Then $f(x) = \sum_{n=1}^{n_0-1} \frac{x_n}{2^{n+1}} + \sum_{n=n_0+1}^{\infty} \frac{x_n}{2^{n+1}} \leq \sum_{n=1}^{n_0-1} \frac{y_n}{2^{n+1}} + \frac{1}{2^{n_0}} < f(y)$ (y has an infinity of non-zero digits).

f is surjective: For every $z \in [0, 1]$, let $z = 0_2, z_1 z_2 \cdots z_n \cdots$ be the writing of z in base 2. If $x = 0_3, x_1 x_2 x_3 \cdots x_n \cdots$, where $x_n = 2z_n$, for all $n \in \mathbb{N}^*$, then $x \in D$ and $f(x) = z$.

Therefore $\text{card}(D) = \text{card}([0, 1]) = \text{card}(\mathbb{R}) = c > \aleph_0 = \text{card}(\mathbb{N})$, and then D is not countable.

The set D is a not countable null set. The set of Cantor will be defined by adding to D the points located at the right end of the intervals J_n: $E = \left\{ \frac{2}{3}, \frac{2}{3^2}, \frac{8}{3^2}, \frac{2}{3^3}, \frac{8}{3^3}, \frac{20}{3^3}, \frac{26}{3^3}, \cdots \right\}$.

So the Cantor set is $C = D \cup E$. D and E are null sets, and then C is a null set. Moreover $\text{card}(C) = \text{card}(D) = c$ (see Exercise (1) of ▶ Sect. 1.5).

Noting that $J \setminus E$ is an open set (countable union of open intervals) and that $C = I \setminus (J \setminus E)$, we note that C is closed and bounded and therefore is a compact set.

Exercise (4) of ▶ Sect. 1.5 gives another example of a null set which has the power of the continuum.

Remarks 1.3.17

(i) Because Lebesgue measure is complete (see (ii) of Remarks 1.3.4), the subsets of C are measurable; therefore $\mathcal{P}(C) \subseteq \mathcal{L} \subseteq \mathcal{P}(\mathbb{R})$. Then $\text{card}(\mathcal{P}(C)) = 2^c \leq \text{card}(\mathcal{L}) \leq \text{card}(\mathcal{P}(\mathbb{R})) = 2^c$, and so $\text{card}(\mathcal{L}) = 2^c$.

(ii) Although having the same cardinality, however, $\mathcal{L}$ is a proper subset of $\mathcal{P}(\mathbb{R})$. Indeed, we note that λ^* is σ-additive on $\mathcal{L}$ (Theorem 1.3.10), but it is not σ-additive on $\mathcal{P}(\mathbb{R})$ (see 1.2.7). The Vitali set (see 1.2.7) is not measurable Lebesgue; indeed, if we assume that $V \in \mathcal{L}$, then $r + V \in \mathcal{L}$, for any $r \in \mathbb{Q}_1$, and therefore, from Theorem 1.3.10 and from (1) of Theorem 1.3.15, $\lambda\left(\bigcup_{r \in \mathbb{Q}_1} (r + V)\right) = \sum_{r \in \mathbb{Q}_1} \lambda(r + V) = \sum_{r \in \mathbb{Q}_1} \lambda(V) = +\infty$ which contradicts $(*)$ of 1.2.7.

We have noticed in 1.2.7 that the existence of V depends essentially on the axiom of choice. There exist models of the theory of sets without axiom of choice, where all the sets of real numbers are measurable in the sense of Lebesgue.

Next we will define an important class of measurable sets, the class of Borel sets.

Definition 1.3.18

Let $\mathcal{A} \subseteq \mathcal{P}(\mathbb{R})$; $\mathcal{A}$ is said to be a σ-algebra on $\mathbb{R}$ if:

(1) $\bigcup_{n=1}^{\infty} A_n \in \mathcal{A}$, for every $(A_n)_n \subseteq \mathcal{A}$;
(2) $A \setminus B \in \mathcal{A}$, for any $A, B \in \mathcal{A}$;
(3) $\mathbb{R} \in \mathcal{A}$.

(Continued)

Definition 1.3.18 (continued)

Let $\mathcal{U} \subseteq \mathcal{P}(\mathbb{R})$; then there is the smallest σ-algebra on $\mathbb{R}$, $\mathcal{A}(\mathcal{U})$, which contains the class $\mathcal{U}$ (see Proposition 1.4.3). $\mathcal{A}(\mathcal{U})$ is said to be the **σ-algebra generated** by $\mathcal{U}$.. The σ-algebra generated by the usual topology τ_u, $\mathcal{A}(\tau_u) = \mathcal{B}_u$, is said to be the family of **Borel sets** of $(\mathbb{R}, \tau_u)$; every $B \in \mathcal{B}_u$ is called a **Borel set**. The open and closed subsets of $(\mathbb{R}, \tau_u)$ are Borel sets. The countable sets are Borel sets (as countable union of closed sets).

Remark 1.3.19 Because $\tau_u \subseteq \mathscr{L}$(see Remark 1.3.2), $\mathscr{L}$contains the σ-algebra generated by τ_u, $\mathcal{B}_u$ (Definition 1.3.18). We can show that $\text{card}(\mathcal{B}_u) = c$ (see Theorem 3.3.18 of [8]). Then $\text{card}(\mathcal{B}_u) = c < 2^c = \text{card}(\mathscr{L})$. A subset of a Borel null set is not necessarily a Borel set; indeed the Cantor set C is a Borel set, but $\mathcal{P}(C) \nsubseteq \mathcal{B}_u$ $(\text{card}(\mathcal{B}_u) = c < 2^c = \text{card}(\mathcal{P}(C)))$. Therefore the restriction of Lebesgue measure on $\mathcal{B}_u$ is not complete.

Although $\mathscr{L}$contains "many more" elements than $\mathcal{B}_u$, the sets of $\mathscr{L}$differ from those of $\mathcal{B}_u$ by null sets.

Theorem 1.3.20

$A \in \mathscr{L}$ if and only if $A = B \cup N$, where $B \in \mathcal{B}_u$ and $\lambda(N) = 0$.

Proof

Let $A \in \mathscr{L}$; then $A^c \in \mathscr{L}$, and so, for any $n \in \mathbb{N}^*$, there exists $D_n \in \tau_u$ such that $A^c \subseteq D_n$ and $\lambda(D_n \setminus A^c) = \lambda(D_n \cap A) < \frac{1}{n}$. The set $F_n = D_n^c \subseteq A$ is closed, and therefore $B = \bigcup_{n=1}^{\infty} F_n \in \mathcal{B}_u$. Moreover

$$\lambda(A \setminus B) = \lambda\left(\bigcap_{n=1}^{\infty}(A \cap D_n)\right) < \frac{1}{n}, \quad \text{for all } n \in \mathbb{N}^*.$$

Hence $N = A \setminus B$ is negligible, and $A = B \cup N$.

Conversely, if $A = B \cup N$ with $B \in \mathcal{B}_u \subseteq \mathscr{L}$and N a null set, it follows that $N \in \mathscr{L}$, and therefore $A \in \mathscr{L}$. ■

The previous result shows us that $\mathscr{L}$ is the smallest complete σ-algebra with respect to λ which contains $\mathcal{B}_u$. Thus $(\mathbb{R}, \mathscr{L}, \lambda)$ is the completion of $(\mathbb{R}, \mathcal{B}_u, \lambda)$—in the last triplet λ is the restriction of Lebesgue measure to the Borel sets (see Exercise (18) of ▶ Sect. 1.5).

We conclude this section with the measurability criterion of Carathéodory. The Carathéodory condition allows us, that in an abstract space on which we have defined an outer measure μ^*, to define the class of μ^*-measurable sets and a measure on this class.

Theorem 1.3.21 (Carathéodory criterion)
Let λ^* be the Lebesgue outer measure on $\mathbb{R}$, and let $A \subseteq \mathbb{R}$. $A \in \mathscr{L}$ if and only if

$$\lambda^*(T) = \lambda^*(T \cap A) + \lambda^*(T \setminus A), \text{ for every } T \subseteq \mathbb{R}. \tag{C}$$

Proof
We note that $\lambda^*(T) = \lambda^*[(T \cap A) \cup (T \setminus A)] \leq$
$\leq \lambda^*(T \cap A) + \lambda^*(T \setminus A)$. So checking the condition (C) is like checking

$$\lambda^*(T) \geq \lambda^*(A \cap T) + \lambda^*(T \setminus A), \text{ for every } T \subseteq \mathbb{R} \text{ with } \lambda^*(T) < +\infty$$

(if $\lambda^*(T) = +\infty$, then the above inequality is obviously verified).

(C) *is necessary*. We suppose that $A \in \mathscr{L}$; let $T \subseteq \mathbb{R}$ with $\lambda^*(T) < +\infty$, and let $\varepsilon > 0$; there exists $D \in \tau_u$ such that $T \subseteq D$ and $\lambda(D) < \lambda^*(T) + \varepsilon$. Because $D, A \in \mathscr{L}$ and λ is a measure on $\mathscr{L}$, $\lambda(D) = \lambda(D \cap A) + \lambda(D \setminus A) \geq \lambda^*(T \cap A) + \lambda^*(T \setminus A)$, from where $\lambda^*(T) + \varepsilon > \lambda^*(T \cap A) + \lambda^*(T \setminus A)$, for any $\varepsilon > 0$.

(C) *is sufficient*. If $\lambda^*(A) < +\infty$, then, for every $\varepsilon > 0$, there exists $D \in \tau_u$ such that $A \subseteq D$ and $\lambda(D) < \lambda^*(A) + \varepsilon$. We use the condition (C) with $T = D$: $\lambda(D) = \lambda^*(D) = \lambda^*(D \cap A) + \lambda^*(D \setminus A) = \lambda^*(A) + \lambda^*(D \setminus A)$, from where $\lambda^*(D \setminus A) < \varepsilon$.

If $\lambda^*(A) = +\infty$, we denote, for any $n \in \mathbb{N}^*$, $A_n = A \cap]-n, n[$. Since $]-n, n[\in \tau_u \subseteq \mathscr{L}$, according to the necessity of (C),

$$\lambda^*(T) = \lambda^*(T \cap]-n, n[) + \lambda^*(T \setminus]-n, n[), \text{ for every } T \subseteq \mathbb{R}. \tag{1.3}$$

We write (C) for A with $T_n = T \cap]-n, n[$:

$$\lambda^*(T_n) = \lambda^*(T_n \cap A) + \lambda^*(T_n \setminus A). \tag{1.4}$$

From 1.3 and 1.4, we obtain

$$\lambda^*(T) = \lambda^*(T_n \cap A) + \lambda^*(T_n \setminus A) + \lambda^*(T \setminus]-n, n[), \text{ for every } T \subseteq \mathbb{R}. \tag{1.5}$$

In 1.5, we substitute T with $T \setminus A_n$:

$$\lambda^*(T \setminus A_n) = \lambda^*(T_n \setminus A) + \lambda^*(T \setminus]-n, n[). \tag{1.6}$$

Finally, from 1.5 and 1.6,

$$\lambda^*(T) = \lambda^*(T \cap A_n) + \lambda^*(T \setminus A_n), \text{ for every } T \subseteq \mathbb{R}.$$

Therefore, $\lambda^*(A_n) < +\infty$ and A_n satisfy the condition (C), for all $n \in \mathbb{N}^*$. After the first part of the sufficiency, $A_n \in \mathscr{L}$, for any $n \in \mathbb{N}^*$. Then $A = \bigcup_{n=1}^{\infty} A_n \in \mathscr{L}$. ∎

1.4 Abstract Setting

So far we have built a measure on $\mathbb{R}$—the Lebesgue measure. In this section, we will approach a more abstract point: we will consider a general measure defined on a class of the subsets of an abstract space. Since the construction that we present generalizes that of the Lebesgue measure, only a part of the properties will be found in the general case.

Definition 1.4.1

Let X be an abstract set, let $\mathcal{P}(X)$ be the family of all subsets of X, and let $\mathcal{A} \subseteq \mathcal{P}(X)$; $\mathcal{A}$ is said to be a σ-**algebra** on X if:

(1) $\bigcup_{n=1}^{\infty} A_n \in \mathcal{A}$, for every $(A_n)_n \subseteq \mathcal{A}$;
(2) $A \setminus B \in \mathcal{A}$, for any $A, B \in \mathcal{A}$;
(3) $X \in \mathcal{A}$.

If $\mathcal{A}$ is a σ-algebra on X, then $(X, \mathcal{A})$ is a **measurable space**.

Remarks 1.4.2

(i) Let $\mathcal{A}$ be a σ-algebra on X; then
 (a) $\emptyset = X \setminus X \in \mathcal{A}$.
 (b) $A \in \mathcal{A}$ if and only if $A^c \in \mathcal{A}$.
 (c) $A \cup B = A \cup B \cup \emptyset \cup \cdots \cup \emptyset \cup \cdots \in \mathcal{A}$, for any $A, B \in \mathcal{A}$.
 (d) $A \cap B = (A^c \cup B^c)^c \in \mathcal{A}$, for any $A, B \in \mathcal{A}$.
 (e) $\bigcap_{n=1}^{\infty} A_n = (\bigcup_{n=1}^{\infty} A_n^c)^c \in \mathcal{A}$, for every $(A_n)_n \subseteq \mathcal{A}$.
 (f) $\liminf_n A_n = \bigcup_{n=1}^{\infty} \bigcap_{k=n}^{\infty} A_k \in \mathcal{A}$, $\limsup_n A_n = \bigcap_{n=1}^{\infty} \bigcup_{k=n}^{\infty} A_k \in \mathcal{A}$, for every $(A_n)_n \subseteq \mathcal{A}$.

(ii) From Theorems 1.3.3 and 1.3.6 and Corollary 1.3.8, it follows that the family of measurable sets in the sense of Lebesgue is a σ-algebra on $\mathbb{R}$ and that $(\mathbb{R}, \mathcal{L})$ is a measurable space.

Proposition 1.4.3 *Let $\mathcal{U} \subseteq \mathcal{P}(X)$; then there is a smallest σ-algebra on X, $\mathcal{A}(\mathcal{U})$, which contains the class $\mathcal{U}$.*

Proof

It is easy to prove that any intersection (finite or infinite) of σ-algebras is a σ-algebra. Then $\mathcal{A}(\mathcal{U})$ is the intersection of all σ-algebras which contains the class $\mathcal{U}$ (at least $\mathcal{P}(X)$ is such σ-algebra). $\mathcal{A}(\mathcal{U})$ is the smallest σ-algebra containing $\mathcal{U}$. ■

Definition 1.4.4

The σ-algebra $\mathcal{A}(\mathcal{U})$ is called the **σ-algebra generated** by $\mathcal{U}$. If τ is a topology on X, then the σ-algebra generated by τ, $\mathcal{A}(\tau) = \mathcal{B}$, is said to be the family of **Borel sets** of (X, τ); every $B \in \mathcal{B}$ is called a **Borel set**. The open sets and the closed sets of (X, τ) are Borel sets.

Definition 1.4.5

Let $\mathcal{A}$ be a σ-algebra on X, and let $\gamma : \mathcal{A} \to \bar{\mathbb{R}}_+$ be a set function; γ is called a **measure** on X and the triple $(X, \mathcal{A}, \gamma)$ a **measure space** if:

(1) $\gamma(\emptyset) = 0$.

(2) $\gamma(\bigcup_{n=1}^{\infty} A_n) = \sum_{n=1}^{\infty} \gamma(A_n)$, for every $(A_n)_n \subseteq \mathcal{A}$, $A_n \cap A_m = \emptyset$ and every $n \neq m$.

If $\gamma(X) < +\infty$, then γ is a **finite** measure on X; if $\gamma(X) = 1$, γ is said to be a **probability** on X.
If $X = \bigcup_{n=1}^{\infty} A_n$ and, for any $n \in \mathbb{N}$, $A_n \in \mathcal{A}$ and $\gamma(A_n) < +\infty$, then γ is called **σ-finite**.
The measure γ is **complete** on $\mathcal{A}$ (or the σ-algebra $\mathcal{A}$ is complete with respect to γ) if, for every $A \in \mathcal{A}$ with $\gamma(A) = 0$ and every $B \subseteq A$, it follows that $B \in \mathcal{A}$ (obviously, from (2) of Theorem 1.4.7 $\gamma(B) = 0$).
If τ is a topology on the measure space $(X, \mathcal{A}, \gamma)$ such that $\tau \subseteq \mathcal{A}$ (and so $\mathcal{A}(\tau) = \mathcal{B} \subseteq \mathcal{A}$), then γ is called **regular** when, for every $A \in \mathcal{A}$ and every $\varepsilon > 0$, there exist an open set D and a closed set F such that $F \subseteq A \subseteq D$ and $\gamma(D \setminus F) < \varepsilon$.

Examples 1.4.6

(i) Let $x \in X$ and $\delta_x : \mathcal{P}(X) \to \mathbb{R}_+$, $\delta_x(A) = \begin{cases} 1, x \in A, \\ 0, x \notin A. \end{cases}$ δ_x is a probability complete on X; it is called the Dirac measure (the unit mass concentrated in x).

(ii) Let $\gamma : \mathcal{P}(\mathbb{N}) \to \bar{\mathbb{R}}_+$, defined by $\gamma(A) = \begin{cases} \text{card}(A)\ , A = \text{ finite}, \\ +\infty\ , A = \text{ infinite}. \end{cases}$ γ is a complete σ-finite measure on $\mathbb{N}$; it is called the **counting measure**.

The proof of the following theorem is similar to that of Theorem 1.3.11.

Theorem 1.4.7

Let $\gamma : \mathcal{A} \to \bar{\mathbb{R}}_+$ be a measure on $(X, \mathcal{A})$; then:

(1) $\gamma\left(\bigcup_{k=1}^{n} A_k\right) = \sum_{k=1}^{n} \gamma(A_k)$, *for every* $(A_k)_{k=1}^{n} \subseteq \mathcal{A}$, $A_k \cap A_l = \emptyset$ *and every* $k \neq l$.

(Continued)

Theorem 1.4.7 (continued)

(2) $\gamma(A) \le \gamma(B)$, *for every* $A, B \in \mathcal{A}$ *with* $A \subseteq B$.

(3) $\gamma(B \setminus A) = \gamma(B) - \gamma(A)$, *for every* $A, B \in \mathcal{A}$, $A \subseteq B$, $\gamma(A) < +\infty$.

(4) $\gamma(A \cup B) + \gamma(A \cap B) = \gamma(A) + \gamma(B)$, *for every* $A, B \in \mathcal{A}$.

(5) $\gamma\left(\bigcup_{n=1}^{\infty} A_n\right) \le \sum_{n=1}^{\infty} \gamma(A_n)$, *for every* $(A_n) \subseteq \mathcal{A}$.

(6) $\gamma\left(\bigcup_{n=1}^{\infty} A_n\right) = \lim_n \gamma(A_n)$, *for every* $(A_n) \subseteq \mathcal{A}$ *with* $A_n \subseteq A_{n+1}$ *for any* $n \in \mathbb{N}$.

(7) $\gamma\left(\bigcap_{n=1}^{\infty} A_n\right) = \lim_n \gamma(A_n)$, *for every* $(A_n) \subseteq \mathcal{A}$ *with* $A_{n+1} \subseteq A_n$, *for any* $n \in \mathbb{N}$, *and* $\gamma(A_1) < +\infty$.

(8) $\gamma(\liminf_n A_n) \le \liminf_n \gamma(A_n)$, *for every* $(A_n) \subseteq \mathcal{A}$.

(9) $\limsup_n \gamma(A_n) \le \gamma(\limsup_n A_n)$, $\forall (A_n) \subseteq \mathcal{A}$, $\gamma\left(\bigcup_{n=1}^{\infty} A_n\right) < +\infty$.

Definition 1.4.8

Property (1) is said to be the property of **finite additivity** of the measure γ; property (2) is the property of **monotonicity**, (5) is the σ**-subadditivity**, and (6) and (7) are the properties of **continuity** of the measure γ from **below**, respectively from **above**.

Remark 1.4.9 The family of Lebesgue measurable subsets of $\mathbb{R}$, $\mathscr{L}$, is a σ-algebra and Lebesgue measure, λ, is a measure σ-finite, complete and regular on $\mathscr{L}$.

If $A \in \mathscr{L}$, then $\mathscr{L}(A)$ is a σ-algebra on A, and the restriction of λ on $\mathscr{L}(A)$ is a measure on A.

1.5 Exercises

(1) Let $A = \{a_0, a_1, \cdots, a_n, \cdots\}$ be a countable set, and let B be an infinite set.

 (a) Show that there is a countable set $C = \{c_0, c_1, \cdots, c_n, \cdots\} \subseteq B$ such that $B \setminus C$ is infinite; deduce from here that $\aleph_0 \overset{\text{definition}}{=\!=} \text{card}(A) \le \text{card}(B)$ ($\aleph_0$ is the smallest transfinite cardinal).

 (b) Suppose that $A \cap B = \emptyset$; show that the function $f : A \cup B \to B$, defined by

$$f(x) = \begin{cases} x, \ x \in B \setminus C, \\ c_{2k-1}, \ x = c_k \in C, \\ c_{2k}, \ x = a_k \in A, \end{cases} \quad \text{is a bijection.}$$

 Deduce from here that $\text{card}(B) = \text{card}(A \cup B) \overset{\text{definition}}{=\!=} \text{card}(A) + \text{card}(B)$.

(2) Show that the following functions are bijections:

 (a) $f :]a, b[\to]c, d[$, $f(x) = \dfrac{c-d}{a-b} \cdot x + \dfrac{ad-bc}{a-b}$.

 (b) $g :]a, b[\to]0, +\infty[$, $g(x) = \dfrac{x-a}{b-x}$.

(c) $h :]0, +\infty[\to \mathbb{R}, h(x) = \ln x$.
From the above, it will be concluded that $\operatorname{card}(]a, b[) = \operatorname{card}(]c, d[) =$

$= \operatorname{card}(]0, +\infty[) = \operatorname{card}(\mathbb{R}) = c$.

(3) Show that:
(a) $d(B, C) \leq d(A, C)$, for every $A, B, C \subseteq \mathbb{R}$ with $A \subseteq B$.
(b) $d(A \cup B, C) = \min\{d(A, C), d(B, C)\}$, for every $A, B, C \subseteq \mathbb{R}$.

(4) Let $A \subseteq]0, 1[$ be the set of numbers which, written in base 10, use only the digits 0 and 1. Show that $\operatorname{card}(A) = c$ $(= \operatorname{card}(\mathbb{R}))$ and that $\lambda^*(A) = 0$.

Indication: $A = \bigcup_{n=1}^{\infty} A_n$, where A_n is the set of elements of A for which, in decimal writing, the digit 1 appears for the first time in the place n. It is shown that $A_1 = \left\{\frac{1}{10}\right\} \cup \left(\frac{1}{10} + (\cup_{n=2}^{\infty} A_n)\right)$ and that, for all $n, p \geq 1$, $d(A_n, A_{n+p}) > \dfrac{8}{10^{n+p}} > 0$ and $A_n = 10 \cdot A_{n+1}$.

(5) Show that, for every bounded set $A \subseteq \mathbb{R}$, $\lambda^*(A) < +\infty$. Does the converse hold true?

(6) Let a continuous function $f : \mathbb{R} \to \mathbb{R}$ such that $A = \{x \in \mathbb{R} : f(x) \neq 0\}$ is a null set ($\lambda^*(A) = 0$). Show that $f(x) = 0$, for every $x \in \mathbb{R}$.

(7) Let $C \subseteq \mathbb{R}$ be a bounded closed set (a compact). Show that $\lambda^*(C) = 0$ if and only if, for every $\varepsilon > 0$, there is $\{I_1, \cdots, I_n\} \subseteq \mathcal{I}$ such that $C \subseteq \bigcup_{k=1}^{n} I_k$ and $\sum_{k=1}^{n} |I_k| < \varepsilon$ (a compact set is a Lebesgue null set if and only if it is a Jordan null set).

Indication: see Lemma 7.1.6

(8) Let $\lambda_* : \mathcal{P}(\mathbb{R}) \to \bar{\mathbb{R}}_+$ be a set function defined by $\lambda_*(A) = \sup\{\lambda(F) : F \subseteq A, F$ closed set$\}$, for every $A \subseteq \mathbb{R}$; λ_* is called the inner Lebesgue measure on $\mathbb{R}$. Show that, for every $A \subseteq \mathbb{R}$, $\lambda_*(A) \leq \lambda^*(A)$; if $A \in \mathcal{L}$, then $\lambda_*(A) = \lambda^*(A)$. Conversely, if $\lambda_*(A) = \lambda^*(A) < +\infty$, then $A \in \mathcal{L}$. Calculate $\lambda_*(\mathbb{R} \setminus V)$, where V is a Vitali set.

(9) Let $A \subseteq \mathbb{R}$; for any $n \in \mathbb{N}^*$, we denote $D_n = \{x \in \mathbb{R} : d(x, A) < \frac{1}{n}\}$. Show that $(D_n)_n \subseteq \tau_u$ and that if A is compact, then $\lambda(A) = \lim_n \lambda(D_n)$.
Show that the compactness hypothesis cannot be removed.

(10) Show that $A \in \mathcal{L}$ if and only if, for every $\varepsilon > 0$, there exist a closed set F and an open set D such that $F \subseteq A \subseteq D$ and $\lambda(D \setminus F) < \varepsilon$.

(11) Let $A \subseteq B \subseteq C$, $A, C \in \mathcal{L}$, and $\lambda(A) = \lambda(C) < +\infty$; show that $B \in \mathcal{L}$.

(12) Show that Lebesgue measure has the Darboux property:
(a) For every $A \in \mathcal{L}$ and every $b \in \mathbb{R}$ with $0 < b < \lambda(A)$, there is $B \in \mathcal{L}$, $B \subseteq A$ such that $\lambda(B) = b$.
(b) For every bounded sets $A, B \in \mathcal{L}$, $A \subseteq B$ and for any $c \in \mathbb{R}$ with $\lambda(A) < c < \lambda(B)$, there is $C \in \mathcal{L}$ such that $A \subseteq C \subseteq B$ and $\lambda(C) = c$.

Indication. a). Let A be bounded from below, let $t_0 = \inf A$, and let $f : [t_0, +\infty) \to \mathbb{R}$ be a function defined by $f(t) = \lambda(A \cap [t_0, t])$. Then f is Lipschitz continuous, $f(t_0) = 0$ and $\lim_{t \to +\infty} f(t) = \lambda(A)$. Since f has the Darboux property and $b \in f((t_0, +\infty))$, there exists t such that $f(t) = b$; we consider $B = A \cap [t_0, t]$. If A is not bounded from below, we reason the same for the set $A_n = A \cap [-n, n]$.

b). If $\lambda(A) < c < \lambda(B)$, then $0 < c - \lambda(A) < \lambda(B \setminus A)$, and the question is reduced to case (a).

(13) Let $A, B \in \mathscr{L}$ with $\lambda(A) < +\infty$ and $\lambda(B) < +\infty$; show that

$$|\lambda(A) - \lambda(B)| \leq \lambda(A \Delta B),$$

where $A \Delta B = (A \setminus B) \cup (B \setminus A)$ is the symmetric difference of A and B.

(14) Let $n \in \mathbb{N}^*$, and let $A_1, A_2, \cdots, A_n \subseteq [0, 1]$ be Lebesgue measurable sets with $\sum_{k=1}^{n} \lambda(A_k) > n - 1$; show that $\lambda(\cap_{k=1}^{n} A_k) > 0$.

Indication. Show that $\lambda(\cap_{k=1}^{n} A_k) = 1 - \lambda(\cup_{k=1}^{n}([0, 1] \setminus A_k))$.

(15) Let $(A_n)_n \subseteq \mathscr{L}$ with $\lambda(A_n \cap A_m) = 0$, for all $n \neq m$. Show that

$$\lambda\left(\bigcup_{n=1}^{\infty} A_n\right) = \sum_{n=1}^{\infty} \lambda(A_n).$$

(16) Let $A \subseteq \mathbb{R}$ be a null set; show that $\mathbb{R} \setminus A$ is dense in $\mathbb{R}$ (for every $x, y \in \mathbb{R}$ with $x < y$, there exists $z \in \mathbb{R} \setminus A$ such that $x < z < y$). Does the converse hold true?

(17) Let $A \subseteq \mathbb{R}$, let $\tau_A = \{D \cap A : D \in \tau_u\}$—the usual topology on A—and let $\mathcal{B}_A$ be the σ-algebra generated by τ_A on A (Definition 1.4.4). Show that $\mathcal{B}_A = \{B \cap A : B \in \mathcal{B}_u\}$. ($\tau_u$ is the usual topology on $\mathbb{R}$ (Definition 1.1.2), and $\mathcal{B}_u$ is the family of Borel sets of $(\mathbb{R}, \tau_u)$ (Definition 1.3.18).)

Indication. Let $\mathcal{A} = \{B \cap A : B \in \mathcal{B}_u\}$; it's easy to verify that $\mathcal{B}_A \subseteq \mathcal{A}$. For reverse inclusion, show that $\mathcal{B}_u = \{B \in \mathcal{B}_u : B \cap A \in \mathcal{B}_A\}$.

(18) Let γ be a measure on a measurable space $(X, \mathcal{A})$, and let $\bar{\mathcal{A}} = \{A \Delta N : A \in \mathcal{A}, N \subseteq B, B \in \mathcal{A}$ and $\gamma(B) = 0\}$. Show that $\bar{\mathcal{A}}$ is a σ-algebra on X and $\bar{\gamma} : \bar{\mathcal{A}} \to \mathbb{R}_+$, $\bar{\gamma}(A \Delta N) = \gamma(A)$ is a measure complete on $\bar{\mathcal{A}}$. $\bar{\gamma}$ is called the **completion** of γ. Show that $\mathcal{A} \subseteq \bar{\mathcal{A}}$ and that the restriction of $\bar{\gamma}$ to $\mathcal{A}$ coincides with γ. The measure $\bar{\gamma}$ is the smallest complete extension of γ.

(19) Let γ be a σ-finite measure on the measurable space $(X, \mathcal{A})$, and let $\{A_i : i \in I\} \subseteq \mathcal{A}$ be a family of pairwise disjoint sets.

Show that $N = \{i \in I : \gamma(A_i) > 0\}$ is a countable set.

Indication. Suppose that $\gamma(X) < +\infty$, and let $\mathcal{C} = \{C \subseteq I : C$ be countable$\}$. For every $C \in \mathcal{C}$, $\sum_{i \in C} \gamma(A_i) = \gamma(\bigcup_{i \in C} A_i) \leq \gamma(X) < +\infty$. Hence $s = \sup_{C \in \mathcal{C}} \sum_{i \in C} \gamma(A_i) < +\infty$. Therefore, $\forall n \in \mathbb{N}^*, \exists C_n \in \mathcal{C}$ such that $s - \frac{1}{n} < \sum_{i \in C_n} \gamma(A_i)$. Show that $C_0 = \bigcup_{n=1}^{\infty} C_n \in \mathcal{C}$ and that $\gamma(A_i) = 0$, for every $i \in I \setminus C_0$.

Let now $X = \bigcup_{n=1}^{\infty} X_n$ with $\gamma(X_n) < +\infty$, for any $n \in \mathbb{N}^*$; we can still assume that $X_n \subseteq X_{n+1}, \forall n \in \mathbb{N}^*$. Let $C_n = \{i \in I : \gamma(A_i \cap X_n) > 0\}, \forall n \in \mathbb{N}^*$; show that $C_0 = \bigcup_{n=1}^{\infty} C_n = \{i \in I : \gamma(A_i) > 0\}$.

Measurable Functions

In this chapter, we will present and study a large class of functions—that of measurable functions. In the following chapter, we will have identified the integrable functions among the measurable functions.

The class of measurable functions contains most of the known functions (continuous, monotonic, integrable Riemann functions); in addition, this class benefits from a number of remarkable properties for the passage to the limit.

2.1 Definitions. Properties

Recall that a function $f : A \subseteq \mathbb{R} \to \mathbb{R}$ is continuous at a point $x \in A$ if, for every open interval $I \in \mathcal{I}$ with $f(x) \in I$, there exists $J \in \mathcal{I}$ such that $x \in A \cap J \subseteq f^{-1}(I)$ or, equivalent, for every sequence $(x_n)_n \subseteq A, x_n \to x$, it follows that $f(x_n) \to f(x)$. The function f is continuous on A if it is continuous at every point of set A. A simple characterization of the global continuity (which we will demonstrate below) asserts that a function is continuous on A if and only if the inverse image of any open set of $\mathbb{R}$ is an open set of A (a set of form $A \cap G$, with $G \in \tau_u$).

Proposition 2.1.1 *A function $f : A \subseteq \mathbb{R} \to \mathbb{R}$ is continuous on A if and only if, for every $D \in \tau_u$, there exists $G \in \tau_u$ such that $f^{-1}(D) = A \cap G$.*

Proof
Let us assume that f is continuous on A, and let $D \in \tau_u$; if $f^{-1}(D) = \emptyset$, then $G = \emptyset$ is the desired set. If $f^{-1}(D) \neq \emptyset$, then, for every $x \in f^{-1}(D)$, $f(x) \in D$, and then there exists $I \in \mathcal{I}$ such that $f(x) \in I \subseteq D$; f is continuous at x, and then there exists an open interval $J_x \in \mathcal{I}$ such that $x \in A \cap J_x \subseteq f^{-1}(I)$; the set $G = \bigcup_{x \in f^{-1}(D)} J_x$ meets the conditions of the proposition.

Conversely, we assume that the inverse image by f of any open set of $\mathbb{R}$ is an open set of A. Let $x \in A$ be an arbitrary point, and let $I \in \mathcal{I}$ such that $f(x) \in I$; since $I \in \tau_u$, there exists $G \in \tau_u$ with $x \in f^{-1}(I) = A \cap G$. Let $J \in \mathcal{I}$ such that $x \in J \subseteq G$; then $x \in A \cap J \subseteq A \cap G = f^{-1}(I)$. Therefore f is continuous at x, for every $x \in A$. ■

L. C. Florescu, *Lebesgue Integral*, Compact Textbooks in Mathematics,
https://doi.org/10.1007/978-3-030-60163-8_2

In the previous proposition, the open set D can be replaced by open intervals; so the function $f : A \subseteq \mathbb{R} \to \mathbb{R}$ is continuous on A if and only if, for all $I \in \mathcal{I}$, there is $G \in \tau_u$ such that $f^{-1}(I) = A \cap G$.

Definition 2.1.2

Let $A \in \mathscr{L}$ and let $f : A \to \mathbb{R}$; f is said to be a **measurable function** on A in the Lebesgue sense if $f^{-1}(]-\infty, a[) \in \mathscr{L}(A)$, for any $a \in \mathbb{R}$.
The set of all the functions measurable on A will be denoted with $\mathfrak{L}(A)$; let $\mathfrak{L}_+(A)$ be the measurable and positive functions on A.

Remarks 2.1.3

(i) If A is a null set, then $f \in \mathfrak{L}(A)$, for every $f : A \to \mathbb{R}$ (see (ii) of Examples 1.2.11 and (1) of Theorem 1.3.3).

(ii) For every $B \in \mathscr{L}(A)$ ($B \in \mathscr{L}$, $B \subseteq A$) and for every $f \in \mathfrak{L}(A)$, the restriction of f on B, $f|_B \in \mathfrak{L}(B)$; indeed, $(f|_B)^{-1}(]-\infty, a[) = f^{-1}(]-\infty, a[) \cap B \in \mathscr{L}(B)$, for any $a \in \mathbb{R}$.

(iii) Let $A, B, C \in \mathscr{L}$, $A = B \cup C$. $f \in \mathfrak{L}(A)$ if and only if $f|_B \in \mathfrak{L}(B)$ and $f|_C \in \mathfrak{L}(C)$ $(f^{-1}(]-\infty, a[) = (f|_B)^{-1}(]-\infty, a[) \cup (f|_C)^{-1}(]-\infty, a[)$, for any $a \in \mathbb{R}$).

(iv) The above definition can be extended to the case where the function f can also take the value $+\infty$ ($f : A \to]-\infty, +\infty]$). As in Definition 2.1.2, the function f will be called **measurable** on $A \in \mathscr{L}$ if $f^{-1}(]-\infty, a[) \in \mathscr{L}(A)$, for any $a \in \mathbb{R}$. We remark that f is measurable on A if and only if $Z = f^{-1}(+\infty) \in \mathscr{L}(A)$ and $f|_{A\setminus Z}$ is measurable on $A \setminus Z \in \mathscr{L}$. Indeed, this results immediately if we notice that $Z = A \setminus \bigcup_{n=1}^{\infty}(f < n)$ and that, for any $a \in \mathbb{R}$, $(f|_{A\setminus Z})^{-1}(]\infty, a[) = f^{-1}(]-\infty, a[) = f^{-1}(]-\infty, a[) \cap (A \setminus Z)$.

Definition 2.1.4

Let $A \in \mathscr{L}$, and let $\mathcal{B}_A = \{B \cap A : B \in \mathcal{B}_u\} \subseteq \mathscr{L}(A)$ be the family of Borel sets on A (see Exercise (17) of ▶ Sect. 1.5); it is obvious that $\mathcal{B}_{\mathbb{R}} = \mathcal{B}_u$. A function $f : A \to \mathbb{R}$ is said to be a **Borel function** on A if, for any $a \in \mathbb{R}$, $f^{-1}(]-\infty, a[) \in \mathcal{B}_A$. If $A = \mathbb{R}$, we say that f is a **Borel function**. It is obvious that every Borel function on A is measurable on A.

Proposition 2.1.5 *Let $A \in \mathscr{L}$.*

(1) Every continuous function on A is a Borel function on A.

(2) Every monotonic function on A is a Borel function on A.

Proof

(1) Let f be a continuous function on A. According to Proposition 2.1.1, for any $a \in \mathbb{R}$, there exists $G \in \tau_u \subseteq \mathcal{B}_u$ such that $f^{-1}(]-\infty, a[) = A \cap G \in \mathcal{B}_A$.

(2) Let us assume that $f : A \to \mathbb{R}$ is an increasing function on A, and let $a \in \mathbb{R}$. If $\text{card}(f^{-1}(]-\infty, a[)) \leq 1$, then $f^{-1}(]-\infty, a[) \in \mathcal{B}_A$ (see Definition 1.3.18). If

$\text{card}(f^{-1}(]-\infty, a[)) \geq 2$, then,

$$A \cap]-\infty, x_0[\subseteq f^{-1}(]-\infty, a[) \subseteq A \cap]-\infty, x_0]$$

where $x_0 = \sup f^{-1}(]-\infty, a[) \in]-\infty, +\infty]$. Indeed, for every $x \in A \cap]-\infty, x_0[$, there exists $y \in f^{-1}(]-\infty, a[) \subseteq A$ such that $x < y$. It follows that $f(x) \leq f(y) < a$ and then $x \in f^{-1}(]-\infty, a[)$.

The second inclusion is obvious. From the two inclusions, it follows that $f^{-1}(]-\infty, a[)$ coincides with the left set or with the right set of the above inclusions (the two sets do not differ only with one point); but the two sets are Borel sets of A. ■

Remark 2.1.6 Because there are continuous functions which are not monotonic and monotonic functions which are not continuous, it follows that the two classes (the class of continuous functions and that of monotonic functions) are strictly included in the class of Borel functions.

Since all Borel functions are measurable, we can formulate the following corollary.

Corollary 2.1.7 *Let $A \in \mathscr{L}$.*
(1) Every continuous function on A is measurable on A.
(2) Every monotonic function on A is measurable on A.

Proposition 2.1.8 *For every $A \subseteq \mathbb{R}$, let $\chi_A : \mathbb{R} \to \mathbb{R}$, $\chi_A(x) = \begin{cases} 1, x \in A \\ 0, x \notin A \end{cases}$, the characteristic function of A. Then $\chi_A \in \mathfrak{L}(\mathbb{R})$ if and only if $A \in \mathscr{L}$; χ_A is a Borel function if and only if $A \in \mathcal{B}_u$.*

Proof
If $\chi_A \in \mathfrak{L}(\mathbb{R})$ (χ_A is a Borel function), then $A^c = \mathbb{R} \setminus A = \chi_A^{-1}(]-\infty, \frac{1}{2}[) \in \mathscr{L}(\in \mathcal{B}_u)$, from where $A \in \mathscr{L}(\in \mathcal{B}_u)$. Conversely, if $A \in \mathscr{L}$, then $\chi_A^{-1}(]-\infty, a[) = \begin{cases} \emptyset & , a \leq 0 \\ A^c & , 0 < a \leq 1 \\ \mathbb{R} & , 1 < a \end{cases}$, and therefore $\chi_A^{-1}(]-\infty, a[) \in \mathscr{L}(\in \mathcal{B}_u)$, for any $a \in \mathbb{R}$, which means that $\chi_A \in \mathfrak{L}(\mathbb{R})$ (χ_A is a Borel function). ■

Remark 2.1.9 Let $\mathbb{Q} \subseteq \mathbb{R}$ be the subset of rational numbers; since $\mathbb{Q} \in \mathcal{B}_u$, it follows that $\chi_{\mathbb{Q}}$ (the Dirichlet function) is a Borel function, and so it is measurable.

In the following theorem, several characterizations of the measurability of a function on a set are presented.

Theorem 2.1.10
Let $A \in \mathcal{L}$ and let $f : A \to \mathbb{R}$; the following are equivalent:
(1) $f \in \mathfrak{L}(A)$.
(2) $f^{-1}(]-\infty, a]) \in \mathcal{L}$, *for any* $a \in \mathbb{R}$.
(3) $f^{-1}(]a, +\infty[) \in \mathcal{L}$, *for any* $a \in \mathbb{R}$.
(4) $f^{-1}([a, +\infty[) \in \mathcal{L}$, *for any* $a \in \mathbb{R}$.
(5) $f^{-1}(I) \in \mathcal{L}$, *for any* $I \in \mathcal{I}$.
(6) $f^{-1}(D) \in \mathcal{L}$, *for every* $D \in \tau_u$.
(7) $f^{-1}(B) \in \mathcal{L}$, *for every* $B \in \mathcal{B}_u$.

Proof
(1) $\Longrightarrow$ (2): $f^{-1}(]-\infty, a]) = \bigcap_{n=1}^{\infty} f^{-1}\left(\left]-\infty, a + \frac{1}{n}\right[\right)$, for any $a \in \mathbb{R}$.

(2) $\Longrightarrow$ (3): $f^{-1}(]a, +\infty[) = A \setminus f^{-1}(]-\infty, a])$, for any $a \in \mathbb{R}$.

(3) $\Longrightarrow$ (4): $f^{-1}([a, +\infty[) = \bigcap_{n=1}^{\infty} f^{-1}\left(\left]a - \frac{1}{n}, +\infty\right[\right)$, for any $a \in \mathbb{R}$.

(4) $\Longrightarrow$ (5): Every open interval $I \in \mathcal{I}$ is of the form $I =]-\infty, b[$, $I =]a, +\infty[$ or $I =]a, b[$ with $a < b$.

$f^{-1}(]-\infty, b[) = A \setminus f^{-1}([b, +\infty[)$, for any $b \in \mathbb{R}$,

$f^{-1}(]a, +\infty[) = \bigcup_{n=1}^{\infty} f^{-1}\left(\left[a + \frac{1}{n}, +\infty\right[\right)$, for any $a \in \mathbb{R}$ and

$f^{-1}(]a, b[) = f^{-1}(]-\infty, b[) \cap f^{-1}(]a, +\infty[)$, for all $a, b \in \mathbb{R}$ with $a < b$. In all cases $f^{-1}(I) \in \mathcal{L}$.

(5) $\Longrightarrow$ (6): From the open set structure theorem (Theorem 1.1.3), for every $D \in \tau_u$, $D = \bigcup_{n=1}^{\infty} I_n$ where $\{I_n : n \geq 1\} \subseteq \mathcal{I}$. Then $f^{-1}(D) = \bigcup_{n=1}^{\infty} f^{-1}(I_n) \in \mathcal{L}$.

(6) $\Longrightarrow$ (7): Let $\mathcal{C} = \{C \subseteq \mathbb{R} : f^{-1}(C) \in \mathcal{L}\}$; it is easy to show that $\mathcal{C}$ is a σ-algebra on $\mathbb{R}$, and according to condition (6), $\tau_u \subseteq \mathcal{C}$. Since $\mathcal{B}_u$ is the smallest σ-algebra that contains τ_u, $\mathcal{B}_u \subseteq \mathcal{C}$ (see Definition 1.3.18).

(7) $\Longrightarrow$ (1): Every open interval $]-\infty, a[$ is an open set and therefore is a Borel set. ■

Remarks 2.1.11
(i) Equivalences (1)–(7) remain valid if we replace "$f \in \mathfrak{L}(A)$" by "f a Borel function on A" and $\mathcal{L}$ by $\mathcal{B}_u$. In particular, f is a Borel function on A if and only if $f^{-1}(B) \in \mathcal{B}_A$, for every $B \in \mathcal{B}_u$.
(ii) In general, the inverse image of a measurable set by a measurable function is not measurable. Let $f : D \to I = [0, 1]$ be the function defined in ▶ Sect. 1.3.16; let us remember that f is strictly increasing and surjective and therefore bijective. Then the inverse function of f, $g = f^{-1} : [0, 1] \to D$, is also strictly increasing, so it is measurable (see (2) of Corollary 2.1.7). In ▶ Sect. 1.2.7, we denoted with $V \subseteq [0, 1]$ the Vitali set; V is not measurable ((ii) of Remark 1.3.17), but $N = g(V) \subseteq D$ is measurable ((ii) of Remark 1.3.4). Then $N \in \mathcal{L}$ and $g^{-1}(N) = V \notin \mathcal{L}$.

(iii) If $f \in \mathfrak{L}(A)$, then $f^{-1}(J) \in \mathscr{L}$, for every $J \in \mathcal{J}$. Indeed, it suffices to note that $\mathcal{J} \subseteq \mathcal{B}_u$.

Notations Let $f, g, f_n : A \to \mathbb{R}$, for any $n \in \mathbb{N}$; for the simplification of the writing, we will use in the current way the following abbreviations:

$$(f =_A g) \equiv \{x \in A : f(x) = g(x)\}$$
$$(f \neq_A g) \equiv \{x \in A : f(x) \neq g(x)\}$$
$$(f_n \to_A f) \equiv \{x \in A : f_n(x) \to f(x)\}$$
$$(f_n \not\to_A f) \equiv \{x \in A : f_n(x) \not\to f(x)\}$$

In the same way, we can clearly see the meaning of the notations $(f >_A 0)$, $(f <_A g)$, etc.

When there is no danger of confusion, set A will no longer appear in the notations above.

Definition 2.1.12

A property P is satisfied **almost everywhere** on the set $A \subseteq \mathbb{R}$ if the set $A_P = \{x \in A : x$ does not have the property $P\}$ is a null set, which means that $\lambda^*(A_P) = 0$; according to (1) of Theorem 1.3.3, A_P is a null set if and only if $A_P \in \mathscr{L}$ and $\lambda(A_P) = 0$. We abbreviate by saying that P occurs a.e. on A. So, we will say that

- $f = g$ a.e. on A if $\lambda^*((f \neq_A g)) = 0$; we will denote this with $f \doteq_A g$.
- f is continuous a.e. on A if $\lambda^*(\{x \in A : f$ is discontinuous at $x\}) = 0$.
- The sequence (f_n) converges a.e. on A to the function f if $\lambda^*((f_n \not\to_A f)) = 0$; we will denote this with $f_n \underset{A}{\dot\to} f$.

If there is no danger of confusion about set A on which this property is satisfied almost everywhere, we can omit it, and we can write $f_n \dot\to f$.

Theorem 2.1.13

Let $A \in \mathscr{L}$, $f, g : A \to \mathbb{R}$.

(1) *If $f \in \mathfrak{L}(A)$ and $f \doteq_A g$, then $g \in \mathfrak{L}(A)$.*
(2) *If f is continuous a.e. on A, then $f \in \mathfrak{L}(A)$.*

Proof

(1) Let $N = (f \neq g)$; then $\lambda(N) = 0$.
For every $a \in \mathbb{R}$, $g^{-1}(]-\infty, a[) \equiv (g < a) = [(g < a) \cap N] \cup [(g < a) \setminus N]$.
Since $[(g < a) \cap N] \subseteq N$ is a null set, it is Lebesgue measurable (see Theorem 1.3.3); $(g < a) \setminus N = (f < a) \setminus N \in \mathscr{L}$.
Therefore $g^{-1}(]-\infty, a[) \in \mathscr{L}$, for any $a \in \mathbb{R}$, and so $g \in \mathfrak{L}(A)$.

(2) Let $N = \{x \in A : f \text{ is discontinuous at } x\}$; then $\lambda(N) = 0$.
For every $a \in \mathbb{R}$, $f^{-1}(]-\infty, a[) \equiv (f < a) = [(f < a) \cap N] \cup [(f < a) \setminus N]$. Since $[(f < a) \cap N] \subseteq N$ is negligible, it is Lebesgue measurable (Theorem 1.3.3).
Because f is continuous on $A \setminus N$, $f|_{A \setminus N} \in \mathfrak{L}(A \setminus N)$ (see Corollary 2.1.7), and then $(f < a) \setminus N = (f|_{A\setminus N})^{-1}(]-\infty, a[) \in \mathscr{L}$. Therefore, $f^{-1}(]-\infty, a[) \in \mathscr{L}$, for any $a \in \mathbb{R}$, and so $f \in \mathfrak{L}(A)$. ■

(2) of the previous theorem and the Lebesgue criterion (see Theorem 7.1.7) lead us to the following corollary.

Corollary 2.1.14 *Any Riemann integrable function on an interval* $[a, b]$ *is Lebesgue measurable on* $[a, b]$ ($\mathcal{R}_{[a,b]} \subseteq \mathfrak{L}([a, b])$).

2.1.1 Operations with Measurable Functions

We will treat the behavior of measurability vis-à-vis the operations of composition and passage to the limit. We will then present the compatibility results of measurability with respect to the algebraic operations of addition, multiplication with scalars, and multiplication of functions.

Theorem 2.1.15
Let $A \in \mathscr{L}$, $f \in \mathfrak{L}(A)$, *and let* $g : C \to \mathbb{R}$ *be a Borel function on* $C \subseteq \mathbb{R}$; *if* $f(A) \subseteq C$, *then* $g \circ f \in \mathfrak{L}(A)$.

Proof
For every $B \in \mathcal{B}_u$, there exists $D \in \mathcal{B}_u$ such that $g^{-1}(B) = D \cap C$ (see Exercise (17) of ▶ Sect. 1.5); then, using point (7) of Theorem 2.1.10, $(g \circ f)^{-1}(B) = f^{-1}(g^{-1}(B)) = f^{-1}(D) \in \mathscr{L}$. ■

Remark 2.1.16 The composition of two measurable functions is not, in general, a measurable function. Indeed, the function $g : [0, 1] \to D$, defined in Remarks 2.1.11, is measurable, and $\chi_N : D \to [0, 1]$ is also measurable (Proposition 2.1.8), but $\chi_N \circ g = \chi_V$ is not measurable.

Corollary 2.1.17 *For every* $f \in \mathfrak{L}(A)$, *we have:*
(1) $f^n \in \mathfrak{L}(A)$, *for any* $n \in \mathbb{N}$;
(2) $e^f \in \mathfrak{L}(A)$;
(3) *If* $f(A) \subseteq]0, +\infty[$, *then* $\ln f$, $f^\alpha \in \mathfrak{L}(A)$, *for any* $\alpha \in \mathbb{R}$.

Proof
It suffices to note that the functions of (1), (2), and (3) are the composition of f by a continuous function (therefore a Borel one) g where $g : \mathbb{R} \to \mathbb{R}$, $g(x) = x^n$ (in the case

of (1)); $g : \mathbb{R} \to \mathbb{R}$, $g(x) = e^x$ (in the case of (2)); and $g :]0, +\infty[\to \mathbb{R}$, $g(x) = \ln x$, or $g(x) = x^\alpha$ (in the case of (3)). ■

Theorem 2.1.18

Let $A \in \mathscr{L}$ and $(f_n) \subseteq \mathfrak{L}(A)$; then:

(1) $f = \sup_n f_n \in \mathfrak{L}(A)$, *if* $\sup_n f_n(x) < +\infty$, *for every* $x \in A$;
(2) $f = \inf_n f_n \in \mathfrak{L}(A)$, *if* $\inf_n f_n(x) > -\infty$, *for every* $x \in A$;
(3) $f = \limsup_n f_n \in \mathfrak{L}(A)$, *if* $\limsup_n f_n(x) \in \mathbb{R}$, *for every* $x \in A$;
(4) $f = \liminf_n f_n \in \mathfrak{L}(A)$, *if* $\liminf_n f_n(x) \in \mathbb{R}$, *for every* $x \in A$;
(5) *If* $f_n(x) \to f(x) \in \mathbb{R}$, *for every* $x \in A$, *then* $f \in \mathfrak{L}(A)$.
(6) *If* $f_n \underset{A}{\dot{\to}} f$, *then* $f \in \mathfrak{L}(A)$.

Proof

(1) For every $a \in \mathbb{R}$, $f^{-1}(]a, +\infty[) = \bigcup_{n\in\mathbb{N}} f_n^{-1}(]a, +\infty[) \in \mathscr{L}$.
(2) For every $a \in \mathbb{R}$, $f^{-1}(]-\infty, a[) = \bigcup_{n\in\mathbb{N}} f_n^{-1}(]-\infty, a[) \in \mathscr{L}$.
(3) $\limsup_n f_n = \inf_n \sup_{k\geq n} f_k \in \mathfrak{L}(A)$ (see (1) and (2)).
(4) $\liminf_n f_n = \sup_n \inf_{k\geq n} f_k \in \mathfrak{L}(A)$ (see (1) and (2)).
(5) If $f_n(x) \to f(x)$, for every $x \in A$, then $f = \liminf_n f_n \in \mathfrak{L}(A)$.
(6) Let $N = (f_n \nrightarrow_A f)$; then $\lambda(N) = 0$, and $f_n(x) \to f(x)$, for every $x \in A \setminus N$. The previous point assures us that $f|_{A\setminus N} \in \mathfrak{L}(A \setminus N)$, and (i) of Remarks 2.1.3 assures us that $f|_N \in \mathfrak{L}(N)$. From (iii) of Remarks 2.1.3, $f \in \mathfrak{L}(A)$. ■

Remark 2.1.19 Properties (1)–(5) remain valid for Borel functions but not property (6)! In particular, if f and g are Borel functions on A, then $\sup\{f, g\}$ is a Borel function on A.

Theorem 2.1.20

Let $f, g \in \mathfrak{L}(A)$ and let $\alpha \in \mathbb{R}$; then $f + g \in \mathfrak{L}(A)$, $\alpha \cdot f \in \mathfrak{L}(A)$, and $f \cdot g \in \mathfrak{L}(A)$.

Proof

For every $a \in \mathbb{R}$,

$$(f + g < a) = \bigcup_{r\in\mathbb{Q}} [(f < r) \cap (g < a - r)] \in \mathscr{L};$$

hence $f + g \in \mathfrak{L}(A)$.

If $\alpha > 0$, then, for any $a \in \mathbb{R}$, $(\alpha \cdot f > a) = \left(f > \frac{a}{\alpha}\right) \in \mathscr{L}$; if $\alpha < 0$, then $(\alpha \cdot f > a) = \left(f < \frac{a}{\alpha}\right) \in \mathscr{L}$.

Finally, if $f, g \in \mathfrak{L}(A)$, then $f^2 \in \mathfrak{L}(A)$ (see Corollary 2.1.17), and therefore $f \cdot g = \frac{1}{4}\left[(f+g)^2 - (f-g)^2\right] \in \mathfrak{L}(A)$. ■

Definition 2.1.21

Let $f : A \to \mathbb{R}$; we will define $f^+, f^- : A \to \mathbb{R}$ by $f^+ = \sup\{f, \underline{0}\}$, $f^- = \sup\{-f, \underline{0}\}$.

f^+ is called the **positive part** and f^- the **negative part** of f.

It is obvious that $f = f^+ - f^-$ and $|f| = f^+ + f^-$.

Proposition 2.1.22 *Let* $A \in \mathscr{L}$.

(1) $f \in \mathfrak{L}(A)$ *if and only if* $f^+ \in \mathfrak{L}(A)$ *and* $f^- \in \mathfrak{L}(A)$.

(2) $f \in \mathfrak{L}(A)$ *implies* $|f| \in \mathfrak{L}(A)$.

Proof

(1) If $f \in \mathfrak{L}(A)$, then, according to Theorem 2.1.18, $f^+ \in \mathfrak{L}(A)$, and $f^- \in \mathfrak{L}(A)$; the converse theorem is provided by the preceding theorem. (2) is a consequence of Theorem 2.1.15. ■

Remark 2.1.23 f is a Borel function on A if and only if f^+ and f^- are Borel functions on A.

2.2 Different Types of Convergence

Let $f_n, f : A \to \mathbb{R}$; the sequence $(f_n)_{n\in\mathbb{N}}$ converges **pointwise** on A to f if, for every $x \in A$, $f_n(x) \to f(x)$. We denote this by $f_n \xrightarrow[A]{p} f$. In the previous paragraph, we introduced convergence almost everywhere; we recall that a sequence (f_n) converges **almost everywhere** to the function f on $A \subseteq \mathbb{R}$ if $\lambda^*((f_n \not\to_A f)) = 0$. We will denote this by $f_n \overset{\cdot}{\underset{A}{\to}} f$. In other words, $f_n \overset{\cdot}{\underset{A}{\to}} f$ if and only if there exists a null set N such that $f_n \xrightarrow[A\setminus N]{p} f$.

We have shown in (6) of Theorem 2.1.18 that if $A \in \mathscr{L}$, $(f_n) \subseteq \mathfrak{L}(A)$ and $f_n \overset{\cdot}{\underset{A}{\to}} f$, then $f \in \mathfrak{L}(A)$.

In this paragraph, we will introduce two other types of convergence for the sequences of measurable functions, and we will analyze the links between these convergences.

If $f_n, f : A \to \mathbb{R}$, then $f_n \xrightarrow[A]{u} f$ denotes the **uniform convergence** on A of $(f_n)_n$ to f: for every $\varepsilon > 0$, there exists $n_0 \in \mathbb{N}$ such that, for any $n \geq n_0$ and every $x \in A$, $|f_n(x) - f(x)| < \varepsilon$.

Definition 2.2.1

Let $A \in \mathscr{L}$, $(f_n) \subseteq \mathfrak{L}(A)$, and let $f \in \mathfrak{L}(A)$:

1. (f_n) converges **almost uniformly** on A to f if, for every $\varepsilon > 0$, there exists $A_\varepsilon \in \mathscr{L}$ such that $\lambda(A_\varepsilon) < \varepsilon$ and $f_n \xrightarrow[A \setminus A_\varepsilon]{\text{u}} f$. We will denote this with $f_n \xrightarrow[A]{\text{a.u.}} f$.
2. (f_n) converges **in measure** on A to f if $\lim_n \lambda((|f_n - f| \geq \varepsilon)) = 0$, for every $\varepsilon > 0$; we will denote this with $f_n \xrightarrow[A]{\lambda} f$.
   The sequence $(f_n)_n$ is **convergent in measure** on A if there exists $f \in \mathfrak{L}(A)$ such that $f_n \xrightarrow[A]{\lambda} f$.
3. (f_n) is **Cauchy in measure** on A if, for every $\varepsilon > 0$,
   $\lim\limits_{m,n \to \infty} \lambda((|f_m - f_n| \geq \varepsilon)) = 0$.

Theorem 2.2.2

Let $(f_n) \subseteq \mathfrak{L}(A)$ and let $f \in \mathfrak{L}(A)$:

(1) *If $f_n \xrightarrow[A]{\text{a.u.}} f$, then $f_n \xrightarrow[A]{\lambda} f$;*

(2) *If $f_n \xrightarrow[A]{\text{a.u.}} f$, then $f_n \xrightarrow[A]{\cdot} f$.*

(3) *Every sequence convergent in measure on A is Cauchy in measure.*

Proof

(1) Since $f_n \xrightarrow[A]{\text{a.u.}} f$, for every $\varepsilon > 0$, there exists $A_\varepsilon \in \mathscr{L}$ with $\lambda(A_\varepsilon) < \varepsilon$ and $f_n \xrightarrow[A \setminus A_\varepsilon]{\text{u}} f$. It follows that for every $\eta > 0$, there exists $n_0 \in \mathbb{N}$ such that, for any $n \geq n_0$ and every $x \in A \setminus A_\varepsilon$, $|f_n(x) - f(x)| < \eta$ or, in other words, $A \setminus A_\varepsilon \subseteq (|f_n - f| < \eta)$. If we go to the complement with respect to A in the last inclusion, we get $(|f_n - f| \geq \eta) \subseteq A_\varepsilon$ from where $\lambda(|f_n - f| \geq \eta) < \varepsilon$, and so $\lim_n \lambda(|f_n - f| \geq \eta) = 0$, for every $\eta > 0$, which implies $f_n \xrightarrow[A]{\lambda} f$.

(2) Since $f_n \xrightarrow[A]{\text{a.u.}} f$, for every $\varepsilon > 0$, there exists $A_\varepsilon \in \mathscr{L}$ with $\lambda(A_\varepsilon) < \varepsilon$ and $f_n \xrightarrow[A \setminus A_\varepsilon]{\text{u}} f$ from where $(f_n)_n$ is pointwise convergent to f on $A \setminus A_\varepsilon$, or $A \setminus A_\varepsilon \subseteq (f_n \xrightarrow[A]{\text{p}} f)$. If we go to the complement with respect to A in the last inclusion, we get $(f_n \nrightarrow f) \subseteq A_\varepsilon$, and therefore $\lambda^*(f_n \nrightarrow f) \leq \varepsilon$, for every $\varepsilon > 0$. It follows that $\lambda^*(f_n \nrightarrow f) = 0$, and then $f_n \xrightarrow[A]{\cdot} f$.

(3) Let $(f_n)_n \subseteq \mathfrak{L}(A)$ be a sequence convergent in measure on A; then there exists $f \in \mathfrak{L}(A)$ such that $f_n \xrightarrow[A]{\lambda} f$.

For every $\varepsilon > 0$, $\lim_n \lambda((|f_n - f| \geq \frac{\varepsilon}{2})) = 0$; therefore, for every $\eta > 0$, there exists $n_0 \in \mathbb{N}$, such that, for any $n \geq n_0$, $\lambda((|f_n - f| \geq \frac{\varepsilon}{2})) < \frac{\eta}{2}$. Now let $m, n \geq n_0$; because

$|f_m - f_n| \leq |f_m - f| + |f - f_n|$,

$$\left(|f_m - f| < \frac{\varepsilon}{2}\right) \bigcap \left(|f_n - f| < \frac{\varepsilon}{2}\right) \subseteq (|f_m - f_n| < \varepsilon),$$

or, if we go to the complement with respect to A,

$$(|f_m - f_n| \geq \varepsilon)) \leq \left(|f_m - f| \geq \frac{\varepsilon}{2}\right) \bigcup \left(|f_n - f| \geq \frac{\varepsilon}{2}\right)$$

and then

$$\lambda((|f_m - f_n| \geq \varepsilon)) \leq \lambda\left(\left(|f_m - f| \geq \frac{\varepsilon}{2}\right)\right) + \lambda\left(\left(|f_n - f| \geq \frac{\varepsilon}{2}\right)\right) < \eta.$$

It follows that $\lim_{m,n\to\infty} \lambda((|f_m - f_n| \geq \varepsilon)) = 0$. ■

The following examples show that the converse of (1) and (2) in the previous theorem is not true.

Examples 2.2.3

(i) Let $f_n : \mathbb{R} \to \mathbb{R}$, $f_n = \chi_{]n,+\infty[}$. Then $f_n \underset{\mathbb{R}}{\overset{\cdot}{\to}} \underline{0}$, but $(f_n)_n$ does not converge almost uniformly to $\underline{0}$.

(ii) For every $n \in \mathbb{N}^*$ and every $k = 1, \ldots, n$, we denote $f_{n,k} = \chi_{\left]\frac{k-1}{n}, \frac{k}{n}\right[}$. Let the sequence (g_p) be defined by

$g_1 = f_{1,1}, g_2 = f_{2,1}, g_3 = f_{2,2}, \ldots, g_{\frac{n(n-1)}{2}+1} = f_{n,1}, \ldots, g_{\frac{n(n-1)}{2}+n} = f_{n,n}, \ldots$

For every $p \in \mathbb{N}^*$, there is a unique n_p such that $\frac{n_p(n_p-1)}{2} < p \leq \frac{n_p(n_p+1)}{2}$, and then $g_p = f_{n_p,k_p}$, where $k_p = p - \frac{n_p(n_p-1)}{2} \in \{1, 2, \ldots, n_p\}$.

For every $\varepsilon > 0$, $\lambda(|g_p| > \varepsilon) \leq \lambda\left(\left]\frac{k_p-1}{n_p}, \frac{k_p}{n_p}\right[\right) = \frac{1}{n_p} \to 0$; hence $g_p \underset{\mathbb{R}}{\overset{\lambda}{\to}} \underline{0}$.

On the other hand, for every $x \in]0, 1[$ and every $n \in \mathbb{N}, n \geq 3$, there exist $k', k'' \in \{1, \ldots, n\}$ such that $x \in \left]\frac{k'-1}{n}, \frac{k'}{n}\right[\setminus \left]\frac{k''-1}{n}, \frac{k''}{n}\right[$, and therefore there exists $p'_n = \frac{n(n-1)}{2} + k'$, $p''_n = \frac{n(n-1)}{2} + k''$ such that $g_{p'_n}(x) = 1$ and $g_{p''_n}(x) = 0$, which shows that $(g_p(x))_{p\in\mathbb{N}}$ is divergent.

So (g_p) is not convergent a.e. on $]0, 1[$ to $\underline{0}$ from where it turns out that (g_p) is not almost uniformly convergent on $]0, 1[$ to $\underline{0}$.

Theorem 2.2.4

Let $A \in \mathcal{L}$, $(f_n) \subseteq \mathfrak{L}(A)$, $f, g \in \mathfrak{L}(A)$.

(1) *Let* $f_n \underset{A}{\overset{\lambda}{\to}} f$; *then* $f_n \underset{A}{\overset{\lambda}{\to}} g$ *if and only if* $f = g$ *a.e.*

(2) *Let* $f_n \underset{A}{\overset{\cdot}{\to}} f$; *then* $f_n \underset{A}{\overset{\cdot}{\to}} g$ *if and only if* $f = g$ *a.e.*

(3) *Let* $f_n \underset{A}{\overset{\text{a.u.}}{\longrightarrow}} f$; *then* $f_n \underset{A}{\overset{\text{a.u.}}{\longrightarrow}} g$ *if and only if* $f = g$ *a.e.*

Proof

(1) ($\Longrightarrow$): We assume that $f_n \xrightarrow[A]{\lambda} f$, $f_n \xrightarrow[A]{\lambda} g$ and let $\varepsilon > 0$; from the inequality $|f - g| \le |f - f_n| + |f_n - g|$, it follows the inclusion $(|f_n - f| < \frac{\varepsilon}{2}) \cap (|f_n - g| < \frac{\varepsilon}{2}) \subseteq (|f - g| < \varepsilon)$. We go to the complement in the last inclusion, and we get $(|f - g| \ge \varepsilon) \subseteq (|f_n - f| \ge \frac{\varepsilon}{2}) \cup (|f_n - g| \ge \frac{\varepsilon}{2})$ from where, using the monotonicity and the finite subadditivity of measure, $\lambda(|f - g| \ge \varepsilon) \le \lambda(|f_n - f| \ge \frac{\varepsilon}{2}) + \lambda(|f_n - g| \ge \frac{\varepsilon}{2})$. Passing to the limit in the previous inequality, we get that, for every $\varepsilon > 0$, $\lambda(|f - g| \ge \varepsilon) = 0$.

On the other hand, $\lambda(f \ne g) = \lambda(|f - g| > 0) = \lambda\left(\bigcup_{p=1}^{\infty}\left(|f - g| \ge \frac{1}{p}\right)\right)$

$\le \sum_{p=1}^{\infty} \lambda\left(|f - g| \ge \frac{1}{p}\right) = 0$, and then $f = g$ a.e.

($\Longleftarrow$): We assume that $f_n \xrightarrow[A]{\lambda} f$ and $f = g$ a.e. For every $\varepsilon > 0$, $(|f_n - g| \ge \varepsilon) \subseteq (|f_n - f| \ge \varepsilon) \cup (f \ne g)$; by applying the properties of monotonicity and finite subadditivity of the measure, we obtain $\lambda(|f_n - g| \ge \varepsilon) \le \lambda(|f_n - f| \ge \varepsilon)$ and, passing to the limit, $\lim_n \lambda(|f_n - g| \ge \varepsilon) = 0$.

(2) ($\Longrightarrow$): From inclusion $(f \ne g) \subseteq (f_n \nrightarrow f) \cup (f_n \nrightarrow g)$ and from the properties of the measure λ, it follows that $\lambda(f \ne g) = 0$.

($\Longleftarrow$): The inclusion $(f_n \nrightarrow g) \subseteq (f_n \nrightarrow f) \cup (f \ne g)$ leads us to $\lambda(f_n \nrightarrow g) = 0$.

(3) ($\Longrightarrow$): If $f_n \xrightarrow[A]{\text{a.u.}} f$ and $f_n \xrightarrow[A]{\text{a.u.}} g$, then, from point (2) of Theorem 2.2.2, $f_n \xrightarrow[A]{\cdot} f$, and $f_n \xrightarrow[A]{\cdot} g$. According to the previous point, $f = g$ a.e.

($\Longleftarrow$): Let us assume that $f_n \xrightarrow[A]{\text{a.u.}} f$ and that $f = g$ a.e.; for every $\varepsilon > 0$, there exists $A_\varepsilon \in \mathscr{L}$ such that $\lambda(A_\varepsilon) < \varepsilon$ and $f_n \xrightarrow[A \setminus A_\varepsilon]{\text{u}} f$. Then $f_n \xrightarrow[A \setminus (A_\varepsilon \cup (f \ne g))]{\text{u}} f$ or $f_n \xrightarrow[A \setminus (A_\varepsilon \cup (f \ne g))]{\text{u}} g$, and, since $\lambda(A_\varepsilon) = \lambda(A_\varepsilon \cup (f \ne g)) < \varepsilon$, it follows that $f_n \xrightarrow[A]{\text{a.u.}} g$. ■

Theorem 2.2.5 (Riesz)

(1) *Any Cauchy in measure sequence on $A \in \mathscr{L}$ has an almost uniform convergent subsequence on A.*

(2) *If $f_n \xrightarrow[A]{\lambda} f$, then there exists $k_n \uparrow +\infty$ such that $f_{k_n} \xrightarrow[A]{\text{a.u.}} f$.*

(3) *Any Cauchy in measure sequence on $A \in \mathscr{L}$ is convergent in measure on A.*

Proof

(1) Let $(f_n)_n \subseteq \mathfrak{L}(A)$ be a Cauchy in measure sequence on A; for every $\varepsilon > 0$, $\lim_{m,n \to +\infty} \lambda(|f_n - f_m| \ge \varepsilon) = 0$. Hence, for every $\varepsilon > 0$, there exists $k_\varepsilon \in \mathbb{N}$ such that, for any $k \ge k_\varepsilon$, $\lambda(|f_k - f_{k_\varepsilon}| \ge \varepsilon) < \varepsilon$. We will give values for ε in the set $\{\frac{1}{2^k} : k \in \mathbb{N}\}$.

(0) $\varepsilon = 1$, there exists k_0, such that $\lambda(|f_k - f_{k_0}| \ge 1) < 1$, $\forall k > k_0$,

(1) $\varepsilon = \frac{1}{2}$, there exists $k_1 > k_0$, such that $\lambda(|f_k - f_{k_1}| \ge \frac{1}{2}) < \frac{1}{2}$, $\forall k > k_1$,

...

(n) $\varepsilon = \frac{1}{2^n}$, there exists $k_n > k_{n-1}$, s.t. $\lambda(|f_k - f_{k_n}| \geq \frac{1}{2^n}) < \frac{1}{2^n}$, $\forall k > k_n$,

...

If we replace $k = k_{n+1} > k_n$ in the relation (n), then we get:
$\lambda(|f_{k_{n+1}} - f_{k_n}| \geq \frac{1}{2^n}) < \frac{1}{2^n}$, for any $n \in \mathbb{N}$.

For every $n \in \mathbb{N}$, we denote $B_n = \bigcup_{i=n}^{\infty}(|f_{k_{i+1}} - f_{k_i}| \geq \frac{1}{2^i})$, and we note that $\lambda(B_n) \leq \sum_{i=n}^{\infty} \lambda(|f_{k_{i+1}} - f_{k_i}| \geq \frac{1}{2^i}) < \sum_{i=n}^{\infty} \frac{1}{2^i} = \frac{1}{2^{n-1}}$.

Let $B = \bigcap_{n=1}^{\infty} B_n$; then, for any $n \in \mathbb{N}$, $\lambda(B) \leq \lambda(B_n) < \frac{1}{2^{n-1}}$, from where it follows that $\lambda(B) = 0$.

For every $x \in A \setminus B = \bigcup_{n=1}^{\infty}(A \setminus B_n)$, there exists $n_0 \in \mathbb{N}$ such that $x \in A \setminus B_{n_0}$; hence, for any $n \geq n_0$, $|f_{k_{n+1}}(x) - f_{k_n}(x)| < \frac{1}{2^n}$. Therefore, for any $n > m \geq n_0$, $|f_{k_n}(x) - f_{k_m}(x)| \leq |f_{k_n}(x) - f_{k_{n-1}}(x)| + \cdots + |f_{k_{m+1}}(x) - f_{k_m}(x)| < \frac{1}{2^{n-1}} + \frac{1}{2^{n-2}} + \cdots + \frac{1}{2^m} < \frac{1}{2^{m-1}}$. It follows that the sequence $(f_{k_n}(x))_n$ is a Cauchy on $\mathbb{R}$, and then there exists $\lim_n f_{k_n}(x) \in \mathbb{R}$, for every $x \in A \setminus B$.

We will define $f : A \to \mathbb{R}$ by $f(x) = \begin{cases} \lim_{n\to\infty} f_{k_n}(x), & x \in A \setminus B \\ 0, & x \in B. \end{cases}$

Then $f_{k_n} \xrightarrow[A]{} f$ and, according to (6) of Theorem 2.1.18, $f \in \mathfrak{L}(A)$.

We will show that $f_{k_n} \xrightarrow[A]{\text{a.u.}} f$.

For every $\varepsilon > 0$, there exists $n_0 \in \mathbb{N}$ such that $\frac{1}{2^{n_0-1}} < \varepsilon$; then $\lambda(B_{n_0}) < \frac{1}{2^{n_0-1}} < \varepsilon$.

We show that $f_{k_n} \xrightarrow[A\setminus B_{n_0}]{\text{u}} f$.

For every $x \in A \setminus B_{n_0} = \bigcap_{n=n_0}^{\infty}(|f_{k_{n+1}} - f_{k_n}| < \frac{1}{2^n})$, $|f_{k_{n+1}}(x) - f_{k_n}(x)| < \frac{1}{2^n}$, for any $n \geq n_0$. So, like above,

$$|f_{k_n}(x) - f_{k_m}(x)| < \frac{1}{2^{m-1}}, \text{ for any } n > m \geq n_0.$$

We remark that $x \in A \setminus B_{n_0} \subseteq A \setminus B$, and so $\lim_n f_{k_n}(x) = f(x)$. If we go to the limit for $n \to \infty$ in the above inequality, we obtain

$$|f(x) - f_{k_m}(x)| \leq \frac{1}{2^{m-1}}, \text{ for every } m \geq n_0 \text{ and every } x \in A \setminus B_{n_0},$$

which assures us that $f_n \xrightarrow[A\setminus B_{n_0}]{\text{u}} f$.

(2) If $f_n \xrightarrow[A]{\lambda} f$, then, from (3) of Theorem 2.2.2, $(f_n)_n$ is Cauchy in measure on A. Above, we have shown that, in this case, $(f_n)_n$ has a subsequence $(f_{k_n})_n$ almost uniformly convergent to a function $g \in \mathfrak{L}(A)$. This subsequence will converge in measure to g (see (1) of Theorem 2.2.2). On the other hand, $(f_{k_n})_n$ converges in measure to f (any subsequence of a sequence convergent in measure converges in measure to the same function). From (1) of Theorem 2.2.4, it results that $f = g$ a.e. and, from (3) of the same theorem, $f_{k_n} \xrightarrow[A]{\text{a.u.}} f$.

(3) From (1), every Cauchy in measure sequence, $(f_n)_n \subseteq \mathfrak{L}(A)$, admits a subsequence $(f_{k_n})_n$ almost uniformly convergent on A to a function $f \in \mathfrak{L}(A)$; then $f_{k_n} \xrightarrow[A]{\lambda} f$. Since

$|f_n - f| \le |f_n - f_{k_n}| + |f_{k_n} - f|$, it follows that, for every $\varepsilon > 0$, $\left(|f_n - f_{k_n}| < \frac{\varepsilon}{2}\right) \cap \left(|f_{k_n} - f| < \frac{\varepsilon}{2}\right) \subseteq (|f_n - f| < \varepsilon)$ from where, passing to the complement,

$$(|f_n - f| \ge \varepsilon) \subseteq \left(|f_n - f_{k_n}| \ge \frac{\varepsilon}{2}\right) \cup \left(|f_{k_n} - f| \ge \frac{\varepsilon}{2}\right)$$

hence

$$\lambda(|f_n - f| \ge \varepsilon) \le \lambda\left(|f_n - f_{k_n}| \ge \frac{\varepsilon}{2}\right) + \lambda\left(|f_{k_n} - f| \ge \frac{\varepsilon}{2}\right).$$

Then $\lim_n \lambda((|f_n - f| \ge \varepsilon)) = 0$, for every $\varepsilon > 0$; therefore, $f_n \underset{A}{\overset{\lambda}{\rightarrow}} f$. ■

Corollary 2.2.6 *If $f_n \underset{A}{\overset{\lambda}{\rightarrow}} f$, then there exists $k_n \uparrow +\infty$ such that $f_{k_n} \underset{A}{\overset{\cdot}{\rightarrow}} f$.*

Remark 2.2.7 From (3) of the previous theorem, the space $\mathfrak{L}(A)$ is complete with respect to the convergence in measure. In fact we can show that this convergence is pseudo-metrizable. Let $d : \mathfrak{L}(A) \times \mathfrak{L}(A) \to \mathbb{R}_+, d(f, g) = \min\{1, \inf_{\alpha>0}\{\alpha + \lambda(|f - g| \ge \alpha\}\}$; d is a pseudo-metric on $\mathfrak{L}(A)$ and $d(f_n, f) \to 0$ if and only if $f_n \underset{A}{\overset{\lambda}{\rightarrow}} f$; therefore, the pseudo-metric space $(\mathfrak{L}(A), d)$ is complete.

The convergence almost everywhere does not imply the convergence almost uniform (see (i) of Examples 2.2.3). On the sets of finite measure this implication works.

Theorem 2.2.8 (Egoroff)
Let $A \in \mathscr{L}$ with $\lambda(A) < +\infty$, and let $(f_n)_n \subseteq \mathfrak{L}(A)$ such that $f_n \underset{A}{\overset{\cdot}{\rightarrow}} f$; then $f_n \underset{A}{\overset{\text{a.u.}}{\longrightarrow}} f$.

Proof
According to (6) of Theorem 2.1.18, $f \in \mathfrak{L}(A)$. Since $f_n \underset{A}{\overset{\cdot}{\rightarrow}} f$, the set $B = (f_n \not\to_A f)$ is a null set which is equivalent to $B \in \mathscr{L}$ and $\lambda(B) = 0$. For all $k, m \in \mathbb{N}^*$, we denote

$$E_{k,m} = \left\{x \in A \setminus B : |f_n(x) - f(x)| < \frac{1}{m}, \text{ for any } n \ge k\right\}.$$

We remark that $E_{k,m} = \bigcap_{n=k}^{\infty} \left(|f_n - f| < \frac{1}{m}\right) \setminus B$. Since $|f_n - f| \in \mathfrak{L}(A)$, it follows that $E_{k,m} \in \mathscr{L}$, for all $k, m \in \mathbb{N}^*$. Moreover,

$$E_{k,m} \subseteq E_{k+1,m} \subseteq A \setminus B, \text{ for all } k, m \in \mathbb{N}^*.$$

For all $x \in A \setminus B$, $f_n(x) \to f(x)$, and then there exists $k_0 \in \mathbb{N}^*$ such that, for any $n \ge k_0$, $|f_n(x) - f(x)| < \frac{1}{m}$ or $x \in E_{k_0,m}$. We have shown that, for any $m \in \mathbb{N}^*$, the sequence

$(E_{k,m})_{k\in\mathbb{N}^*}$ is increasing and that $\bigcup_{k=1}^{\infty} E_{k,m} = A\backslash B$. We now use the propriety of continuity from below of measure (see (6) of Theorem 1.3.11), and to obtain

$$\lambda(A \setminus B) = \lim_{k\to+\infty} \lambda(E_{k,m}), \text{ for any } m \in \mathbb{N}^*.$$

But $\lambda(A \setminus B) = \lambda(A) < +\infty$, and, according to (3) of Theorem 1.3.11, for every $\varepsilon > 0$, and any $m \in \mathbb{N}^*$, there exists $k_m \in \mathbb{N}^*$ such that

$$\lambda(A \setminus E_{k_m,m}) = \lambda(A) - \lambda(E_{k_m,m}) < \frac{\varepsilon}{2^m}.$$

For every $\varepsilon > 0$, let $A_\varepsilon = \bigcup_{m=1}^{\infty}(A \setminus E_{k_m,m})$.

Then $\lambda(A_\varepsilon) \le \sum_{m=1}^{\infty} \lambda(A\backslash E_{k_m,m}) < \sum_{m=1}^{\infty} \frac{\varepsilon}{2^m} = \varepsilon$. Let us show that $(f_n)_n$ is uniformly convergent on $A \setminus A_\varepsilon$ to f.

For every $x \in A \setminus A_\varepsilon = \bigcap_{m=1}^{\infty} E_{k_m,m}$, for any $m \in \mathbb{N}^*$ and for any $n \ge k_m$, $|f_n(x) - f(x)| < \frac{1}{m}$.

Let η be arbitrary positive, and let $m_0 \in \mathbb{N}^*$ such that $\frac{1}{m_0} < \eta$. There exists $n_\varepsilon = k_{m_0} \in \mathbb{N}$ such that, for any $n \ge n_\varepsilon$, and every $x \in A \setminus A_\varepsilon$, $|f_n(x) - f(x)| < \frac{1}{m_0} < \eta$, from where $f_n \xrightarrow[A\backslash A_\varepsilon]{\text{u}} f$. Therefore $f_n \xrightarrow[A]{\text{a.u.}} f$. ∎

Corollary 2.2.9 *If* $f_n \xrightarrow[A]{\cdot} f$ *and* $\lambda(A) < +\infty$, *then* $f_n \xrightarrow[A]{\lambda} f$.

Remark 2.2.10 The condition $\lambda(A) < +\infty$ of Egoroff's theorem is essential. Indeed, if $f_n = \chi_{[n,\,n+1]}$, then (f_n) is pointwise convergent on $\mathbb{R}$ to the function $\underline{0}$, but it does not converge in measure, and even less it does not converge almost uniformly.

In the following diagram, we summarize the relationships between the different types of convergences: the uniform convergence (U), the pointwise convergence (P), the almost uniform convergence (AU), the almost everywhere convergence (AE), and the convergence in measure (M).

The dotted arrows in ◻ Fig. 2.1 represent the convergence of a subsequence.

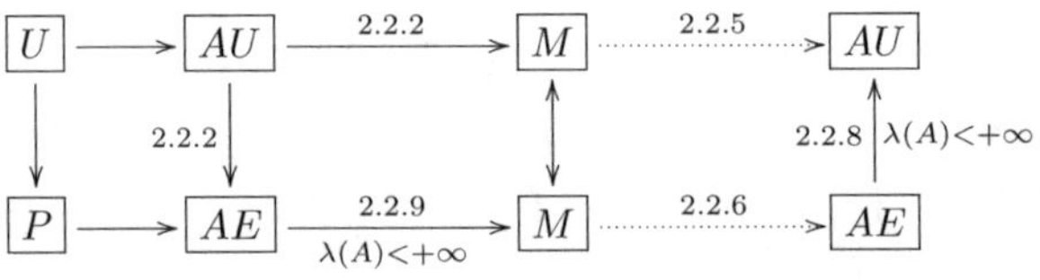

◻ **Fig. 2.1** Relations between different types of convergence on $\mathfrak{L}(A)$

2.3 The Structure of Measurable Functions

Let χ_A be the characteristic function of $A \subseteq \mathbb{R}$: $\chi_A(x) = \begin{cases} 1, & x \in A \\ 0, & x \in \mathbb{R} \setminus A \end{cases}$.
If $B \subseteq A$, then $\chi_B \leq \chi_A$. In what follows, we will identify, without risk of confusion, the function χ_B with the restriction of this function to A: $\chi_B|_A : A \to \{0, 1\}$.

We note that if $\{A_i : i \in I\} \subseteq \mathcal{P}(\mathbb{R})$ is an arbitrary family (finite or countably infinite) of pairwise disjoint sets ($A_i \cap A_j = \emptyset$, for all $i, j \in I, i \neq j$), then $\chi_{\cup_{i\in I} A_i} = \sum_{i \in I} \chi_{A_i}$.

Definition 2.3.1

Let $A \in \mathscr{L}$ and let $f : A \to \mathbb{R}$; the function f is called **simple function** on A if $f(A) = \{a_1, \ldots, a_p\} \subseteq \mathbb{R}$ and, for any $i \in \{1, \ldots, p\}$, $A_i = f^{-1}(\{a_i\}) \in \mathscr{L}(A)$.
In this situation, $f = \sum_{i=1}^{p} a_i \cdot \chi_{A_i}$ (as we mentioned above, the characteristic functions of sets A_i are considered as functions defined on A). Among the values a_i we assume that there is also 0 and that $a_i \neq a_j$, for all $i \neq j$; then the family $\{A_1, \ldots, A_p\}$ is a $\mathscr{L}$-partition of A ($A_i \cap A_j = \emptyset$, for all $i \neq j$ and $\cup_{i=1}^{p} A_i = A$).
Let $\mathcal{E}(A)$ be the set of all simple functions on A.
We remark that $\mathcal{E}(A) \subseteq \mathfrak{L}(A)$; indeed, for every $f = \sum_{i=1}^{p} a_i \cdot \chi_{A_i} \in \mathcal{E}(A)$ and for any $a \in \mathbb{R}$, $(f < a) = f^{-1}(]-\infty, a[) = \cup_{a_i < a} A_i \in \mathscr{L}$; the simple function f is a Borel function on A if and only if $A_i \in \mathcal{B}_A$, for any $i = 1, \cdots, p$.

Proposition 2.3.2 *$\mathcal{E}(A)$ is a vector subspace of $\mathfrak{L}(A)$.*

Proof
Let $f = \sum_{i=1}^{p} a_i \cdot \chi_{A_i}, g = \sum_{j=1}^{q} b_j \cdot \chi_{B_j} \in \mathcal{E}(A)$;
then $f + g = \sum_{i=1}^{p} a_i \cdot (\sum_{j=1}^{q} \chi_{A_i \cap B_j}) + \sum_{j=1}^{q} b_j \cdot (\sum_{i=1}^{p} \chi_{B_j \cap A_i}) =$
$= \sum_{i=1}^{p} \sum_{j=1}^{q} (a_i + b_j) \cdot \chi_{A_i \cap B_j} \in \mathcal{E}(A)$ and $c \cdot f = \sum_{i=1}^{p} (ca_i) \cdot \chi_{A_i} \in \mathcal{E}(A)$, for any $c \in \mathbb{R}$. ■

Theorem 2.3.3 (Simple Functions Approximation Theorem)
Let $A \in \mathscr{L}$.

(1) *For every $f \in \mathfrak{L}_+(A)$, there exists an increasing sequence of simple functions $(f_n) \subseteq \mathcal{E}_+(A)$ such that $f_n \uparrow f$ (throughout this book, the notation $f_n \uparrow f$ means that, for every $x \in A$, the sequence $(f_n(x))_{n\in\mathbb{N}}$ is increasing and $f_n(x) \to f(x)$).*
(2) *For every $f \in \mathfrak{L}(A)$, there exists a sequence of simple functions $(f_n) \subseteq \mathcal{E}(A)$ such that $f_n \xrightarrow[A]{p} f$ and $|f_n| \leq |f|$, for all $n \in \mathbb{N}$.*

(Continued)

Theorem 2.3.3 (continued)

(3) *If $f \in \mathfrak{L}(A)$ is a bounded measurable function, then we can choose the functions of the sequence $(f_n) \subseteq \mathcal{E}(A)$ so that $f_n \overset{u}{\underset{A}{\to}} f$.*

Proof

(1) Firstly, let's assume that $f : A \to \mathbb{R}_+$ is measurable and positive. We note that, for any $n \in \mathbb{N}$,

$$\mathbb{R}_+ = [0, +\infty[= \bigcup_{k=0}^{n2^n-1} \left[\frac{k}{2^n}, \frac{k+1}{2^n}\right[\cup [n, +\infty[.$$

Then

$$A = f^{-1}(\mathbb{R}_+) = \bigcup_{k=0}^{n2^n-1} f^{-1}\left(\left[\frac{k}{2^n}, \frac{k+1}{2^n}\right[\right) \cup f^{-1}\left([n, +\infty[\right).$$

According to (iii) of Remark 2.1.11, for any $n \in \mathbb{N}$ and any $k = 0, \dots, n2^n - 1$,

$$A_{k,n} = f^{-1}\left(\left[\frac{k}{2^n}. \frac{k+1}{2^n}\right[\right) \in \mathscr{L}.$$

For any $n \in \mathbb{N}$, let

$$f_n = \sum_{k=0}^{n2^n-1} \frac{k}{2^n} \cdot \chi_{A_{k,n}}.$$

Then $(f_n)_{n\in\mathbb{N}} \subseteq \mathcal{E}(A)$ and $f_n \geq 0$, for any $n \in \mathbb{N}$. We show that the sequence $(f_n)_{n\in\mathbb{N}}$ is increasing and that it converges pointwise on A to f.

For all $x \in A$, there exists $n_0 \in \mathbb{N}$, such that $f(x) < n$, for any $n \geq n_0$. Then $f(x) \in [0, n)$, and so there exists $k \in \{0, \dots, n2^n - 1\}$ such that $\frac{k}{2^n} \leq f(x) < \frac{k+1}{2^n}$.

On the one hand, $f_n(x) = \frac{k}{2^n}$. On the other hand, $\frac{2k}{2^{n+1}} \leq f(x) < \frac{2k+2}{2^{n+1}}$, from where $f_{n+1}(x) \geq \frac{2k}{2^{n+1}} = f_n(x)$.

Furthermore $0 \leq f(x) - f_n(x) < \frac{1}{2^n}$, for any $n \geq n_0$, and then $f_n(x) \to f(x)$.

Note that if f is additionally bounded on A, then there exists $n_0 \in \mathbb{N}$ such that $f(x) < n_0$, for every $x \in A$. Therefore $|f(x) - f_n(x)| < \frac{1}{2^n}$, for any $n \geq n_0$ and every $x \in A$, from where $f_n \overset{u}{\underset{A}{\to}} f$.

In Fig. 2.2 I have marked with a bold line the graph of the simple function f_n attached to a measurable and positive function f.

(2) Let now $f : A \to \mathbb{R}$ be a measurable function. Then $f^+ = \sup\{f, \underline{0}\}$ and $f^- = \sup\{-f, \underline{0}\}$ are measurable and positive functions (see Proposition 2.1.22). According to the first part of the proof, there exist $(g_n), (h_n) \subseteq \mathcal{E}(A)$ such that $g_n \uparrow f^+$ and

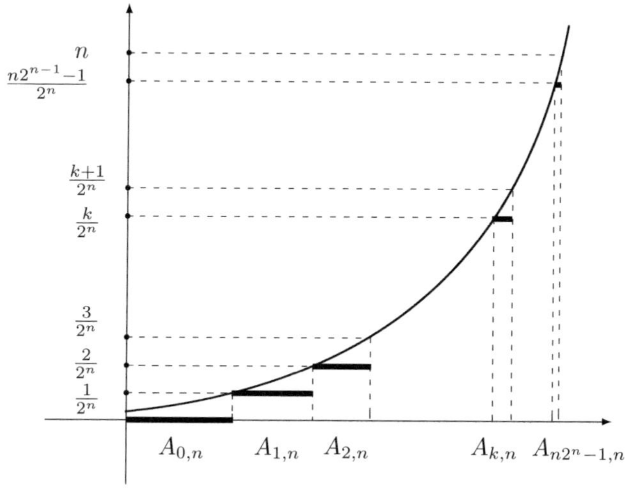

Fig. 2.2 The graph of the function f_n attached to a measurable and positive function f

$h_n \uparrow f^-$. Then $f_n = g_n - h_n \in \mathcal{E}(A)$ and $(f_n) = (g_n - h_n)_n$ is pointwise convergent on A to $f^+ - f^- = f$. Moreover $g_n + h_n \in \mathcal{E}_+(A)$, and $g_n + h_n \uparrow f^+ + f^- = |f|$ so that $|f_n| \le |f|$, for all $n \in \mathbb{N}$.

(3) If $f \in \mathfrak{L}(A)$ is bounded, then f^+ and f^- are bounded, and then, from (1), there exist $(g_n)_n, (h_n)_n \subseteq \mathcal{E}_+(A)$ such that $g_n \xrightarrow[A]{u} f^+$ and $h_n \xrightarrow[A]{u} f^-$, from where $f_n = g_n - h_n \xrightarrow[A]{u} f^+ - f^- = f$. ■

Remarks 2.3.4

(i) The approximation theorem remains valid if, instead of the measurable functions on $A \in \mathscr{L}$, we consider Borel functions on A. Thus, for any positive Borel function f, there is an increasing sequence of simple Borel functions pointwise convergent to f.

(ii) From the previous theorem and from point (5) of Theorem 2.1.18, we observe that $f \in \mathfrak{L}(A)$ if and only if there is $(f_n)_{n\in\mathbb{N}} \subseteq \mathcal{E}(A)$ such that, for all $x \in A$, $f_n(x) \to f(x)$.

(iii) (1) and (2) of the above theorem can be extended to the case where the function f can take the value $+\infty$. So, if $f : A \to]-\infty, +\infty]$ is measurable (see (iv) of Remark 2.1.3), then there exists a sequence of simple functions $(f_n)_n \subseteq \mathcal{E}(A)$ such that $f_n \xrightarrow[A]{p} f$ and $|f_n| \le |f|$, for all $n \in \mathbb{N}$. Indeed, if $Z = f^{-1}(+\infty) \in \mathscr{L}(A)$, then $g = f|_{A\setminus Z} \in \mathscr{L}(A \setminus Z)$, and therefore there exists $(g_n)_n \subseteq \mathcal{E}(A \setminus Z)$ such that $g_n \xrightarrow[A\setminus Z]{p} g$ and $|g_n| \le |g|$ on $A \setminus Z$, for all $n \in \mathbb{N}$. If we define $f_n : A \to \mathbb{R}$ by $f_n(x) = \begin{cases} g_n(x), & x \in A \setminus Z \\ n, & x \in Z \end{cases}$, then $(f_n)_n \subseteq \mathcal{E}(A)$ and, obviously, $f_n \xrightarrow[A]{p} f$ and $|f_n| \le |f|$, for all $n \in \mathbb{N}$. Moreover, if $f : A \to [0, +\infty]$ is positive, then $(f_n) \subseteq \mathcal{E}_+(A)$, and $f_n \uparrow f$.

The limits in measure of simple functions sequences constitute an important subset of the measurable functions set.

Definition 2.3.5

Let $A \in \mathscr{L}$; the function $f \in \mathfrak{L}(A)$ is said to be **totally measurable** on A if there is a sequence of simple functions $(f_n) \subseteq \mathcal{E}(A)$ convergent in measure to f ($f_n \xrightarrow[A]{\lambda} f$). Let $\mathfrak{L}_t(A)$ be the set of all totally measurable functions on A.

Theorem 2.3.6

Let $f \in \mathfrak{L}(A)$; then $f \in \mathfrak{L}_t(A)$ if and only if, for every $\varepsilon > 0$, there exists $k > 0$ such that $\lambda(|f| \geq k) < \varepsilon$.

Proof

Let $f \in \mathfrak{L}_t(A)$ and let $(f_n)_n \subseteq \mathcal{E}(A)$ such that $f_n \xrightarrow[A]{\lambda} f$. According to the Riesz theorem (Theorem 2.2.5), there exists a subsequence $k_n \uparrow +\infty$ such that $f_{k_n} \xrightarrow[A]{\text{a.u.}} f$. Then, for every $\varepsilon > 0$, there exists $A_\varepsilon \in \mathscr{L}$ with $\lambda(A_\varepsilon) < \varepsilon$ such that $f_{k_n} \xrightarrow[A\setminus A_\varepsilon]{u} f$. The functions f_{k_n} are bounded (they are simple functions), and since uniform convergence preserves the property of being bounded, f is bounded on $A \setminus A_\varepsilon$. Therefore there exists $k > 0$ such that, for every $x \in A \setminus A_\varepsilon$, $|f(x)| < k$ or, equivalent, $A \setminus A_\varepsilon \subseteq (|f| < k)$. Passing to the complement set in the last relation, $(|f| \geq k) \subseteq A_\varepsilon$, and then $\lambda(|f| \geq k) < \varepsilon$.

Now let $f \in \mathfrak{L}(A)$ such that, for every $\varepsilon > 0$, there exists $k > 0$ so that $\lambda(|f| \geq k) < \varepsilon$. Then, for all $n \in \mathbb{N}^*$, there exists $k_n > 0$ such that $\lambda(|f| \geq k_n) < \frac{1}{n}$. $(k_n)_n$ can be chosen a strictly increasing sequence of natural numbers, and then $k_n \geq n$, for all $n \in \mathbb{N}^*$. We denote $A_n = (|f| \geq k_n)$. Then $-k_n < f(x) < k_n$, for every $x \in A \setminus A_n$. Therefore

$$A \setminus A_n \subseteq f^{-1}([-k_n, k_n[) = \bigcup_{k=-nk_n}^{nk_n-1} f^{-1}\left(\left[\frac{k}{n}, \frac{k+1}{n}\right[\right).$$

Since $f \in \mathfrak{L}(A)$, $A_{k,n} = f^{-1}\left(\left[\frac{k}{n}, \frac{k+1}{n}\right[\right) \in \mathscr{L}(A)$, and then

$$f_n = \sum_{-nk_n}^{nk_n-1} \frac{k}{n} \cdot \chi_{A_{k,n}} \in \mathcal{E}(A).$$

Let us show that $(f_n)_n$ converges in measure on A to f. We can easily see that, for all $n \in \mathbb{N}^*$, $(|f_n - f| > \frac{1}{n}) \subseteq A_n$.

For every $\varepsilon > 0$, there exists $n_0 \in \mathbb{N}^*$ such that $\frac{1}{n} < \varepsilon$, for all $n \geq n_0$. Then

$$\lambda((|f_n - f| \geq \varepsilon)) \leq \lambda\left(\left(|f_n - f| > \frac{1}{n}\right)\right) \leq \lambda(A_n) < \frac{1}{n}, \text{ for any } n \geq n_0,$$

from where $\lim_n \lambda((|f_n - f| \geq \varepsilon)) = 0$, for every $\varepsilon > 0$; hence $f_n \xrightarrow[A]{\lambda} f$, and so $f \in \mathfrak{L}_t(A)$.
■

Remarks 2.3.7

(i) The previous theorem states that a function is totally measurable on A if and only if it is measurable and asymptotically bounded on A: For every $\varepsilon > 0$, there exists $A_\varepsilon \in \mathscr{L}$ with $\lambda(A_\varepsilon) < \varepsilon$, and f is bounded on $A \setminus A_\varepsilon$ ($A_\varepsilon = (|f| > k)$).

Obviously, a bounded function on a measurable set is totally measurable if and only if it is measurable.

(ii) If $\lambda(A) < +\infty$, then $\mathfrak{L}_t(A) = \mathfrak{L}(A)$. Indeed, let $f \in \mathfrak{L}(A)$; then $\bigcap_{n=0}^{\infty}(|f| \geq n) = \emptyset$, and since $\lambda(A) < +\infty$, we can use the property of continuity from above of λ ((7) of Theorem 1.3.11). So $\lim_n \lambda(|f| \geq n) = 0$. It follows that, for every $\varepsilon > 0$, there exists $n_0 \in \mathbb{N}$ such that $\lambda(|f| \geq n_0) < \varepsilon$. The function f is then asymptotically bounded on A, and so $f \in \mathfrak{L}_t(A)$.

(iii) Let $f \in \mathfrak{L}(A)$ and $B \in \mathscr{L}(A)$ with $\lambda(B) < +\infty$; then $f \cdot \chi_B \in \mathfrak{L}_t(A)$. Indeed, $f \cdot \chi_B \in \mathfrak{L}(A)$ (see Exercise (4) of ▶ Sect. 2.5). From (ii), $f|_B \in \mathfrak{L}_t(B)$. According to Theorem 2.3.6, for every $\varepsilon > 0$, there exists $k > 0$ such that $\lambda\left(\left|f|_B\right| \geq k\right) = \lambda(\{x \in B : |f(x)| \geq k\}) = \lambda(\{x \in A : |f \cdot \chi_B| \geq k\}) < \varepsilon$, from where $f \cdot \chi_B \in \mathfrak{L}_t(A)$.

(iv) The function $f : \mathbb{R} \to \mathbb{R}$, $f(x) = x$ is continuous everywhere on $\mathbb{R}$, and then it is measurable on $\mathbb{R}$; however f is not asymptotically bounded on $\mathbb{R}$ (for any $k > 0$, $\lambda(|f| > k) = +\infty$), and therefore it is not totally measurable. The suite of simple functions (f_n), $f_n = \sum_{k=-n.2^n}^{n.2^n} \frac{k}{2^n} \cdot \chi_{\left[\frac{k}{2^n}, \frac{k+1}{2^n}\right[}$, is pointwise convergent to f on $\mathbb{R}$.

We have noticed that the almost everywhere continuous functions are measurable (see (2) of Theorem 2.1.13). The Dirichlet function (see Remark 2.1.9) is an example of a measurable function which is discontinuous at each point of $\mathbb{R}$. The following theorem shows us that, although the measurable functions can be discontinuous everywhere, they are continuous on sets whose complement has the arbitrarily small measure (continuity related to the topology induced on these sets).

Theorem 2.3.8 (Lusin)
Let $A \in \mathscr{L}$; then $f \in \mathfrak{L}(A)$ if and only if, for every $\varepsilon > 0$, there exists a closed set $F_\varepsilon \subseteq A$, such that $\lambda(A \setminus F_\varepsilon) < \varepsilon$ and $f|_{F_\varepsilon}$ is continuous (for the relative topology of τ_u on F_ε).

Proof

(1) *The condition is necessary.* We will prove this part in three steps.

(I) Let $f = \sum_{k=1}^{n} a_k \chi_{A_k} \in \mathcal{E}(A)$. Then, for any $k \neq l$, $a_k \neq a_l$, $A_k \cap A_l = \emptyset$, $A_k \in \mathscr{L}(A)$, and $\bigcup_{k=1}^{n} A_k = A$. Corollary 1.3.9 assures us that, for every $\varepsilon > 0$ and any $k = 1, \ldots, n$, there exists a closed set $F_k \subseteq A_k$ such that $\lambda(A_k \setminus F_k) < \frac{\varepsilon}{2^k}$. Then $F_\varepsilon = \bigcup_{k=1}^{n} F_k \subseteq A$ is also closed, and

$$\lambda(A \setminus F_\varepsilon) = \lambda\left(\bigcup_{k=1}^{n} A_k \setminus F_\varepsilon\right) = \sum_{k=1}^{n} \lambda(A_k \setminus F_\varepsilon) \leq$$

$$\leq \sum_{k=1}^{n} \lambda(A_k \setminus F_k) < \sum_{k=1}^{n} \frac{\varepsilon}{2^k} < \sum_{k=1}^{\infty} \frac{\varepsilon}{2^k} = \varepsilon.$$

Let $g = f|_{F_\varepsilon}$, $x_0 \in F_\varepsilon$, and $(x_n)_n \subseteq F_\varepsilon$, $x_n \to x_0$. There is $i \in \{1, \cdots, n\}$ so that $x_0 \in F_i$. It follows that there exists $n_0 \in \mathbb{N}$ such that $x_n \in F_i$, for all $n \geq n_0$ (if we suppose an infinity between the terms of the sequence $(x_n)_n$ would be in another set F_j, then, since F_j is closed, it would result from this that $x_0 \in F_j$, which is impossible because $F_i \cap F_j = \emptyset$).

But on F_i, the function g is constant, and so, for any $n \geq n_0$, $g(x_n) = a_i \to a_i = g(x_0)$. It follows that g is continuous everywhere on F_ε.

(II) $f \in \mathfrak{L}_t(A)$. Then the function f is a limit in measure of a sequence of simple functions. According to Riesz theorem (Theorem 2.2.5), this sequence admits a subsequence, let's say $(f_n)_n \subseteq \mathcal{E}(A)$, almost uniformly convergent to f. For every $\varepsilon > 0$, there exists $A_\varepsilon \in \mathscr{L}(A)$ with $\lambda(A_\varepsilon) < \frac{\varepsilon}{4}$ and $f_n \xrightarrow[A \setminus A_\varepsilon]{u} f$. $A \setminus A_\varepsilon \in \mathscr{L}$; according to Corollary 1.3.9, there exists a closed set $G_\varepsilon \subseteq A \setminus A_\varepsilon$ such that $\lambda((A \setminus A_\varepsilon) \setminus G_\varepsilon) < \frac{\varepsilon}{4}$. From the first step of the demonstration, for any $n \in \mathbb{N}$, there exists a closed set $F_n^\varepsilon \subseteq A$, such that $f_n|_{F_n^\varepsilon}$ is continuous and $\lambda(A \setminus F_n^\varepsilon) < \frac{\varepsilon}{2^{n+1}}$.

Let $F_\varepsilon = G_\varepsilon \cap \bigcap_{n=1}^{\infty} F_n^\varepsilon \subseteq A$. The set F_ε is closed, and $\lambda(A \setminus F_\varepsilon) \leq \lambda(A \setminus G_\varepsilon) + \sum_{n=1}^{\infty} \lambda(A \setminus F_n^\varepsilon) < \frac{\varepsilon}{2} + \sum_{n=1}^{\infty} \frac{\varepsilon}{2^{n+1}} = \frac{\varepsilon}{2} + \frac{\varepsilon}{2} = \varepsilon$.

According to the definition, $F_\varepsilon \subseteq G_\varepsilon \subseteq A \setminus A_\varepsilon$, and then $f_n \xrightarrow[F_\varepsilon]{u} f$, and for any $n \in \mathbb{N}$, $f_n|_{F_\varepsilon}$ is continuous for the relative topology on F_ε.

The preserving continuity by uniform convergence then assures us that $f|_{F_\varepsilon}$ is continuous.

(III) $f \in \mathfrak{L}(A)$. For all $n \in \mathbb{N}$, let $A_n = A \cap]-n, n[\in \mathscr{L}(A)$; then $A = \bigcup_{n=1}^{\infty} A_n$. Since $\lambda(A_n) < +\infty$, $f \cdot \chi_{A_n} \in \mathfrak{L}_t(A)$ (see (iii) of Remark 2.3.7). From step II, for every $\varepsilon > 0$ and any $n \in \mathbb{N}^*$, there exists a closed set $F_n^\varepsilon \subseteq A$ such that $(f \cdot \chi_{A_n})|_{F_n^\varepsilon}$ is continuous and $\lambda(A \setminus F_n^\varepsilon) < \frac{\varepsilon}{2^n}$.

Let $F_\varepsilon = \bigcap_{n=1}^{\infty} F_n^\varepsilon$; it follows that $F_\varepsilon \subseteq A$ is a closed set and

$$\lambda(A \setminus F_\varepsilon) = \lambda\left(\bigcup_{n=1}^{\infty} (A \setminus F_n^\varepsilon)\right) \leq \sum_{n=1}^{\infty} \lambda(A \setminus F_n^\varepsilon) < \varepsilon.$$

Let us show that $f|_{F_\varepsilon}$ is continuous.

For every $x \in F_\varepsilon \subseteq A$, there exists $n_0 \in \mathbb{N}^*$ such that $x \in A_{n_0}$. Let $(x_p)_p \subseteq F_\varepsilon, x_p \to x$. There exists $p_0 \in \mathbb{N}$ so that $x_p \in A\cap]-n_0, n_0[= A_{n_0}$, for any $p \geq p_0$. Since $(x_p)_p \subseteq F^\varepsilon_{n_0}$ and $x \in F^\varepsilon_{n_0}$, for any $p \geq p_0$,

$$f(x_p) = (f \cdot \chi_{A_{n_0}})(x_p) = (f \cdot \chi_{A_{n_0}})|_{F^\varepsilon_{n_0}}(x_p) \to (f \cdot \chi_{A_{n_0}})|_{F^\varepsilon_{n_0}}(x) = f(x).$$

Consequently $f|_{F_\varepsilon}$ is continuous.

(2) *The condition is sufficient.* According to the hypothesis, for all $n \in \mathbb{N}^*$, there exists a closed set $F_n \subseteq A$ such that $\lambda(A \setminus F_n) < \frac{1}{n}$ and $f|_{F_n}$ is continuous. Let $F = \bigcup_{n=1}^\infty F_n \in \mathscr{L}(A)$; then $\lambda(A \setminus F) = \lambda(\bigcap_{n=1}^\infty (A \setminus F_n)) \leq \frac{1}{n}$, for any $n \in \mathbb{N}^*$, from where $\lambda(A \setminus F) = 0$. Therefore $N = A \setminus F$ is a null set. For any $n \in \mathbb{N}^*$, $f|_{F_n}$ is continuous; it follows that $f|_{F_n} \in \mathfrak{L}(F_n)$ (see (i) of Corollary 2.1.7). Then, for any $a \in \mathbb{R}$ and any $n \in \mathbb{N}^*$, $A_n = \{x \in F_n : f(x) < a\} \in \mathscr{L}$. The set $A_0 = \{x \in N : f(x) < a\} \subseteq N$ is a null set so that $A_0 \in \mathscr{L}$. Thus $\{x \in A : f(x) < a\} = A_0 \cup \bigcup_{n=1}^\infty A_n \in \mathscr{L}$, from where $f \in \mathfrak{L}(A)$. ■

Corollary 2.3.9 *Let $A \in \mathscr{L}$ such that $\lambda(A) < +\infty$; then $f \in \mathfrak{L}(A)$ if and only if, for every $\varepsilon > 0$, there exists a compact set $K_\varepsilon \subseteq A$ such that $\lambda(A \setminus K_\varepsilon) < \varepsilon$ and $f|_{K_\varepsilon}$ is continuous.*

Proof

We need only to prove that the condition is necessary. For any $n \in \mathbb{N}$, let $A_n = A\cap]-n, n[\in \mathscr{L}(A)$; then $A = \bigcup_{n=1}^\infty A_n$; hence $\lambda(A) = \lim_n \lambda(A_n)$. For every $\varepsilon > 0$, there exists $n_0 \in \mathbb{N}$ such that $\lambda(A \setminus A_{n_0}) = \lambda(A) - \lambda(A_{n_0}) < \frac{\varepsilon}{2}$ (see (3) of Theorem 1.3.11). From the previous theorem, there is a closed set $F_\varepsilon \subseteq A_{n_0}$ such that $\lambda(A_{n_0} \setminus F_\varepsilon) < \frac{\varepsilon}{2}$ and $f|_{F_\varepsilon}$ is continuous. Then $\lambda(A \setminus F_\varepsilon) < \varepsilon$ and F_ε is a bounded closed set and hence compact. ■

We will present in the following another theorem to approximate measurable functions with continuous functions.

Theorem 2.3.10 (Borel)
Let $A \in \mathscr{L}$; for every $f \in \mathfrak{L}(A)$ and every $\varepsilon > 0$, there exists a continuous function $f_\varepsilon : A \to \mathbb{R}$, such that $\lambda(f \neq f_\varepsilon) < \varepsilon$. Moreover, we can choose f_ϵ so that $\sup_{x\in A} |f_\varepsilon(x)| \leq \sup_{x\in A} |f(x)|$.

Proof

Let $f \in \mathfrak{L}(A)$ and let $\alpha = \sup_{x\in A} |f(x)| \in [0, +\infty]$; then $f(A) \subseteq I = [-\alpha, \alpha]$. According to the Lusin theorem, for every $\varepsilon > 0$, there exists a closed set $F_\varepsilon \subseteq A$ such that $f|_{F_\varepsilon}$ is continuous and $\lambda(A \setminus F_\varepsilon) < \varepsilon$; $f(F_\varepsilon) \subseteq [-\alpha, \alpha]$. We will now use a classic theorem of topology, the Tietze extension theorem: *For every closed set $F \subseteq \mathbb{R}$ and every continuous mapping $f : F \to \mathbb{R}$, there exists a continuous extension $\bar{f} : \mathbb{R} \to \mathbb{R}$ of f. Moreover, if*

$f(F) \subseteq [-\alpha, +\alpha]$, *then* $\bar{f}$ *can be chosen so that* $\bar{f}(\mathbb{R}) \subseteq [-\alpha, +\alpha]$ (for a demonstration on normal spaces, see Theorem 5.1, page 149 in [4]).

If $\alpha < +\infty$, let $\bar{f} : \mathbb{R} \to [-\alpha, +\alpha]$ be a continuous function on $\mathbb{R}$ such that $\bar{f}|_{F_\varepsilon} = f$, and let $f_\varepsilon = \bar{f}|_A : A \to \mathbb{R}$. It's obvious that $\lambda(f \neq f_\varepsilon) \leq \lambda(A \setminus F_\varepsilon) < \varepsilon$ and $\sup_{x\in\mathbb{R}} |f_\varepsilon(x)| \leq \alpha = \sup_{x\in\mathbb{R}} |f(x)|$.

If $\alpha = +\infty$, then $\sup_{x\in A} |f_\varepsilon(x)| \leq +\infty = \sup_{x\in F} |f(x)|$. ■

Example 2.3.11

Let $\mathbb{Q} = \{q_1, q_2, \ldots, q_n, \ldots\} \subseteq \mathbb{R}$ be the set of rational numbers; then $\mathbb{Q} \in \mathcal{B}_u$, and so $\chi_{\mathbb{Q}} \in \mathfrak{L}(\mathbb{R})$ (see Remark 2.1.9). For every $\varepsilon > 0$ and any $n \in \mathbb{N}^*$, let

$$I_n^\varepsilon = \left]q_n - \frac{\varepsilon}{2^{n+1}}, q_n + \frac{\varepsilon}{2^{n+1}}\right[.$$

Then $\mathbb{Q} \subseteq \cup_{n=1}^\infty I_n^\varepsilon = D_\varepsilon \in \tau_u$. $F_\varepsilon = \mathbb{R} \setminus D_\varepsilon \subseteq \mathbb{R} \setminus \mathbb{Q}$ is a closed set. $\lambda(\mathbb{R} \setminus F_\varepsilon) = \lambda(D_\varepsilon) \leq \sum_{n=1}^\infty \frac{\varepsilon}{2^n} = \varepsilon$. $\chi_{\mathbb{Q}}(x) = 0$, for every $x \in F_\varepsilon$, then $\chi_{\mathbb{Q}}$ is continuous on this set. We note that the function $\chi_{\mathbb{Q}}$ is not continuous in any point of $\mathbb{R}$.

The function $f_\varepsilon = \underline{0}$ is continuous on $\mathbb{R}$ and $\lambda^*(\chi_{\mathbb{Q}} \neq f_\varepsilon) = 0 < \varepsilon$.

2.4 Abstract Setting

Just like at the end of ▶ Chap. 1, we will present the measurable functions and their properties on an abstract measure space.

Definition 2.4.1

Let $(X, \mathcal{A}, \gamma)$ be a measure space where γ is a measure σ-finite and complete; the function $f : X \to \mathbb{R}$ is $\mathcal{A}$-**measurable** if, for any $a \in \mathbb{R}$, $f^{-1}(]-\infty, a[) \in \mathcal{A}$.
We denote $\mathcal{M}(X)$ (or simply with $\mathcal{M}$ when there is no danger of confusion) the set of all measurable functions on X.

For every $A \subseteq X$, $\chi_A \in \mathcal{M}$ if and only if $A \in \mathcal{A}$.

A function $\mathcal{A}$-**simple** (or simple, if there is no risk of confusion) is a function $f : X \to \mathbb{R}$ for which $f(X) = \{a_1, \ldots, a_p\} \subseteq \mathbb{R}$ and $A_i = f^{-1}(\{a_i\}) \in \mathcal{A}$, for any $i = 1, \ldots, p$. We denote with $\mathcal{E}(X)$ the set of all simple functions; for every $f \in \mathcal{E}(X)$, $f = \sum_{i=1}^p a_i \cdot \chi_{A_i}$, where $\{A_1, \ldots, A_p\}$ is a partition $\mathcal{A}$-measurable of X. Obviously, $\mathcal{E}(X) \subseteq \mathcal{M}(X)$.

If X is provided with a topology τ and if $\tau \subseteq \mathcal{A}$, then the real continuous functions on (X, τ) are measurable ($C(X) \subseteq \mathcal{M}(X)$).

If we replace correctly A by X, $\mathscr{L}$ by $\mathcal{A}$, and $\mathscr{L}(A)$ by $\mathcal{M}(X)$, then the results from Theorems 2.1.10, 2.1.13, 2.1.18, 2.1.20, and Proposition 2.1.22 are preserved. A set $A \subseteq X$ is a null set if $A \in \mathcal{A}$ and $\gamma(A) = 0$. Since γ is a complete measure, every subset of a null set is a null set. A property P is satisfied γ-almost everywhere (almost everywhere or a.e.) on X if $\{x \in X : x$ does not have the property $P\}$ is a null set.

Theorem 2.4.2
Let $f : X \to \mathbb{R}$; the following are equivalent:
(1) $f \in \mathcal{M}(X)$.
(2) $f^{-1}(]-\infty, a]) \in \mathcal{A}$, *for any* $a \in \mathbb{R}$.
(3) $f^{-1}(]a, +\infty[) \in \mathcal{A}$, *for any* $a \in \mathbb{R}$.
(4) $f^{-1}([a, +\infty[) \in \mathcal{A}$, *for any* $a \in \mathbb{R}$.
(5) $f^{-1}(I) \in \mathcal{A}$, *for every* $I \in \mathcal{I}$.
(6) $f^{-1}(D) \in \mathcal{A}$, *for every* $D \in \tau_u$.
(7) $f^{-1}(B) \in \mathcal{A}$, *for every* $B \in \mathcal{B}_u$.

Theorem 2.4.3
Let $(f_n) \subseteq \mathcal{M}(X)$*; then:*
(1) $f = \sup_n f_n \in \mathcal{M}(X)$, *if* $\sup_n f_n(x) < +\infty$, *for every* $x \in X$.
(2) $f = \inf_n f_n \in \mathcal{M}(X)$, *if* $\inf_n f_n(x) > -\infty$, *for every* $x \in X$.
(3) $f = \limsup_n f_n \in \mathcal{M}(X)$, *if* $\limsup_n f_n(x) \in \mathbb{R}$, *for every* $x \in X$.
(4) $f = \liminf_n f_n \in \mathcal{M}(X)$, *if* $\liminf_n f_n(x) \in \mathbb{R}$, *for every* $x \in X$.
(5) *If* $f_n(x) \to f(x) \in \mathbb{R}$, *for every* $x \in X$, *then* $f \in \mathcal{M}(X)$.
(6) *If* $f_n \underset{X}{\dot{\to}} f$, *then* $f \in \mathcal{M}(X)$.

Theorem 2.4.4
Let $f, g : X \to \mathbb{R}$.
(1) *If* $f \in \mathcal{M}(X)$ *and* $f = g$ *a.e., then* $g \in \mathcal{M}(X)$.
(2) *If* f *is continuous a.e. on* X, *then* $f \in \mathcal{M}(X)$.

Theorem 2.4.5
Let $f, g \in \mathcal{M}(X)$ *and let* $\alpha \in \mathbb{R}$*; then* $f + g \in \mathcal{M}(X)$, $\alpha \cdot f \in \mathcal{M}(X)$ *and* $f \cdot g \in \mathcal{M}(X)$.

Proposition 2.4.6
(1) $f \in \mathcal{M}(X)$ *if and only if* $f^+ \in \mathcal{M}(X)$ *and* $f^- \in \mathcal{M}(X)$.
(2) *If* $f \in \mathcal{M}(X)$, *then* $|f| \in \mathcal{M}(X)$.

We can define, as in Definition 2.2.1, the convergence in measure and almost uniform convergence, and we find the results Theorems 2.2.2, 2.2.4, 2.2.5, 2.2.8, Corollary 2.2.6, Remark 2.2.7 and Corollary 2.2.9.

Definition 2.4.7

Let $(f_n) \subseteq \mathcal{M}(X)$ and $f \in \mathcal{M}(X)$.

1. (f_n) converges **almost uniformly** to f on X ($f_n \xrightarrow[X]{\text{a.u.}} f$) if, for every $\varepsilon > 0$, there exists $A_\varepsilon \in \mathcal{A}$ such that $\gamma(A_\varepsilon) < \varepsilon$ and $f_n \xrightarrow[X \setminus A_\varepsilon]{\text{u}} f$.
2. (f_n) converges **in measure** to f on X ($f_n \xrightarrow[X]{\gamma} f$) if, for every $\varepsilon > 0$, $\lim_n \gamma((|f_n - f| \geq \varepsilon)) = 0$.
   The sequence $(f_n)_n$ is **convergent in measure** on X if there is $f \in \mathcal{M}(X)$ such that $f_n \xrightarrow[X]{\gamma} f$.
3. (f_n) is **Cauchy in measure** if, for every $\varepsilon > 0$, $\lim_{m,n\to\infty} \gamma((|f_m - f_n| \geq \varepsilon)) = 0$.

Theorem 2.4.8

Let $(f_n) \subseteq \mathcal{M}(X)$ and $f \in \mathcal{M}(X)$; then:

(1) *If $f_n \xrightarrow[X]{\text{a.u.}} f$, then $f_n \xrightarrow[X]{\gamma} f$.*

(2) *If $f_n \xrightarrow[X]{\text{a.u.}} f$, then $f_n \xrightarrow[X]{\cdot} f$.*

(3) *Every sequence convergent in measure is Cauchy in measure on X.*

Theorem 2.4.9

Let $(f_n) \subseteq \mathcal{M}(X)$, $f, g \in \mathcal{M}(X)$.

(1) *Let $f_n \xrightarrow[X]{\gamma} f$; then $f_n \xrightarrow[X]{\gamma} g$ if and only if $f = g$ a.e.*

(2) *Let $f_n \xrightarrow[X]{\cdot} f$; then $f_n \xrightarrow[X]{\cdot} g$ if and only if $f = g$ a.e.*

(3) *Let $f_n \xrightarrow[X]{\text{a.u.}} f$; then $f_n \xrightarrow[X]{\text{a.u.}} g$ if and only if $f = g$ a.e.*

Theorem 2.4.10 (Riesz)

(1) *Every sequence Cauchy in measure on X has a subsequence almost uniformly convergent on X.*

(2) *If $f_n \xrightarrow[X]{\gamma} f$, then there exists $k_n \uparrow +\infty$ such that $f_{k_n} \xrightarrow[X]{\text{a.u.}} f$.*

(3) *Every sequence Cauchy in measure on X is convergent in measure.*

Corollary 2.4.11 *If $f_n \xrightarrow[X]{\gamma} f$, then there exists $k_n \uparrow +\infty$ such that $f_{k_n} \xrightarrow[X]{\cdot} f$.*

Theorem 2.4.12 (Egoroff)
If $\gamma(X) < +\infty$, $(f_n)_n \subseteq \mathcal{M}(X)$, *and* $f_n \underset{X}{\overset{\cdot}{\to}} f$, *then* $f_n \underset{X}{\overset{\text{a.u.}}{\longrightarrow}} f$.

Corollary 2.4.13 *If* $f_n \underset{X}{\overset{\cdot}{\to}} f$ *and* $\gamma(X) < +\infty$, *then* $f_n \underset{X}{\overset{\gamma}{\to}} f$.

In this abstract framework, we can also prove the theorem of approximation of measurable functions with simple functions (see Theorem 2.3.3).

Theorem 2.4.14
(1) *If* $f : X \to \mathbb{R}_+$, $f \in \mathcal{M}_+(X)$, *there exists* $(f_n) \subseteq \mathcal{E}_+(X)$, $f_n \uparrow f$.
(2) *If* $f \in \mathcal{M}(X)$, *there exists* $(f_n) \subseteq \mathcal{E}(X)$, $f_n \underset{X}{\overset{p}{\to}} f$.
(3) *If* $f \in \mathcal{M}(X)$ *is bounded, there exists* $(f_n) \subseteq \mathcal{E}(X)$, $f_n \underset{X}{\overset{u}{\to}} f$.

Let (X, τ) be a normal topological space (see [4, page 144]; particularly, X can be a metric space or even a normed space), let $\mathcal{A}$ be a σ-algebra on X such that $\tau \subseteq \mathcal{A}$, and let γ be a σ-finite complete regular measure on $(X, \mathcal{A})$; in this context we can formulate the generalizations of the theorems of Lusin and Borel.

Theorem 2.4.15 (Lusin)
$f \in \mathcal{M}(X)$ *if and only if, for every* $\varepsilon > 0$, *there exists a closed set* $F_\varepsilon \subseteq X$ *such that* $\gamma(X \setminus F_\varepsilon) < \varepsilon$ *and* $f|_{F_\varepsilon}$ *is continuous.*

Theorem 2.4.16 (Borel)
For every $f \in \mathcal{M}(X)$ *and every* $\varepsilon > 0$, *there exists a continuous function* $f_\varepsilon : X \to \mathbb{R}$ *such that* $\gamma(f \neq f_\varepsilon) < \varepsilon$.
Moreover, we can choose f_ϵ *so that* $\sup_{x\in X} |f_\varepsilon(x)| \leq \sup_{x\in X} |f(x)|$.

2.5 Exercises

(1) Let $(A_n)_n \subseteq \mathscr{L}(\mathbb{R})$ be a sequence of pairwise disjoint measurable sets ($A_n \cap A_m = \emptyset$, for all $n \neq m$), and let $(a_n)_n \subseteq \mathbb{R}$. Show that the function $f : \mathbb{R} \to \mathbb{R}$, defined by $f(x) = \sum_{n=0}^{\infty} a_n \cdot \chi_{A_n}(x)$, for every $x \in \mathbb{R}$, is measurable.

(2) Let $A \subseteq \mathbb{R}$ be a non-measurable Lebesgue set ($A \notin \mathscr{L}(\mathbb{R})$), and let $f : \mathbb{R} \to \mathbb{R}$, defined by $f(x) = \begin{cases} 1, & x \in A \\ -1, & x \notin A \end{cases}$. Show that $|f| \in \mathfrak{L}(\mathbb{R})$ but $f \notin \mathfrak{L}(\mathbb{R})$.

(3) Check if the Riemann function: $f : [0, 1] \to \mathbb{R}$,
$f(x) = \begin{cases} 0, & x \in (\mathbb{R} \setminus \mathbb{Q}) \cap [0, 1] \cup \{0\} \\ \frac{1}{q}, & x = \frac{p}{q} \in [0, 1],\ p, q \in \mathbb{N}^*,\ p, q \text{ are relatively prime} \end{cases}$,
is Lebesgue measurable.

Indication. Show that f is continuous at all irrational points and in 0 and discontinuous at rational points of $]0, 1]$.

(4) Let $f : A \to \mathbb{R}$, $f \in \mathfrak{L}(A)$ and let $B \in \mathscr{L}(A)$; show that the function $g : A \to \mathbb{R}$, defined by $g(x) = \begin{cases} 0, & x \in A \setminus B \\ f(x), & x \in B \end{cases}$, is Lebesgue measurable.

(5) Let $A \in \mathscr{L}, a, b \in \mathbb{R}, a \neq 0$, $f : aA+b \to \mathbb{R}$ and $g : A \to \mathbb{R}$, $g(x) = f(ax+b)$, for every $x \in A$. Show that $f \in \mathfrak{L}(aA + b)$ if and only if $g \in \mathfrak{L}(A)$.

Indication. Show that, for any $\alpha \in \mathbb{R}$, $g^{-1}(]-\infty, \alpha[) = -\frac{b}{a} + \frac{1}{a} f^{-1}(]-\infty, \alpha[)$, and use Theorem 1.3.15.

(6) Let $I \in \mathcal{I}$, and let $f : I \to \mathbb{R}$ be a differentiable function on open interval I; show that the derivative of f, f', is Lebesgue measurable.

Indication. Show that, for every $x \in \mathbb{R}$, $f'(x) = \lim_{n\to\infty} n \cdot \left[f\left(x + \frac{1}{n}\right) - f(x) \right]$.

(7) Let $A \in \mathscr{L}$, $f \in \mathfrak{L}(A)$ and let $a \in \mathbb{R}$; show that $(f = \underline{a}) = \{x \in A : f(x) = a\} \in \mathscr{L}$.

(8) Let $A \in \mathscr{L}$ and let $f, g \in \mathfrak{L}(A)$; show that $(f < g) = \{x \in A : f(x) < g(x)\} \in \mathscr{L}$.

(9) Let $A \in \mathscr{L}$, $f \in \mathfrak{L}(A)$, and let $\|f\| = \inf\{\alpha + \lambda(|f| > \alpha) : \alpha > 0\}$.
Show that:
(a) $\|f\| = 0$ if and only if $f = \underline{0}$ a.e.,
(b) $\|-f\| = \|f\|$, for every $f \in \mathfrak{L}(A)$,
(c) $\|f + g\| \leq \|f\| + \|g\|$, for every $f, g \in \mathfrak{L}(A)$,
(d) $f_n \xrightarrow[A]{\lambda} f$ if and only if $\|f_n - f\| \to 0$.

Indication. See Remark 2.2.7

(10) Check if the following sequences are convergent a.e., almost uniform, or in measure on their definition sets:
(a) $f_n : \mathbb{R} \to \mathbb{R}$, $f_n = \frac{1}{n} \cdot \chi_{[0, n]}$.
(b) $f_n : \mathbb{R} \to \mathbb{R}$, $f_n = n \cdot \chi_{[0, \frac{1}{n}]}$.

(11) Show that the series $\sum_{n=1}^{\infty} \frac{\sin nx}{n}$ converges on $\mathbb{R}$ to a measurable function.

(12) Show that the sequence $(f_n)_n \subseteq \mathbb{R}$, $f_n(x) = \sin nx$, is not convergent in measure to $\underline{0}$.

(13) Let $A \in \mathscr{L}$ with $\lambda(A) < +\infty$.
(a) $f \in \mathscr{L}(A) \Longrightarrow \forall \varepsilon > 0, \exists k > 0$ such that $\lambda(|f| \geq k) < \varepsilon$.
(b) $f_n \xrightarrow[A]{\lambda} f \Longrightarrow \forall \varepsilon > 0, \exists k > 0$ such that $\lambda(|f_n| \geq k) < \varepsilon, \forall n \in \mathbb{N}$.

(c) $f_n \underset{A}{\overset{\lambda}{\to}} f \Longrightarrow f_n^2 \underset{A}{\overset{\lambda}{\to}} f^2$.

(d) $f_n \underset{A}{\overset{\lambda}{\to}} f,\ g_n \underset{A}{\overset{\lambda}{\to}} g \Longrightarrow f_n g_n \underset{A}{\overset{\lambda}{\to}} fg$.

Indications.

(a) See (ii) of Remarks 2.3.7.

(b) We will be using inclusion $(|f_n| \geq k+1) \subseteq (|f_n - f| \geq 1) \cup (|f| \geq k)$ and point (a).

(c) We will be using inclusion $(|f_n^2 - f^2| \geq \varepsilon) \subseteq (|f_n - f| \geq \frac{\varepsilon}{2k}) \cup (|f_n| \geq k) \cup (|f| \geq k)$ and point (b).

(d) We will take into account the relation $f_n g_n = \frac{1}{4}[(f_n + g_n)^2 - (f_n - g_n)^2]$ and point (c).

(14) Show that the sequence $(f_n)_n \subseteq \mathfrak{L}(\mathbb{R})$, defined by $f_n(x) = x + \frac{1}{n}$, is convergent in measure to the function $f \in \mathfrak{L}(\mathbb{R})$, $f(x) = x$, but $(f_n^2)_n$ is not convergent in measure to f^2. Explain the result.

(15) Show that, for every $A \in \mathscr{L}$ and every $f \in \mathfrak{L}_+(A)$, $f = \sup\{\varphi : \varphi \in \mathcal{E}_+(A), \varphi \leq f\}$.

Lebesgue Integral

In this chapter, we will build the Lebesgue integral, first for the measurable and positive functions and then for the measurable functions in general.

We will show that the family of Lebesgue integrable functions on a set $A \in \mathscr{L}$ is organized as a vector subspace of the space $\mathfrak{L}(A)$ and that the integral is a linear operator on this space.

We will present the main properties of the class of integrable functions and of the integral; among these stand out the properties of passing to the limit under the integral. At the end of the chapter, we will carry out a comparative study of the two integrals, Riemann and Lebesgue.

3.1 Integrals of Nonnegative Measurable Functions

Let $A \in \mathscr{L}$ and let $\mathcal{E}_+(A)$ be the set of all simple and positive functions on A; if $f \in \mathcal{E}_+(A)$, then $f = \sum_{i=1}^{p} a_i \chi_{A_i}$, where $\{a_i : i = 1, \ldots, p\} \subseteq \mathbb{R}_+$, $a_i \neq a_j$, for any $i \neq j$ and $A_i = f^{-1}(\{a_i\}) \in \mathscr{L}(A)$, for every $i = 1, \ldots, p$. Suppose that, among the values a_i, there is also the value 0; then $\{A_i : i = 1, \ldots, p\}$ is a $\mathscr{L}$-measurable partition of the set A.

Definition 3.1.1

Let $f \in \mathcal{E}_+(A)$, $f = \sum_{i=1}^{p} a_i \chi_{A_i}$; we say that $\displaystyle\int_A f d\lambda = \sum_{i=1}^{p} a_i \lambda(A_i) \in [0, +\infty]$ is the **integral** of f over the set A. The function f is **integrable** over A if $\int_A f d\lambda < +\infty$. We denote with $\mathcal{E}_+^1(A)$ the set of all simple positive integrable functions over A.

We remark that $\sum_{i=1}^{p} a_i \lambda(A_i) < +\infty$ if and only if, for those i for whom $\lambda(A_i) = +\infty$, we have $a_i = 0$ (we will agree that $0 \cdot (+\infty) = 0$).

(Continued)

L. C. Florescu, *Lebesgue Integral*, Compact Textbooks in Mathematics,
https://doi.org/10.1007/978-3-030-60163-8_3

Definition 3.1.1 (continued)

If $B \in \mathscr{L}$, $B \subseteq A$, then the restriction of f on the set B is $f|_B = \sum_{i=1}^{p} a_i \chi_{A_i \cap B} \in \mathcal{E}_+(B)$ (we recall that we identify $\chi_{A_i \cap B}$ with the restrictions of these characteristic functions to the set B).
It is obvious that $\int_B f|_B d\lambda = \sum_{i=1}^{p} a_i \lambda(A_i \cap B) = \int_A f \cdot \chi_B d\lambda \leq \int_A f d\lambda$; we will denote this integral with $\int_B f d\lambda$. If $f \in \mathcal{E}_+^1(A)$, then $f|_B \in \mathcal{E}_+^1(B)$.

If $B \in \mathscr{L}(A)$, then $\chi_B = \chi_B + 0 \cdot \chi_{A \setminus B} \in \mathcal{E}_+(A)$ and $\int_A \chi_B d\lambda = \lambda(B)$; $\chi_B \in \mathcal{E}_+^1(A)$ if and only if $\lambda(B) < +\infty$.

Remarks 3.1.2

(i) If $f = \sum_{i=1}^{p} a_i \chi_{A_i} = \sum_{j=1}^{q} b_j \chi_{B_j}$, where $\{A_i : i = 1, \ldots, p\}$ and $\{B_j : j = 1, \ldots, q\}$ are measurable partitions of A, then

$$\sum_{i=1}^{p} a_i \lambda(A_i) = \sum_{i=1}^{p} \sum_{j=1}^{q} a_i \lambda(A_i \cap B_j) = \sum_{i=1}^{p} \sum_{j=1}^{q} b_j \lambda(A_i \cap B_j) = \sum_{j=1}^{q} b_j \lambda(B_j).$$

The equality in the middle occurs because, for all (i, j) for which $A_i \cap B_j \neq \emptyset$, $a_i = b_j$. Thus, the integral of the function f is well defined.

(ii) If $f = \sum_{i=1}^{p} a_i \chi_{A_i}, g = \sum_{j=1}^{q} b_j \chi_{B_j} : A \to \mathbb{R}$ and $f = g$ a.e., then $\int_A f d\lambda = \int_A g d\lambda$. Indeed, since $f = g$ a.e., $\lambda(A_i \cap B_j) = 0$, for all i, j for which $a_i \neq b_j$. Then $\int_A f d\lambda = \sum_{a_i = b_j} a_i \lambda(A_i \cap B_j) + \sum_{a_i \neq b_j} a_i \lambda(A_i \cap B_j) = \sum_{a_i = b_j} b_j \lambda(A_i \cap B_j) = \int_B g d\lambda$.

Proposition 3.1.3 *Let* $A \in \mathscr{L}$, $c \geq 0$, *and* $f, g \in \mathcal{E}_+(A)$; *then*

(1) $cf \in \mathcal{E}_+(A)$ *and* $\int_A cf d\lambda = c \int_A f d\lambda$.
(2) $f + g \in \mathcal{E}_+(A)$ *and* $\int_A (f + g) d\lambda = \int_A f d\lambda + \int_A g d\lambda$.
(3) $\int_A f d\lambda \leq \int_A g d\lambda$ *if* $f \leq g$ *a.e.*
(4) $\int_B f d\lambda \leq \int_C f d\lambda$, *for every* $B, C \in \mathscr{L}$, $B \subseteq C \subseteq A$.
(5) *For every* $\varepsilon > 0$, *there exists* $\delta > 0$ *such that* $\int_B f d\lambda < \varepsilon$, *for any* $B \in \mathscr{L}(A)$ *with* $\lambda(B) < \delta$.

Proof

If $f = \sum_{i=1}^{p} a_i \chi_{A_i}$, then $cf = \sum_{i=1}^{p} c a_i \chi_{A_i}$, and therefore (1) is obvious.

(2) Let $f = \sum_{i=1}^{p} a_i \chi_{A_i}, g = \sum_{j=1}^{q} b_j \chi_{B_j}$, where $\{A_i : i = 1, \ldots, p\}$ and $\{B_j : j = 1, \ldots, q\}$ are measurable partitions of A; then

$$f + g = \sum_{i=1}^{p} \sum_{j=1}^{q} a_i \chi_{A_i \cap B_j} + \sum_{i=1}^{p} \sum_{j=1}^{q} b_j \chi_{A_i \cap B_j} = \sum_{i=1}^{p} \sum_{j=1}^{q} (a_i + b_j) \chi_{A_i \cap B_j};$$

therefore $f + g \in \mathcal{E}_+(A)$ and

$$\int_A (f+g)d\lambda = \sum_{i=1}^{p}\sum_{j=1}^{q}(a_i+b_j)\lambda(A_i\cap B_j) =$$

$$= \sum_{i=1}^{p} a_i\left(\sum_{j=1}^{q}\lambda(A_i\cap B_j)\right) + \sum_{j=1}^{q} b_j\left(\sum_{i=1}^{p}\lambda(A_i\cap B_j)\right) =$$

$$= \sum_{i=1}^{p} a_i\lambda(A_i) + \sum_{j=1}^{q} b_j\lambda(B_j) = \int_A f d\lambda + \int_A g d\lambda.$$

(3) Let $f = \sum_{i=1}^{p} a_i\chi_{A_i}$, $g = \sum_{j=1}^{q} b_j\chi_{B_j}$, where $\{A_i : i = 1,\dots,p\}$ and $\{B_j : j = 1,\dots,q\}$ are measurable partitions of A. Since $f \le g$ a.e., $\sum_{i=1}^{p}\sum_{j=1}^{q} a_i\chi_{A_i\cap B_j} \le \sum_{i=1}^{p}\sum_{j=1}^{q} b_j\chi_{A_i\cap B_j}$ a.e.; we notice that, for any pair (i,j) for which $a_i > b_j$, $\lambda(A_i\cap B_j) = 0$. Therefore

$$\int_A f d\lambda = \sum_{i=1}^{p}\sum_{j=1}^{q} a_i\lambda(A_i\cap B_j) = \sum_{(a_i\le b_j)} a_i\lambda(A_i\cap B_j) \le$$

$$\le \sum_{(a_i<b_j)} b_j\lambda(A_i\cap B_j) = \sum_{i=1}^{p}\sum_{j=1}^{q} b_j\lambda(A_i\cap B_j) = \int_A g d\lambda.$$

(4) If we remark that $\int_B f d\lambda = \int_A f\cdot\chi_B d\lambda$, $\int_C f d\lambda = \int_A f\cdot\chi_C d\lambda$ and that $f\chi_B \le f\chi_C$, then (4) results from (3).

5). Let $f = \sum_{i=1}^{p} a_i\chi_{A_i}$; if $f = \underline{0}$, then the condition is obviously verified.

If $f \ne \underline{0}$, then let $M = \max\{a_i : i = 1,\dots,p\} > 0$. For every $\varepsilon > 0$, there exists $\delta = \frac{\varepsilon}{M} > 0$ such that, for every $B \in \mathscr{L}(A)$ with $\lambda(B) < \delta$, $\int_B f d\lambda = \sum_{i=1}^{p} a_i\lambda(A_i\cap B) \le M\cdot\sum_{i=1}^{p}\lambda(A_i\cap B) = M\cdot\lambda(B) < M\cdot\delta = \varepsilon$. ■

For every $A \in \mathscr{L}$ and every $f \in \mathfrak{L}_+(A)$, $f = \sup\{\varphi : \varphi \in \mathcal{E}_+(A), \varphi \le f\}$ (see Exercise 15) of 2.5). Therefore, the following definition of the integral for the measurable and positive functions is as natural as possible:

Definition 3.1.4

For every $f \in \mathfrak{L}_+(A)$ we say that

$$\int_A f d\lambda = \sup\left\{\int_A \varphi d\lambda : \varphi \in \mathcal{E}_+(A), \varphi \le f\right\} \in [0, +\infty]$$

(Continued)

Definition 3.1.4 (continued)

is the **integral** of f over the measurable set A. The function f is **integrable** over A if $\int_A f d\lambda < +\infty$. $\mathfrak{L}^1_+(A)$ denotes the set of all positive and integrable functions over A.

If $B \in \mathscr{L}(A)$, then $f\chi_B \in \mathfrak{L}_+(A)$ and $\int_A f\chi_B d\lambda = \int_B f d\lambda$. The restriction of f to B, $f|_B$, is positive and measurable on B (see Definition 2.1.2). If $f \in \mathfrak{L}^1_+(A)$, then $f|_B \in \mathfrak{L}^1_+(B)$ and $\int_B f|_B d\lambda = \int_B f d\lambda$.

Remarks 3.1.5

(i) The definition of the integral for the measurable and positive functions does not contradict that of the integral for simple and positive functions. Indeed, if $f \in \mathcal{E}_+(A)$, then $\int_A f d\lambda$ is the biggest element of the set $\left\{\int_A \varphi d\lambda : \varphi \in \mathcal{E}_+(A), \varphi \leq f\right\}$ and then the upper bound for this set. One can notice that $\mathcal{E}^1_+(A) \subseteq \mathfrak{L}^1_+(A)$.

(ii) $\displaystyle\int_A f d\lambda = \sup\left\{\int_A \varphi d\lambda : \varphi \in \mathcal{E}_+(A), \varphi \leq f \text{ a.e.}\right\}$. Indeed, if $\varphi \in \mathcal{E}_+(A)$ and $\varphi \leq f$ a.e., then $\lambda((\varphi > f)) = 0$. Therefore $\varphi_1 = \varphi \cdot \chi_{(\varphi \leq f)} \in \mathcal{E}_+(A), \varphi_1 \leq f$ and $\int_A \varphi_1 d\lambda = \int_A \varphi d\lambda$.

In the following proposition, we highlight certain immediate properties of the integral of the measurable and positive functions.

Proposition 3.1.6 *Let $f, g \in \mathfrak{L}_+(A)$ and $c \geq 0$; then*

(1) *$cf \in \mathfrak{L}_+(A)$ and $\int_A cf d\lambda = c\int_A f d\lambda$.*

(2) *$\int_A f d\lambda \leq \int_A g d\lambda$, if $f \leq g$ a.e..*

(3) *$\int_B f d\lambda \leq \int_C f d\lambda$, for every $B, C \in \mathscr{L}, B \subseteq C \subseteq A$.*

Proof

(1) If $c = 0$, then $cf = \underline{0}$ and so $\int_A cf d\lambda = c\int_A f d\lambda$.
If $c > 0$, then

$$\int_A cf d\lambda = \sup\left\{\int_A \varphi d\lambda : \varphi \in \mathcal{E}_+(A), \varphi \leq cf\right\} =$$

$$= \sup\left\{c\int_A \frac{1}{c}\varphi d\lambda : \varphi \in \mathcal{E}_+(A), \frac{1}{c}\varphi \leq f\right\} =$$

$$= \sup\left\{c\int_A \psi d\lambda : \psi \in \mathcal{E}_+(A), \psi \leq f\right\} = c\int_A f d\lambda.$$

(2) We can see that if $\varphi \leq f$ a.e., then $\varphi \leq g$ a.e., which leads to

$$\left\{\int_A \varphi d\lambda : \varphi \in \mathcal{E}_+(A), \varphi \leq f \text{ a.e.}\right\} \subseteq \left\{\int_A \varphi d\lambda : \varphi \in \mathcal{E}_+(A), \varphi \leq g \text{ a.e.}\right\}.$$

The inequality results from (ii) of Remark 3.1.5.

(3) We remark that $f\chi_B \leq f\chi_C$ and then we use (2). ■

The following theorem plays an extremely important role in Lebesgue integral theory.

Theorem 3.1.7 (Monotone Convergence Theorem)
Let $A \in \mathcal{L}$, $f : A \to \mathbb{R}_+$, *and let* $(f_n) \subseteq \mathfrak{L}_+(A)$ *be an increasing sequence such that* $f_n \uparrow f$; *then* $f \in \mathfrak{L}_+(A)$ *and* $\int_A f_n d\lambda \uparrow \int_A f d\lambda$.

Proof
It is obvious that $f \in \mathfrak{L}_+(A)$ (see 5) of 2.1.18) and, since $f_n \leq f$, for any $n \in \mathbb{N}$, $\int_A f_n d\lambda \leq \int_A f d\lambda$. In addition, from (2) of the previous proposition, the sequence $\left(\int_A f_n d\lambda\right)_{n\in\mathbb{N}}$ is increasing in $[0, +\infty]$, and therefore there exists $\lim_n \int_A f_n d\lambda \in [0, +\infty]$; then

$$\lim_n \int_A f_n d\lambda \leq \int_A f d\lambda \tag{1}$$

Let $t \in]0, 1[$ be an arbitrary number, and let $\varphi = \sum_{i=1}^p a_i \chi_{A_i} \in \mathcal{E}_+(A)$, $\varphi \leq f$, an arbitrary simple function but, for the time being, fixed. For any $n \in \mathbb{N}$, let $B_n = \{x \in A : f_n(x) \geq t\varphi(x)\} \in \mathcal{L}$. Then

$$B_n \subseteq B_{n+1}, \text{ for any } n \in \mathbb{N} \text{ and } \bigcup_{n=1}^{\infty} B_n = A. \tag{2}$$

The inclusion $B_n \subseteq B_{n+1}$ is a consequence of the fact that the sequence $(f_n)_n$ is increasing. The equality is shown by double inclusion; the inclusion $\subseteq$ is true because $B_n \subseteq A$, for any $n \in \mathbb{N}$. Let's show the inclusion $\supseteq$. Let $x \in A$; if $\varphi(x) = 0$, then $x \in B_n$, for any $n \in \mathbb{N}$, because the functions f_n are positive. If $\varphi(x) > 0$, then $t\varphi(x) < \varphi(x) \leq f(x)$; since $f_n(x) \uparrow f(x)$, there exists $n \in \mathbb{N}$ such that $t\varphi(x) < f_n(x)$ and so $x \in B_n$.

Now, using the measure property of continuity from below (see the property 6) of Theorem 1.3.11) and the relations (2), it follows that

$$\int_A t\varphi d\lambda = \sum_{i=1}^p ta_i\lambda(A_i) = \sum_{i=1}^p ta_i\lambda\left(\bigcup_{n=1}^{\infty}(A_i \cap B_n)\right) = \tag{3}$$

$$= \lim_n \sum_{i=1}^p ta_i\lambda(A_i \cap B_n) = \lim_n \int_{B_n} t\varphi d\lambda \leq \lim_n \int_{B_n} f_n d\lambda \leq \lim_n \int_A f_n d\lambda.$$

From (3), $\int_A \varphi d\lambda \leq \frac{1}{t} \cdot \lim_n \int_A f_n d\lambda$, and, since the function $\varphi \leq f$ is arbitrary, $\int_A f d\lambda \leq \frac{1}{t} \cdot \lim_n \int_A f_n d\lambda$. If in the previous relations, $t \to 1$, then

$$\int_A f d\lambda \leq \lim_n \int_A f_n d\lambda. \tag{4}$$

The inequalities (1) and (4) end the proof. ■

The above theorem can be extended to the case where $f : A \to [0, +\infty]$ (the function f can also take the value $+\infty$).

Corollary 3.1.8 *Let $A \in \mathscr{L}$ and let $(f_n)_n \subseteq \mathfrak{L}_+(A)$ be an increasing sequence such that $f_n \uparrow f$, where $f : A \to [0, +\infty]$. Then $Z = f^{-1}(+\infty) \in \mathscr{L}(A)$, f is measurable on A, and if we denote by*

$$\int_A f d\lambda = \begin{cases} \int_{A\setminus Z} f d\lambda, & \text{if } \lambda(Z) = 0 \\ +\infty, & \text{if } \lambda(Z) > 0 \end{cases}, \text{ then}$$

$$\lim_n \int_A f_n d\lambda = \int_A f d\lambda.$$

Proof
For every $p, n \in \mathbb{N}$, we denote $Z_n^p = \{x \in A : f_n(x) \geq p\}$; since f_n are measurable functions, $Z_n^p \in \mathscr{L}(A)$. If we remark that $Z = \bigcap_{p=1}^{\infty} \bigcup_{n=1}^{\infty} Z_n^p$, then $Z \in \mathscr{L}(A)$. Let $g_n = f_n|_{A\setminus Z}$ and $g = f|_{A\setminus Z}$. Since $g_n \uparrow g$ it follows from the previous theorem that $g \in \mathfrak{L}_+(A \setminus Z)$ and $\int_{A\setminus Z} g_n d\lambda \uparrow \int_{A\setminus Z} g d\lambda$. According to (iv) of Remark 2.1.3, f is measurable on A.

If $\lambda(Z) = 0$, then

$$\int_A f_n d\lambda = \int_{A\setminus Z} f_n d\lambda = \int_{A\setminus Z} g_n d\lambda \uparrow \int_{A\setminus Z} g d\lambda = \int_{A\setminus Z} f d\lambda.$$

If $\lambda(Z) > 0$, then, for all $n, p \in \mathbb{N}$,

$$\int_A f_n d\lambda \geq \int_Z f_n d\lambda \geq \int_{Z\cap Z_n^p} f_n d\lambda \geq p \cdot \lambda(Z \cap Z_n^p).$$

Then, using the property of continuity from below of λ,

$$\lim_n \int_A f_n d\lambda \geq p \cdot \lambda\left(Z \cap \bigcup_{n=1}^{\infty} Z_n^p\right) = p \cdot \lambda(Z), \text{ for any } p \in \mathbb{N}.$$

Therefore $\lim_n \int_A f_n d\lambda = +\infty$. ■

Corollary 3.1.9 $\int_A (f + g) d\lambda = \int_A f d\lambda + \int_A g d\lambda$, *for every* $f, g \in \mathfrak{L}_+(A)$.

Proof
Let $(f_n)_n, (g_n)_n \subseteq \mathcal{E}_+(A)$ such that $f_n \uparrow f$ and $g_n \uparrow g$ (see Theorem 2.3.3); then $f_n + g_n \uparrow f + g$, and, according to the previous theorem and Proposition 3.1.3,

$$\int_A (f+g)d\lambda = \lim_n \int_A (f_n + g_n)d\lambda = \lim_n \int_A f_n d\lambda + \lim_n \int_A g_n d\lambda =$$

$$= \int_A f d\lambda + \int_A g d\lambda.$$ ■

Corollary 3.1.10 (Beppo Levi) *Let $(f_n)_n \subseteq \mathfrak{L}_+(A)$ be a sequence such that the function series $\sum_{n=1}^{\infty} f_n$ is pointwise convergent on A, and let $f = \sum_{n=1}^{\infty} f_n$; then $f \in \mathfrak{L}_+(A)$ and*

$$\int_A f d\lambda = \sum_{n=1}^{\infty} \int_A f_n d\lambda.$$

Proof
The sequence of partial sums of the series $(s_n)_n$, defined by $s_n = \sum_{k=1}^{n} f_k$, consists of measurable functions and $s_n \uparrow f$. It follows that $f \in \mathfrak{L}_+(A)$ and $\int_A s_n d\lambda \uparrow \int_A f d\lambda$.

On the other hand, from the previous corollary,

$$\int_A s_n d\lambda = \sum_{k=1}^{n} \int_A f_k d\lambda \xrightarrow[n\to+\infty]{} \sum_{k=1}^{\infty} \int_A f_k d\lambda,$$

and therefore $\int_A f d\lambda = \sum_{n=1}^{\infty} \int_A f_n d\lambda$. ■

Corollary 3.1.11 *Let $f \in \mathfrak{L}_+(A)$ and let $(A_n)_n \subseteq \mathscr{L}(A)$ be a sequence of pairwise disjoint sets; then*

$$\int_{\bigcup_{n=1}^{\infty} A_n} f d\lambda = \sum_{n=1}^{\infty} \int_{A_n} f d\lambda.$$

Proof
For any $n \in \mathbb{N}^*$, let $f_n = f \cdot \chi_{A_n} \subseteq \mathfrak{L}_+(A)$; since $\sum_{n=1}^{\infty} f_n = f \cdot \chi_{\bigcup_{n=1}^{\infty} A_n} \le f$, the series $\sum_{n=1}^{\infty} f_n$ converges punctually on A.

We are in the hypotheses of the previous corollary and we therefore obtain

$$\int_{\bigcup_{n=1}^{\infty} A_n} f d\lambda = \int_A f \cdot \chi_{\bigcup_{n=1}^{\infty} A_n} d\lambda = \int_A \sum_{n=1}^{\infty} f \cdot \chi_{A_n} d\lambda = \sum_{n=1}^{\infty} \int_{A_n} f d\lambda.$$ ■

Corollary 3.1.12 (Fatou's Lemma) *Let $(f_n) \subseteq \mathfrak{L}_+(A)$ such that $f(x) = \liminf_n f_n(x) < +\infty$, for every $x \in A$; then*

$$\int_A \liminf_n f_n d\lambda \le \liminf_n \int_A f_n d\lambda.$$

Proof

We recall that $\liminf_n f_n = \sup_{n\in\mathbb{N}} \inf_{k\geq n} f_k$. If, for any $n \in \mathbb{N}$, $g_n = \inf_{k\geq n} f_k$, then $(g_n)_n \subseteq \mathfrak{L}_+(A)$ is an increasing sequence, and $f = \liminf_n f_n = \sup_{n\in\mathbb{N}} g_n = \lim_n g_n$. Therefore $f \in \mathfrak{L}_+(A)$, and, according to the monotone convergence theorem (Theorem 3.1.7), $\int_A g_n d\lambda \uparrow \int_A f d\lambda$.

On the other hand, since $g_n \leq f_n$, $\int_A g_n d\lambda \leq \int_A f_n d\lambda$, for any $n \in \mathbb{N}$, from where $\lim_n \int_A g_n d\lambda = \liminf_n \int_A g_n d\lambda \leq \liminf_n \int_A f_n d\lambda$. So $\int_A f d\lambda \leq \liminf_n \int_A f_n d\lambda$. ■

Proposition 3.1.13 *Let $A \in \mathscr{L}$ and let $f \in \mathscr{L}_+(A)$; then*

$$\int_A f d\lambda = 0 \textit{ if and only if } f = \underline{0} \textit{ a.e.}$$

Proof

($\Longrightarrow$): We suppose that $\int_A f d\lambda = 0$ and let, for any $n \in \mathbb{N}^*$, $A_n = (f \geq \frac{1}{n}) \in \mathscr{L}$. Then $A_n \subseteq A_{n+1}$, for any $n \in \mathbb{N}^*$, and $\bigcup_{n=1}^{\infty} A_n = (f > 0)$, from where $\lambda(f \neq 0) = \lambda(f > 0) = \lim_n \lambda(A_n)$.

On the other hand, for any $n \in \mathbb{N}^*$, $\int_{A_n} f d\lambda \leq \int_A f d\lambda = 0$, from where $\int_{A_n} f d\lambda = 0$. Since $\int_{A_n} f d\lambda \geq \frac{1}{n}\lambda(A_n)$, for any $n \in \mathbb{N}^*$, $\lambda(A_n) = 0$. It follows that $\lambda(f \neq 0) = 0$, and therefore $f = \underline{0}$ a.e.

($\Longleftarrow$): If $f = \underline{0}$ a.e., then, for every $\varphi \in \mathcal{E}_+(A)$ with $\varphi \leq f, \varphi = \underline{0}$ a.e. If $\varphi = \sum_{i=1}^{p} a_i \chi_{A_i}$, then, for any i for which $\lambda(A_i) \neq 0, a_i = 0$, and so $\int_A \varphi d\lambda = 0$. Since the function φ is arbitrary, $\int_A f d\lambda = 0$. ■

Corollary 3.1.14

(1) Let $A \in \mathscr{L}$, $B \in \mathscr{L}(A)$ with $\lambda(B) = 0$ and let $f \in \mathfrak{L}_+(A)$; then $\displaystyle\int_B f d\lambda = 0$.

(2) Let $A \in \mathscr{L}$ and let $f, g \in \mathfrak{L}_+(A)$ with $f \doteq g$ on A; then $\displaystyle\int_A f d\lambda = \int_A g d\lambda$.

Proof

(1) We remark that $\displaystyle\int_B f d\lambda = \int_A f \chi_B d\lambda$ and that $f\chi_B = \underline{0}$ a.e.

(2) According to Corollary 3.1.11 and the previous point 1),

$$\int_A f d\lambda = \int_{(f=g)} f d\lambda + \int_{(f\neq g)} f d\lambda = \int_{(f=g)} f d\lambda = \int_{(f=g)} g d\lambda = \int_A g d\lambda. \blacksquare$$

Remark 3.1.15 The Dirichlet function, $\chi_{\mathbb{Q}}$, is zero a.e. on $\mathbb{R}$. It follows from Proposition 3.1.13 that this function is integrable and that its integral is 0. From (3) of Proposition 3.1.6, this function is integrable over every measurable set, and its integral is 0.

Note that the Dirichlet function is everywhere discontinuous, so it is not Riemann integrable over closed intervals (see Lebesgue Criterion 7.1.7).

Theorem 3.1.16
Let $f \in \mathfrak{L}^1_+(A)$, let $\mathscr{L}(A)$ be the σ-algebra of all measurable subsets of A (see Definition 1.3.1), and let $\gamma : \mathscr{L}(A) \to \mathbb{R}_+$, $\gamma(B) = \int_B f d\lambda$, for every $B \in \mathscr{L}(A)$. Then γ is a finite measure on $\mathscr{L}(A)$ which checks the following two conditions:
(1) *For every $\varepsilon > 0$, there exists $\delta > 0$ such that $\gamma(B) = \int_B f d\lambda < \varepsilon$, for every $B \in \mathscr{L}(A)$ with $\lambda(B) < \delta$.*
(2) *For every $\varepsilon > 0$, there exists $A_0 \in \mathscr{L}(A)$ with $\lambda(A_0) < +\infty$ such that $\gamma(A \setminus A_0) = \int_{A\setminus A_0} f d\lambda < \varepsilon$.*

Proof
$\mathscr{L}(A) = \{B \in \mathscr{L} : B \subseteq A\}$ is a σ-algebra on A. We will show that γ verifies the conditions of Definition 1.4.5:

$\gamma(\emptyset) = \int_\emptyset f d\lambda = 0$ because $\lambda(\emptyset) = 0$.

Let $(B_n)_n \subseteq \mathscr{L}(A)$, $B_n \cap B_m = \emptyset$, for all $n \neq m$,and let $B = \bigcup_{n=1}^\infty B_n \in \mathscr{L}(A)$; by applying the Beppo Levi theorem (see Corollary 3.1.11),

$$\gamma(B) = \int_B f d\lambda = \int_B \left(\sum_{n=1}^\infty f\chi_{B_n}\right) d\lambda = \sum_{n=1}^\infty \int_{B_n} f d\lambda = \sum_{n=1}^\infty \gamma(B_n).$$

Therefore γ is a measure on A, and, since $\gamma(A) = \int_A f d\lambda < +\infty$, γ is finite.

(1) For any $n \in \mathbb{N}$, let $A_n = (f > n) \in \mathscr{L}(A)$; then $A_n \supseteq A_{n+1}$, for any $n \in \mathbb{N}$ and $\bigcap_{n=0}^\infty A_n = (f = +\infty) = \emptyset$. Then we remark that $f\chi_{A \setminus A_n} \uparrow f$ and, according to monotone convergence theorem, $\int_{A\setminus A_n} f d\lambda \uparrow \int_A f d\lambda$. Since $f = f\chi_{A_n} + f\chi_{A \setminus A_n}$, it follows that $\int_{A_n} f d\lambda \downarrow 0$. Then, for every $\varepsilon > 0$, there exists $n_0 \in \mathbb{N}$ such that $\int_{A_{n_0}} f d\lambda < \frac{\varepsilon}{2}$.

Let $\delta = \frac{\varepsilon}{2n_0} > 0$ and let $B \in \mathscr{L}(A)$ be an arbitrary set with $\lambda(B) < \delta$; then

$$\gamma(B) = \int_B f d\lambda = \int_{B\cap A_{n_0}} f d\lambda + \int_{B\setminus A_{n_0}} f d\lambda \le \int_{A_{n_0}} f d\lambda + \int_{B\cap(f\le n_0)} f d\lambda <$$

$$< \frac{\varepsilon}{2} + n_0 \cdot \lambda(B) < \frac{\varepsilon}{2} + n_0 \cdot \frac{\varepsilon}{2n_0} = \varepsilon.$$

(2) For any $n \in \mathbb{N}$, let $A_n = A \cap [-n, n] \in \mathscr{L}(A)$; then $A_n \subseteq A_{n+1}$, for any $n \in \mathbb{N}$ and $\bigcup_{n=1}^\infty A_n = A$. Then $f\chi_{A_n} \uparrow f$, from where $\int_{A\setminus A_n} f d\lambda \downarrow 0$. Hence, for every $\varepsilon > 0$, there exists $n_0 \in \mathbb{N}$ such that $\int_{A\setminus A_{n_0}} f d\lambda < \varepsilon$.

Let $A_0 = A_{n_0}$; $\lambda(A_0) \le 2n_0 < +\infty$ and $\int_{A\setminus A_0} f d\lambda = \gamma(A \setminus A_0) < \varepsilon$. ■

Remarks 3.1.17

(i) The property (1) is the property of **absolute continuity** of the Lebesgue integral. In fact, this property says that the measure γ is absolutely continuous with respect to the Lebesgue measure λ, $\gamma \ll \lambda$ (see Definition 6.2.1 and Proposition 6.2.3).

(ii) The property (2) shows that, for an integrable function, the integral depends on the behavior of that function on sets of finite measures.

3.2 Integrable Functions. Lebesgue Integral

For every real function $f : A \to \overline{\mathbb{R}}$, let $f^+ = \sup\{f, \underline{0}\}$ be the positive part of f and $f^- = \sup\{-f, \underline{0}\}$ be the negative part of f; $f = f^+ - f^-$ and $|f| = f^+ + f^-$ (see Definition 2.1.21). Let $A \in \mathscr{L}$; then $f \in \mathfrak{L}(A)$ if and only if $f^+, f^- \in \mathfrak{L}_+(A)$ (see Proposition 2.1.22).

Definition 3.2.1

Let $f \in \mathfrak{L}(A)$; then

(1) f has an **integral** over A if $\int_A f^+ d\lambda < +\infty$ **or** $\int_A f^- d\lambda < +\infty$, and, in this case, the integral of f over A is

$$\int_A f d\lambda = \int_A f^+ d\lambda - \int_A f^- d\lambda \in [-\infty, +\infty].$$

(2) f is **Lebesgue integrable** over A if $\int_A f^+ d\lambda < +\infty$ **and** $\int_A f^- d\lambda < +\infty$; in this case, the Lebesgue integral of f over A is

$$\int_A f d\lambda = \int_A f^+ d\lambda - \int_A f^- d\lambda \in \mathbb{R}.$$

When it is necessary to specify the variable after which the integration is done, we will also denote the integral with $\int_A f(x) d\lambda(x)$.

If there is no risk of confusion (as will be the case when we are also discussing Riemann integral or integrability), it is simply said that f has the integral over A, respectively f is integrable over A.

On denote $\mathfrak{L}^1(A)$ the set of all integrable functions over A; by definition, $f \in \mathfrak{L}^1(A)$ if and only if $f^+, f^- \in \mathfrak{L}^1_+(A)$. Of course $\mathcal{E}^1_+(A) \subseteq \mathfrak{L}^1_+(A) \subseteq \mathfrak{L}^1(A)$.

Let $B \in \mathscr{L}(A)$ and let $f \in \mathfrak{L}(A)$; f is integrable over B if and only if $f \cdot \chi_B$ is integrable over A and $\int_B f d\lambda = \int_A f \chi_B d\lambda$.

Theorem 3.2.2
Let $f \in \mathfrak{L}(A)$; $f \in \mathfrak{L}^1(A)$ if and only if $|f| \in \mathfrak{L}^1_+(A)$ and, in this case,

$$\left|\int_A f d\lambda\right| \leq \int_A |f| d\lambda.$$

Proof
($\Longrightarrow$): We suppose that f is integrable over A; then $f^+, f^- \in \mathfrak{L}^1_+(A)$, and, according to Corollary 3.1.9, $|f| = f^+ + f^- \in \mathfrak{L}^1_+(A)$.

($\Longleftarrow$): Let $|f| \in \mathfrak{L}^1_+(A)$; since $f^+, f^- \leq |f|$, $\int_A f^+ d\lambda \leq \int_A |f| d\lambda < +\infty$ and $\int_A f^- d\lambda \leq \int_A |f| d\lambda < +\infty$ (see 2) of Proposition 3.1.6). It follows that $f \in \mathfrak{L}^1(A)$. Also, using Corollary 3.1.9,

$$\left|\int_A f d\lambda\right| = \left|\int_A f^+ d\lambda - \int_A f^- d\lambda\right| \leq \int_A f^+ d\lambda + \int_A f^- d\lambda = \int_A |f| d\lambda. \qquad \blacksquare$$

We observe that, for the Lebesgue integral, the integrability of the function is equivalent with the integrability of its module; here one does not meet, as in the case of the Riemann integral, the semi-convergence phenomenon (the case where f is integrable and $|f|$ is not).

Since the module of a measurable function is a positive and measurable function, we obtain a simple condition of integrability.

Corollary 3.2.3 *Let $A \in \mathscr{L}$ and let $f : A \to \mathbb{R}$.*
(1) *If $f = \underline{0}$ a.e. over A, then $f \in \mathfrak{L}^1(A)$ and $\int_A f d\lambda = 0$.*
(2) *If $\lambda(A) = 0$, then $f \in \mathfrak{L}^1(A)$ and $\int_A f d\lambda = 0$.*

Proof
(1) Because the constant function $\underline{0}$ is continuous on A, it is measurable and, because of point (1) of Theorem 2.1.13, f is measurable. Then $|f| \in \mathfrak{L}_+(A)$; from the Proposition 3.1.13, $\int_A |f| d\lambda = 0$ ($|f| = \underline{0}$ a.e.). Therefore $|f| \in \mathfrak{L}^1_+(A)$, and, according to Theorem 3.2.2, $f \in \mathfrak{L}^1(A)$. The inequality of the same Theorem 3.2.2 shows that $\int_A f d\lambda = 0$. ■
(2) If $\lambda(A) = 0$, then $f = \underline{0}$ a.e. on A and we can apply (1).

Remark 3.2.4 Note that for functions that do not keep a constant sign, only one implication of the equivalence of Proposition 3.1.13 works: if $f \in \mathfrak{L}^1(A)$ and $\int_A f d\lambda = 0$, it does not turn out that $f = \underline{0}$ a.e. Indeed, let $f : [-1, 1] \to \mathbb{R}$, $f(x) = \begin{cases} -1, & x \in [-1, 0[, \\ 1, & x \in [0, 1]. \end{cases}$

Then $f^- = \chi_{[-1,0[}$, $f^+ = \chi_{[0,1]}$ and so $\int_{[-1,1]} f^+ d\lambda = 1 = \int_{[0,1]} f^- d\lambda$. It follows that $\int_{[-1,1]} f d\lambda = 0$ but f is not zero a.e.

Under more stringent conditions, we can always formulate a reciprocal: if the integral of a function f is zero on all the measurable subsets of A, then f is zero a.e. on A.

Theorem 3.2.5

Let $A \in \mathscr{L}$ and $f \in \mathfrak{L}^1(A)$.

(a) If $\int_B f d\lambda \geq 0$, for every $B \in \mathscr{L}(A)$, then $f \underline{\geq 0}$ a.e. on A.

(b) If $\int_B f d\lambda \leq 0$, for every $B \in \mathscr{L}(A)$, then $f \underline{\leq 0}$ a.e. on A.

(c) If $\int_B f d\lambda = 0$, for every $B \in \mathscr{L}(A)$, then $f \underline{= 0}$ a.e. on A.

Proof

(a) Let $(f < 0) = \{x \in A : f(x) < 0\}$; then $(f < 0) = \bigcup_{n=1}^{\infty} \left(f \leq -\frac{1}{n}\right)$. The sequence of sets $\left(\left(f \leq -\frac{1}{n}\right)\right)_{n\in\mathbb{N}}$ is increasing and so, according to Theorem 1.3.11),

$$\lambda(f < 0) = \lim_n \lambda\left(f \leq -\frac{1}{n}\right). \tag{*}$$

On the other hand, from the hypothesis, $\int_{(f\leq -\frac{1}{n})} f d\lambda \geq 0$ and therefore

$$0 \leq \int_{(f\leq -\frac{1}{n})} f d\lambda \leq -\frac{1}{n} \cdot \lambda\left(f \leq -\frac{1}{n}\right),$$

from where $\lambda\left(f \leq -\frac{1}{n}\right) \leq 0$ and then $\lambda\left(f \leq -\frac{1}{n}\right) = 0$, for any $n \in \mathbb{N}^*$. From $(*)$ it follows that $\lambda(f < 0) = 0$ and so $f \geq 0$ a.e. on A.

(b) is likewise proven and (c) is a consequence of points (a) and (b). ■

The following theorem provides a simple integrability condition for measurable functions.

Theorem 3.2.6

Let $f \in \mathfrak{L}(A)$ and $g \in \mathfrak{L}_+^1(A)$ such that $|f| \leq g$ a.e.; then $f \in \mathfrak{L}^1(A)$.

Proof

Since $f \in \mathfrak{L}(A)$, $|f| \in \mathfrak{L}_+(A)$, and, from (2) of Proposition 3.1.6, $\int_A |f| d\lambda \leq \int_A g d\lambda < +\infty$. Therefore $|f| \in \mathfrak{L}_+^1(A)$, and then, from Theorem 3.2.2, $f \in \mathfrak{L}^1(A)$. ■

Theorem 3.2.6 has several consequences.

Corollary 3.2.7 *Let $A \in \mathscr{L}$ and $B \in \mathscr{L}(A)$; if $f \in \mathfrak{L}^1(A)$, then $f \in \mathfrak{L}^1(B)$.*

Proof From hypothesis $f \cdot \chi_B \in \mathfrak{L}(A)$ and $|f \cdot \chi_B| \le |f|$; since $f \in \mathfrak{L}^1(A)$, Theorem 3.2.6 assures us that $f \cdot \chi_B \in \mathfrak{L}^1(A)$, which is equivalent to $f \in \mathfrak{L}^1(B)$.

Corollary 3.2.8 *Any measurable and bounded function is integrable over any measurable set of finite measure.*

Proof Let $A \in \mathscr{L}$ with $\lambda(A) < +\infty$, let $f \in \mathfrak{L}(A)$ be a measurable bounded function, and let $k > 0$ such that $|f(x)| \le k$, for every $x \in A$. The integral of the constant function $\underline{k} \in \mathcal{E}_+(A)$ is $\int_A \underline{k} d\lambda = k \cdot \lambda(A) < +\infty$. Therefore $\underline{k} \in \mathcal{E}^1_+(A) \subseteq \mathfrak{L}^1_+(A)$. Theorem 3.2.6 assures us that $f \in \mathfrak{L}^1(A)$.

Corollary 3.2.9 *Any function Riemann integrable over a bounded and closed interval $[a, b]$ is Lebesgue integrable over $[a, b]$: $\mathcal{R}_{[a,b]} \subsetneq \mathfrak{L}^1([a, b])$.*

Proof Any Riemann integrable function over $[a, b]$ is bounded and measurable (see Corollary 2.1.14); therefore, according to the previous corollary, it is Lebesgue integrable.

We noted in Remark 3.1.15 that the Dirichlet function is Lebesgue integrable on any closed interval $[a, b]$ but it is not Riemann integrable on $[a, b]$. Therefore, the inclusion in the previous corollary is strict.

As we will show later in Theorem 3.3.11, $\mathcal{R}_{[a,b]}$ is a dense subspace of $\mathfrak{L}^1([a, b])$.

Theorem 3.2.10

For every $A \in \mathscr{L}$, $\mathfrak{L}^1(A)$ is a real vector space, and the mapping $I : \mathfrak{L}^1(A) \to \mathbb{R}$, $I(f) = \int_A f d\lambda$, is a linear operator:

(1) $\int_A (f + g) d\lambda = \int_A f d\lambda + \int_A g d\lambda$, *for every $f, g \in \mathfrak{L}^1(A)$;*

(2) $\int_A c f d\lambda = c \int_A f d\lambda$, *for every $f \in \mathfrak{L}^1(A)$, and every $c \in \mathbb{R}$.*

Proof

To show that $\mathfrak{L}^1(A)$ is a vector space, it was enough to show that it is closed for sum and multiplication with scalars ($\mathfrak{L}^1(A)$ is a subset of the vector space of $\mathfrak{L}(A)$).

Let then $f, g \in \mathfrak{L}^1(A)$; from Theorem 3.2.2 it follows that $|f|, |g| \in \mathfrak{L}^1_+(A)$, and so Corollary 3.1.9 tells us that $h = |f| + |g| \in \mathfrak{L}^1_+(A)$ ($\int_A h d\lambda = \int_A |f| d\lambda + \int_A |g| d\lambda < +\infty$).

On the other hand, $|f + g| \le |f| + |g| = h$, and then Theorem 3.2.6 assures us that $f + g$ is integrable. From the relation

$$(f^+ - f^-) + (g^+ - g^-) = f + g = (f + g)^+ - (f + g)^-$$

we obtain that

$$(f+g)^+ + f^- + g^- = (f+g)^- + f^+ + g^+.$$

By integrating the previous equality (we observe that all the intervening functions are integrable and positive) and by using again Corollary 3.1.9, we obtain

$$\int_A (f+g)^+ d\lambda + \int_A f^- d\lambda + \int_A g^- d\lambda = \int_A (f+g)^- d\lambda + \int_A f^+ d\lambda + \int_A g^+ d\lambda,$$

hence, all the terms being finite,

$$\int_A (f+g) d\lambda = \int_A f d\lambda + \int_A g d\lambda.$$

Let now $f \in \mathfrak{L}^1(A)$ and let $c \in \mathbb{R}$; then $|c \cdot f| = |c| \cdot |f| \in \mathfrak{L}^1_+(A)$, from where $c \cdot f \in \mathfrak{L}^1(A)$ and

$$\int_A (c \cdot f) d\lambda = \int_A (c \cdot f)^+ d\lambda - \int_A (c \cdot f)^- d\lambda.$$

If $c > 0$, then $(c \cdot f)^+ = c \cdot f^+$ and $(c \cdot f)^- = c \cdot f^-$, from where, using (1) of Proposition 3.1.6,

$$\int_A (c \cdot f) d\lambda = c \int_A f^+ d\lambda - c \int_A f^- d\lambda = c \int_A f d\lambda.$$

If $c < 0$, then the proof is similar by noting that $(c \cdot f)^+ = -c \cdot f^-$ and $(c \cdot f)^- = -c \cdot f^+$.
■

Theorem 3.2.11
Let $A \in \mathscr{L}$, $\lambda(A) > 0$ and let $f, g \in \mathfrak{L}(A)$.
(1) If $f \geq \underline{0}$ a.e., then f has an integral over A and $\int_A f d\lambda \geq 0$.
(2) If $f \leq g$ a.e. and f, g have integrals, then $\int_A f d\lambda \leq \int_A g d\lambda$.
(3) If $f = g$, a.e. and if f has an integral over A, then g has also an integral over A and $\int_A f d\lambda = \int_A g d\lambda$; $f \in \mathfrak{L}^1(A)$ if and only if $g \in \mathfrak{L}^1(A)$.

Proof
(1) If $f \geq \underline{0}$ a.e., then $f^- = \underline{0}$ a.e., and so $\int_A f^- d\lambda = 0$ (see 1) of Corollary 3.2.3). Therefore f has an integral, and $\int_A f d\lambda = \int_A f^+ d\lambda \geq 0$.
(2) If $f \leq g$ a.e., then $f^+ \leq g^+$ and $f^- \geq g^-$ a.e. on A. Then $\int_A f d\lambda = \int_A f^+ d\lambda - \int_A f^- d\lambda \leq \int_A g^+ d\lambda - \int_A g^- d\lambda = \int g d\lambda$.

(3) Since f has an integral over A, $\int_A f^+ d\lambda < +\infty$ or $\int_A f^- d\lambda < +\infty$. Let us suppose that $\int_A f^+ d\lambda < +\infty$; then, since $f^+ - g^+ = \underline{0}$ a.e., $f^+ - g^+ \in \mathfrak{L}^1(A)$ and $\int_A (f^+ - g^+) d\lambda = 0$ (see Corollary 3.2.3). It follows that $\int_A g^+ d\lambda = \int_A f^+ d\lambda < +\infty$; therefore g has an integral.
Similarly, if $\int_A f^- d\lambda < +\infty$, then $\int_A g^- d\lambda = \int_A f^- d\lambda$.
According to (2), it is obvious that $\int_A f d\lambda = \int_A g d\lambda$. ∎

Remark 3.2.12 Let $f \in \mathfrak{L}^1(A)$ and let $B \in \mathscr{L}(A)$ such that $\lambda(A \setminus B) = 0$. Then $f = f \cdot \chi_B$ a.e. and so $\int_A f d\lambda = \int_B f d\lambda$.

3.3 The Space of Integrable Functions

In this section, we present some important properties of the integral (absolute continuity, countable additivity with respect to the integration domain, etc.). We also show that the space of integrable functions is organized as a complete seminormed space.

Theorem 3.3.1
Let $A \in \mathscr{L}$ and $f, g \in \mathfrak{L}^1(A)$; then
(1) $\int_A f d\lambda \leq \int_A g d\lambda$, *if* $f \leq g$ *a.e.*
(2) *For every* $\varepsilon > 0$, *there exists* $\delta > 0$, *such that* $\int_B |f| d\lambda < \varepsilon$, *for every* $B \in \mathscr{L}(A)$ *with* $\lambda(B) < \delta$.
(3) *For every* $\varepsilon > 0$, *there exists* $A_0 \in \mathscr{L}(A)$ *with* $\lambda(A_0) < +\infty$, *such that* $\int_{A \setminus A_0} |f| d\lambda < \varepsilon$;
(4) $\displaystyle\int_{\bigcup_{n=1}^{\infty} A_n} f d\lambda = \sum_{n=1}^{\infty} \int_{A_n} f d\lambda$, *for every* $(A_n)_n \subseteq \mathscr{L}(A)$, $A_n \cap A_m = \emptyset$, *for* $n \neq m$.

Proof
(1) If $f \leq g$ a.e., then from (1) of Theorem 3.2.5 and from Theorem 3.2.10 it follows that $g - f \in \mathfrak{L}^1_+(A)$ and $\int_A g d\lambda - \int_A f = \int_A (g - f) d\lambda \geq 0$, hence the inequality required.
(2) and (3) are immediate consequences of Theorem 3.1.16. ∎
(4) Apply Corollary 3.1.11 to positive functions f^+ and f^-.

Remarks 3.3.2
(i) Property (2) is the property of **absolute continuity** of the Lebesgue integral. In fact, this property says that the signed measure $\gamma : \mathscr{L}(A) \to \mathbb{R}$, defined by $\gamma(B) = \int_B f d\lambda$, is absolutely continuous with respect to the Lebesgue measure λ, $\gamma \ll \lambda$ (see Definition 6.2.1 and Proposition 6.2.3).
(ii) Property (3) shows that, for an integrable function, the integral depends on the behavior of this function on the sets of finite measures.

Theorem 3.3.3
Let $A \in \mathscr{L}$ and $\|\cdot\|_1 : \mathfrak{L}^1(A) \to \mathbb{R}_+$, defined by
$\|f\|_1 = \int_A |f| d\lambda$, *for every* $f \in \mathfrak{L}^1(A)$*; then*
(1) $\|f\|_1 = 0$ *if and only if* $f = \underline{0}$ *a.e.,*
(2) $\|cf\|_1 = |c| \cdot \|f\|_1$, *for every* $f \in \mathfrak{L}^1(A)$ *and any* $c \in \mathbb{R}$,
(3) $\|f+g\|_1 \le \|f\|_1 + \|g\|_1$, *for every* $f, g \in \mathfrak{L}^1(A)$.

Proof
(1) $\|f\|_1 = \int_A |f| d\lambda = 0$ if and only if $|f| = \underline{0}$ a.e. (see Proposition 3.1.13), and this happens if and only if $f = \underline{0}$ a.e.
(2) is the consequence of (1) of Proposition 3.1.6.
(3) Using (2) of Proposition 3.1.6 and Corollary 3.1.9, we obtain:

$$\|f+g\|_1 = \int_A |f+g| d\lambda \le \int_A |f| d\lambda + \int_A |g| d\lambda = \|f\|_1 + \|g\|_1. \qquad \blacksquare$$

Remark 3.3.4 It follows from the previous theorem that $\|\cdot\|_1$ is a seminorm on the vector space $\mathfrak{L}^1(A)$. $\|\cdot\|_1$ is not a norm on $\mathfrak{L}^1(A)$ because there are positive functions whose integral is zero and which are not zero a.e. (see Remark 3.1.15).

The relation $\doteq$ is an equivalence relation on $\mathfrak{L}^1(A)$ (it is reflexive, symmetric, and transitive). The quotient space, $\mathfrak{L}^1(A)|_{\doteq}$, is $L^1(A) = \{[f] : f \in \mathfrak{L}^1(A)\}$, where $[f] = \{g \in \mathfrak{L}^1(A) : f \doteq g\}$ is the equivalence class to which belongs f. Since the Lebesgue integral is the same for two functions equal almost everywhere (see Theorem 3.2.11), we can define coherently $\|[f]\|_1 = \|f\|_1$, for every $[f] \in L^1(A)$. The application thus defined is a norm on $L^1(A)$.

Definition 3.3.5

A sequence $(f_n) \subseteq \mathfrak{L}^1(A)$ is **L^1-convergent** to $f \in \mathfrak{L}^1(A)$ if $\|f_n - f\|_1 \to 0$ (for every $\varepsilon > 0$, there exists $n_0 \in \mathbb{N}$ such that, for any $n \ge n_0$, $\|f_n - f\|_1 = \int_A |f_n - f| d\lambda < \varepsilon$). We denote this situation by $f_n \xrightarrow[A]{\|\cdot\|_1} f$. The L^1-seminorm $\|\cdot\|_1$ is said to be the **seminorm of L^1-convergence** on $\mathfrak{L}^1(A)$.
A sequence $(f_n) \subseteq \mathfrak{L}^1(A)$ is **L^1-Cauchy** if, for every $\varepsilon > 0$, there is $n_0 \in \mathbb{N}$, such that, for any $m, n \ge n_0$, $\|f_m - f_n\|_1 < \varepsilon$.
If $F \subseteq \mathfrak{L}^1(A)$, then $f \in \mathfrak{L}^1(A)$ is a **L^1-adherent point** for F if there exists $(f_n) \subseteq F$ such that $f_n \xrightarrow[A]{\|\cdot\|_1} f$; we denote $f \in \overline{F}^1$.

Theorem 3.3.6
Let $(f_n) \subseteq \mathfrak{L}^1(A)$ *and* $f \in \mathfrak{L}^1(A)$.
1). *If* $f_n \xrightarrow[A]{\|\cdot\|_1} f$, *then* $f_n \overset{\lambda}{\underset{A}{\to}} f$.
2). *If* $(f_n)_n$ *is* L^1*-Cauchy, then* $(f_n)_n$ *is Cauchy in measure.*

Proof
(1) If $(f_n)_n \subseteq \mathfrak{L}^1(A)$ is L^1-convergent to $f \in \mathfrak{L}^1(A)$, then, for every $\varepsilon > 0$, $\|f_n - f\|_1 = \int_A |f_n - f| d\lambda \geq \int_{(|f_n - f| \geq \varepsilon)} |f_n - f| d\lambda \geq \varepsilon \cdot \lambda(|f_n - f| \geq \varepsilon)$, from where $\lambda(|f_n - f| \geq \varepsilon) \leq \frac{1}{\varepsilon} \cdot \|f_n - f\|_1$ and then $\lim_n \lambda(|f_n - f| \geq \varepsilon) = 0$. ε being positive arbitrary, $f_n \overset{\lambda}{\underset{A}{\to}} f$.
(2) Similarly, we notice that, for all $\varepsilon > 0$, $\lambda(|f_n - f_m| \geq \varepsilon) \leq \frac{1}{\varepsilon} \cdot \|f_n - f_m\|_1$. If $(f_n)_n$ is L^1-Cauchy, then $\|f_n - f_m\|_1 \xrightarrow[m,n \to \infty]{} 0$, and so $\lim_{n,m} \lambda(|f_n - f_m| \geq \varepsilon) = 0$, for every $\varepsilon > 0$.

■

Remark 3.3.7 The converses of the previous theorem are not true. For example, the sequence $(f_n)_n$, $f_n = n \cdot \chi_{[0, \frac{1}{n}]} \in \mathfrak{L}^1([0, 1])$, is convergent in measure to $\underline{0}$ (why?) but is not L^1-convergent (why?).

Theorem 3.3.6 allows us to show that the seminormed space $(\mathfrak{L}^1(A), \|\cdot\|_1)$ is complete.

Theorem 3.3.8
The seminormed space $(\mathfrak{L}^1(A), \|\cdot\|_1)$ *is complete (every* L^1*-Cauchy sequence is* L^1*-convergent).*

Proof
From (2) of Theorem 3.3.6, it follows that any sequence L^1-Cauchy, $(f_n)_n \subseteq \mathfrak{L}^1(A)$, is Cauchy in measure. From (1) of Riesz theorem (Theorem 2.2.5), there exists a subsequence $(f_{k_n})_n$ of $(f_n)_n$ almost uniform convergent to a function $f : A \to \mathbb{R}$. Point (2) of Theorem 2.2.2 tells us that $f_{k_n} \overset{\cdot}{\underset{A}{\to}} f$ and then $f \in \mathfrak{L}(A)$ (see 3) of Theorem 2.1.13).

Let's fix a $m \in \mathbb{N}$; for any $n \in \mathbb{N}$, let $g_n = |f_m - f_{k_n}| \in \mathfrak{L}_+(A)$. We apply to the sequence $(g_n)_n$ Fatou's lemma (Corollary 3.1.12):

$$\int_A \liminf_n g_n d\lambda \leq \liminf_n \int_A g_n d\lambda.$$

Since $\liminf_n g_n = |f_m - f|$ a.e., it follows that

$$\int_A |f_m - f| d\lambda \le \liminf_n \|f_m - f_{k_n}\|_1$$

and, since $\lim_{m,n\to+\infty} \|f_m - f_{k_n}\|_1 = 0$, it results, on the one hand, that $f_m - f \in \mathfrak{L}^1(A)$ and then $f = f_m - (f_m - f) \in \mathfrak{L}^1(A)$ and, on the other hand, that $\|f_m - f\|_1 \to 0$.

The following result shows a very practical condition to pass to the limit under the Lebesgue integral.

Theorem 3.3.9 (Dominated Convergence Theorem)
Let $(f_n) \subseteq \mathfrak{L}(A)$ and $g \in \mathfrak{L}^1(A)$ such that
(1) $f_n \underset{A}{\overset{\cdot}{\to}} f$ and
(2) $|f_n| \le g$, a.e., for any $n \in \mathbb{N}$.
Then $(f_n) \subseteq \mathfrak{L}^1(A)$, $f \in \mathfrak{L}^1(A)$, $f_n \xrightarrow[A]{\|\cdot\|_1} f$, and $\int_A f_n d\lambda \to \int_A f d\lambda$.

Proof
From Theorem 3.2.6, it follows that $(f_n)_n \subseteq \mathfrak{L}^1(A)$; (3) of Theorem 2.1.13 assures us that $f \in \mathfrak{L}(A)$. If we go to the limit in inequality (2), we get $|f| \le g$ a.e.; again Theorem 3.2.6 leads us to $f \in \mathfrak{L}^1(A)$.

Now we remark that $|f_n - f| \le |f_n| + |f| \le 2g$. If $h_n = 2g - |f_n - f|$, then $(h_n)_n \subseteq \mathfrak{L}_+(A)$. Then, we apply Fatou's lemma (Corollary 3.1.12) to the sequence $(h_n)_n$:

$$\int_A \liminf_n h_n d\lambda \le \liminf_n \int_A h_n d\lambda.$$

Because $\liminf_n h_n = 2g$ a.e., we obtain

$$2\int_A g d\lambda \le 2\int_A g d\lambda + \liminf_n \left(-\int_A |f_n - f| d\lambda \right) =$$

$$= 2\int_A g d\lambda - \limsup_n \int_A |f_n - f| d\lambda,$$

from where $\limsup_n \|f_n - f\|_1 \le 0$. It follows that $\limsup_n \|f_n - f\|_1 = 0$; therefore, there is $\lim_n \|f_n - f\|_1 = 0$.

Since $\left| \int_A f_n d\lambda - \int_A f d\lambda \right| \le \int_A |f_n - f| d\lambda = \|f_n - f\|_1$, it results that $\int_A f_n d\lambda \to \int_A f d\lambda$. ■

Corollary 3.3.10 (Bounded Convergence Theorem)

Let $A \in \mathscr{L}$ with $\lambda(A) < +\infty$, let $(f_n) \subseteq \mathfrak{L}(A)$ and let $c \in \mathbb{R}_+$ such that

(1) $f_n \underset{A}{\overset{\cdot}{\to}} f$ *and*

(2) $|f_n| \leq \underline{c}$, *for any* $n \in \mathbb{N}$.

Then $(f_n) \subseteq \mathfrak{L}^1(A)$, $f \in \mathfrak{L}^1(A)$, $f_n \xrightarrow[A]{\|\cdot\|_1} f$ *and* $\int_A f_n d\lambda \to \int_A f d\lambda$.

Proof

The theorem results from the dominated convergence theorem if we notice that, on sets of finite measure, the constant functions are integrable (see Corollary 3.2.8); then we can take $g = \underline{c}$. ■

In what follows, we highlight two dense subsets in $\mathfrak{L}^1(A)$.

Theorem 3.3.11

Let $\mathcal{E}^1(A) = \mathcal{E}(A) \cap \mathfrak{L}^1(A)$ be the set of all integrable simple functions on A, and let $C_1(A) = C(A) \cap \mathfrak{L}^1(A)$ be the set of all integrable continuous functions on A; then

(1) $\overline{\mathcal{E}^1(A)}^1 = \mathfrak{L}^1(A)$ *and*

(2) $\overline{C_1(A)}^1 = \mathfrak{L}^1(A)$.

Proof

By definition $\overline{\mathcal{E}^1(A)}^1 \subseteq \mathfrak{L}^1(A)$ and $\overline{C_1(A)}^1 \subseteq \mathfrak{L}^1(A)$. We have to show the reverse inclusions.

(1) For every $f \in \mathfrak{L}^1(A)$, $f = f^+ - f^-$ and $f^+, f^- \in \mathfrak{L}^1_+(A)$. Taking into account (1) of approximation theorem of the measurable functions (see Theorem 2.3.3), there exist two sequences $(g_n)_n, (h_n)_n \subseteq \mathcal{E}_+(A)$ such that $g_n \uparrow f^+$ and $h_n \uparrow f^-$. Then $f_n = g_n - h_n \to f$ and $(f_n)_n \subseteq \mathcal{E}(A)$.

For any $n \in \mathbb{N}$, $|f_n| \leq g_n + h_n \leq f^+ + f^- = |f| \in \mathfrak{L}^1_+(A)$. From the dominated convergence theorem (Theorem 3.3.9), $(f_n)_n \subseteq \mathfrak{L}^1(A)$; then $(f_n)_n \subseteq \mathcal{E}(A) \cap \mathfrak{L}^1(A) = \mathcal{E}^1(A)$, and, more, $\|f_n - f\|_1 \to 0$. It follows that $f \in \overline{\mathcal{E}^1(A)}^1$.

(2) We will first show that $\mathcal{E}^1(A) \subseteq \overline{C_1(A)}^1$.

For every $f = \sum_{i=1}^{p} a_i \cdot \chi_{A_i} \in \mathcal{E}^1(A)$, let $M = \sup_{x \in A} |f(x)| =$ $= \max\{|a_1|, \cdots, |a_p|\}$. From Borel theorem (Theorem 2.3.10), for every $\varepsilon > 0$, there exists $f_\varepsilon \in C(A)$ such that $\lambda(f \neq f_\varepsilon) < \frac{\varepsilon}{2M}$ and $\sup_{x \in A} |f_\varepsilon(x)| \leq M$. Let $B = (f \neq f_\varepsilon) \in \mathscr{L}(A)$. Then $f_\varepsilon = f_\varepsilon \cdot \chi_B + f_\varepsilon \cdot \chi_{A \setminus B} = f_\varepsilon \cdot \chi_B + f \cdot \chi_{A \setminus B}$, and so $|f_\varepsilon| \leq M \cdot \chi_B + |f| \in \mathcal{E}^1_+(A) \subseteq \mathfrak{L}^1(A)$. From Theorem 3.2.6, $f_\varepsilon \in \mathfrak{L}^1(A)$ and then $f_\varepsilon \in C_1(A)$. Moreover

$$\|f - f_\varepsilon\|_1 = \int_A |f - f_\varepsilon| d\lambda = \int_B |f - f_\varepsilon| d\lambda \leq \int_B (|f| + |f_\varepsilon|) d\lambda \leq 2M \cdot \lambda(B) < \varepsilon.$$

If ε takes the values $\frac{1}{n}$, $n \in \mathbb{N}^*$, one by one, we can find the sequence $(f_n)_n \subseteq C_1(A)$ such that $\|f - f_n\|_1 < \frac{1}{n}$, for any $n \in \mathbb{N}^*$. It follows that $f_n \xrightarrow[A]{\|\cdot\|_1} f$ and therefore $f \in \overline{C_1(A)}^1$, which shows the inclusion $\mathcal{E}^1(A) \subseteq \overline{C_1(A)}^1$.

If we use point (1) and the properties of monotonicity and idempotency of the closure operator in the last inclusion, we obtain

$$\mathfrak{L}^1(A) = \overline{\mathcal{E}^1(A)}^1 \subseteq \overline{\overline{C_1(A)}^1}^1 = \overline{C_1(A)}^1 \subseteq \mathfrak{L}^1(A)$$

hence the second density property. ■

Remarks 3.3.12

(i) If $\lambda(A) < +\infty$, then $\mathcal{E}^1(A) = \mathcal{E}(A)$ (see Exercise 12) of (3.7). Therefore, in this case, $\mathcal{E}(A)$ is dense in $\mathfrak{L}^1(A)$.

(ii) If A is compact, then $C_1(A) = C(A)$. Indeed, in this case, any continuous function on A is bounded (Weierstrass theorem). The compact sets are bounded and therefore have a finite measure. On the sets of finite measure, any measurable and bounded function is integrable (see Corollary 3.2.8). It follows that $C(A)$ is dense in $\mathfrak{L}^1(A)$.

(iii) $\overline{\mathcal{R}_{[a,b]}}^1 = \mathfrak{L}^1([a,b])$. Indeed, in this case, $C([a,b]) \subsetneq \mathcal{R}_{[a,b]}$, and, since $[a,b]$ is compact, we can use the previous remark,

$$\mathfrak{L}^1([a,b]) = \overline{C([a,b])}^1 \subseteq \overline{\mathcal{R}_{[a,b]}}^1 \subseteq \mathfrak{L}^1([a,b].$$

(iv) According to Theorems 3.3.8 and 3.3.11, $\mathfrak{L}^1(A)$ is a **completion** of $(\mathcal{E}^1(A), \|\cdot\|_1)$ and also a completion of $(C_1(A), \|\cdot\|_1)$. From the above (iii), $\mathfrak{L}^1([a,b])$ is a completion of $(\mathcal{R}_{[a,b]}, \|\cdot\|_1)$.

3.4 Comparison with the Riemann Integral

In this section, we will compare the Lebesgue integral to the Riemann integral on both compact and non-compact intervals.

For some reminders on the Riemann integral, one can consult ▶ Sect. 7.1.

Let $[a,b]$ be a closed bounded interval, and let $\mathcal{R}_{[a,b]}$ be the set of all Riemann integrable functions on $[a,b]$.

Theorem 3.4.1

$\mathcal{R}_{[a,b]} \subsetneq \mathfrak{L}^1([a,b])$ and $\int_a^b f(x)dx = \int_{[a,b]} f d\lambda$, for every $f \in \mathcal{R}_{[a,b]}$.

Proof

From Corollary 3.2.9, $\mathcal{R}_{[a,b]} \subsetneq \mathfrak{L}^1([a,b])$; now show that the Riemann integral is the restriction of the Lebesgue integral to space $\mathcal{R}_{[a,b]}$.

Let $f \in \mathcal{R}_{[a,b]}$. For every partition of the compact interval $[a,b]$, $\Delta = \{x_0, x_1, \cdots, x_n\} \in \mathcal{D}([a,b])$, and for any $k = 0, 1, \cdots, n$, let

$$m_k = \inf_{x \in [x_k, x_{k+1}]} f(x) \text{ and } M_k = \sup_{x \in [x_k, x_{k+1}]} f(x);$$

then

$$s_\Delta = \sum_{k=0}^{n-1} m_k \cdot (x_{k+1} - x_k) \text{ and } S_\Delta = \sum_{k=0}^{n-1} M_k \cdot (x_{k+1} - x_k)$$

are the lower and upper Darboux sums corresponding to Δ on $[a,b]$.

$\underline{I} = \sup_{\Delta \in \mathcal{D}([a,b])} s_\Delta$ is lower Darboux integral and $\overline{I} = \inf_{\Delta \in \mathcal{D}([a,b])} S_\Delta$ is upper Darboux integral. Since f is Riemann integrable, $\underline{I} = \overline{I} = \int_a^b f(x)dx$.

On the other hand, we consider the simple functions $f_\Delta, F_\Delta : [a,b] \to \mathbb{R}$ defined by $f_\Delta = \sum_{k=0}^{n-1} m_k \cdot \chi_{[x_k, x_{k+1}[}$, respectively $F_\Delta = \sum_{k=0}^{n-1} M_k \cdot \chi_{[x_k, x_{k+1}[}$. Then $f_\Delta, F_\Delta \in \mathcal{E}([a,b]) = \mathcal{E}^1([a,b]) \subseteq \mathfrak{L}^1([a,b])$ (see (i) of Remarks 3.3.12), and, since $f_\Delta \le f \le F_\Delta$ a.e., $s_\Delta = \int_{[a,b]} f_\Delta d\lambda \le \int_{[a,b]} f d\lambda \le \int_{[a,b]} F_\Delta d\lambda = S_\Delta$.

The partition Δ is arbitrary in $\mathcal{D}([a,b])$ so that $\underline{I} \le \int_{[a,b]} f d\lambda \le \overline{I}$, and then $\int_a^b f(x)dx = \int_{[a,b]} f d\lambda$. ■

Let $a \in \mathbb{R}$, $b \le +\infty$, $a < b$ and $f : [a,b[\to \mathbb{R}$ be a function with the property that $f \in \mathcal{R}_{[a,u]}$, for any $u \in [a,b[$. If there exists $\lim_{u \uparrow b} \int_a^u f(x)dx \in \bar{\mathbb{R}}$, then this limit is called the **generalized (or improper) Riemann integral** of the function f over the interval $[a,b[$, and it is denoted with $\int_a^{b-0} f(x)dx$ or $\int_a^{+\infty} f(x)dx$ if $b = +\infty$. If this limit is finite, then we say that f is **Riemann integrable in the generalized sense** on the non-compact interval $[a,b[$ or that the improper integral **converges**. If the limit does not exist or if it is infinite, then we say that the improper integral **diverges**. Let $\mathcal{R}_{[a,b[}$ be the set of all functions Riemann integrable in the generalized sense over $[a,b[$. If $|f| \in \mathcal{R}_{[a,b[}$, it is said that the improper integral $\int_a^{b-0} f(x)dx$ is **absolutely convergent**; an absolutely convergent integral is convergent but the converse is not true. It is possible for a generalized Riemann integral to be **simply convergent** when it converges, but it is not absolutely convergent (see the example given in point (v) of Remark 3.4.3). The integral of Lebesgue knows no such phenomenon.

The following theorem states that, on a non-compact interval, a function is Lebesgue integrable if and only if its improper integral is absolutely convergent.

Theorem 3.4.2
Let $a \in \mathbb{R}$, $b \leq +\infty$, $a < b$, and let $f : [a, b[\to \mathbb{R}$ be a function such that $f \in \mathcal{R}_{[a,u]}$, for any $u \in [a, b[$; then

$$f \in \mathfrak{L}^1([a, b[) \text{ if and only if } |f| \in \mathcal{R}_{[a,b[}$$

and, in this case, $\int_{[a,b[} f d\lambda = \int_a^{b-0} f(x)dx$.

Proof
($\Longrightarrow$): Let us suppose that $f \in \mathfrak{L}^1([a, b[)$; then $|f| \in \mathfrak{L}^1([a, b[)$ and $\gamma : \mathscr{L}([a, b[) \to \mathbb{R}_+$, $\gamma(A) = \int_A |f| d\lambda$, is a finite measure on $\mathscr{L}([a, b[)$ (see Theorem 3.1.16).

For every sequence $(u_n)_n \subseteq [a, b[$, $u_n \uparrow b$, the sets $A_n = [a, u_n]$ form an increasing sequence in $\mathscr{L}([a, b[)$ which converges to $\bigcup_{n=1}^{\infty} A_n = [a, b[$. From the property of continuity from below of measure γ (see the property 6) of Theorem 1.4.7), $\gamma([a, b[) = \lim_n \gamma(A_n)$ or $\int_{[a,b[} |f| d\lambda = \lim_n \int_{[a,u_n]} |f| d\lambda$. But from the previous theorem, for any $n \in \mathbb{N}$, $\int_{[a,u_n]} |f| d\lambda = \int_a^{u_n} |f(x)| dx$. It follows that there exists $\lim_n \int_a^{u_n} |f(x)| dx = \int_{[a,b[} |f| d\lambda$. The sequence $(u_n)_n$ is arbitrary with the properties mentioned, and so there exists $\lim_{u \uparrow b} \int_a^u |f(x)| dx = \int_{[a,b[} |f| d\lambda < +\infty$. Therefore $|f| \in \mathcal{R}_{[a,b[}$ and

$$\int_a^{b-0} |f(x)| dx = \int_{[a,b[} |f| d\lambda.$$

Note that the above relation is satisfied for any function $g \in \mathfrak{L}^1_+([a, b[)$ which is Riemann integrable on any compact subinterval of $[a, b[$. Apply this relation to the positive and negative parts of f: since $f^+ = \frac{1}{2} \cdot (|f| + f)$, $f^- = \frac{1}{2} \cdot (|f| - f) \in \mathfrak{L}^1_+([a, b[)$,

$$\int_a^{b-0} f^+(x)dx = \int_{[a,b[} f^+ d\lambda \text{ and } \int_a^{b-0} f^-(x)dx = \int_{[a,b[} f^- d\lambda.$$

It follows that $\int_{[a,b[} f d\lambda = \int_a^{b-0} f(x)dx$.

($\Longleftarrow$): Suppose now that $|f| \in \mathcal{R}_{[a,b[}$.

If $b < +\infty$, for any $n \in \mathbb{N}^*$, let $f_n = f \cdot \chi_{[a,\, b - \frac{1}{n}]} \in \mathfrak{L}([a, b[)$ ($f \in \mathcal{R}_{[a,b-\frac{1}{n}]} \subseteq \mathfrak{L}([a, b - \frac{1}{n}])$ and $\underline{0} \in \mathfrak{L}(]b - \frac{1}{n}, b[)$—see Exercise 4) of 2.5). Since $f_n \xrightarrow[{[a,b[}]{p} f$, it follows that $f \in \mathfrak{L}([a, b[)$ (point 5) of Theorem 2.1.18). Then $|f| \in \mathfrak{L}_+([a, b[)$ (point 2) of Proposition 2.1.22). Therefore, for any $n \in \mathbb{N}^*$, $|f_n| = |f| \cdot \chi_{[a,\, b - \frac{1}{n}]} \in \mathfrak{L}_+([a, b[)$, and, since $|f_n| \uparrow |f|$, the monotone convergence theorem (Theorem 3.1.7) assures us that

$$\int_a^{b-0} |f(x)| dx = \lim_n \int_a^{b-\frac{1}{n}} |f(x)| dx = \lim_n \int_{[a,b[} |f_n| d\lambda = \int_{[a,b[} |f| d\lambda.$$

Hence $\int_{[a,b[} |f| d\lambda < +\infty$, from where $|f| \in \mathfrak{L}^1_+([a,b[)$ and then $f \in \mathfrak{L}^1([a,b[)$.
If $b = +\infty$, we choose $f_n = f \cdot \chi_{[a,n]}$. ■

The following observations highlight certain comparisons that can be made between the two types of integrals: the Riemann integral and the Lebesgue integral.

Remarks 3.4.3

(i) The Riemann integral is defined only on intervals, while the Lebesgue integral is calculated on measurable sets which form a much larger class.
(ii) The Riemann integral is sensitive to the modification of function values on a set of zero measures, while the Lebesgue integral is invariant to such changes.
(iii) For the Lebesgue integral, we have much easier criteria for passage to the limit under the integral (dominated convergence theorem, bounded convergence theorem), while for the Riemann integral, we need of uniform convergence for such an operation.
(iv) The Lebesgue integral is σ-additive relating to the set of integration (see 4) of Theorem 3.3.1); the Riemann integral is just finite additive.
(v) On the compact intervals, the Lebesgue integral is more general than the Riemann integral: any function integrable Riemann is integrable Lebesgue, but there are integrable Lebesgue functions which are not integrable Riemann (e.g., the Dirichlet function).

On the non-compact intervals, the Riemann integral distinguishes between simple convergence and absolute convergence. So the function $f : [1, +\infty[\to \mathbb{R}$, $f(x) = \frac{\sin x}{x}$, is Riemann integrable on $[1, +\infty[$ (Dirichlet's test can be applied) but $|f|$ is not Riemann integrable: $|f(x)| \geq \frac{(\sin x)^2}{x} = \frac{1}{2x} - \frac{\cos 2x}{2x}$. The first function of the previous difference is not integrable on $[1, +\infty[$, and the second function is. Therefore the generalized Riemann integral $\int_1^\infty \frac{\sin x}{x} dx$ is simply convergent, and so f is not Lebesgue integrable.

3.5 Properties of the Lebesgue Integral

In this section, we will deal with some useful properties for Lebesgue integrals: change of variables, properties of integrals depending on a parameter, and Jensen inequality.

3.5.1 Change of Variables

In this subsection, we will present a change of variables formula for the Lebesgue integral, similar to that of the Riemann integral. One of the problems to be solved in this case consists in finding conditions which a function must satisfy in order to transport measurable sets into measurable sets. In general, a measurable function does not have such a property. Indeed, let $V \subseteq [0, 1]$ be the Vitali set (see 1.2.7). In (ii) of Remarks 1.3.17, we noted that $V \in \mathcal{P}(\mathbb{R}) \setminus \mathscr{L}(\mathbb{R})$. The function $f : D \to [0, 1]$, defined in 1.3.16, is strictly increasing and surjective; according to (2) of Corollary 2.1.7,

$f \in \mathfrak{L}(D)$. If $A = f^{-1}(V) \subseteq D$, then, since $\lambda(D) = 0$, $A \in \mathscr{L}$ (see (ii) of Remarks 1.3.4). Therefore $A \in \mathscr{L}$ and $f(A) = V \notin \mathscr{L}$.

Definition 3.5.1

A function $g : \mathbb{R} \to \mathbb{R}$ is said to be **Lipschitz function** on $A \subseteq \mathbb{R}$ if there exists a constant $L > 0$ such that g is L-Lipschitz, that is to say: $|g(x) - g(y)| \leq L \cdot |x - y|$, for every $x, y \in A$.

g is **locally Lipschitz** if for each $x \in \mathbb{R}$, there exists an open set $D \subseteq \mathbb{R}$ such that $x \in D$ and g is Lipschitz on D (for a certain constant L which can depend on x).

Remark 3.5.2

(i) Every locally Lipschitz function $g : \mathbb{R} \to \mathbb{R}$ is continuous on $\mathbb{R}$. Indeed, for every $x \in \mathbb{R}$ and every $(x_n)_n \subseteq \mathbb{R}$, $x_n \to x$, there exists an open set $D \subseteq \mathbb{R}$ and $L > 0$, such that $x \in D$ and g is L-Lipschitz on D. Let $n_0 \in \mathbb{N}$ such that $x_n \in D$, for any $n \geq n_0$. Then $|g(x_n) - g(x)| \leq L \cdot |x_n - x|$, for any $n \geq n_0$, from where $g(x_n) \to g(x)$.

(ii) Let $g : \mathbb{R} \to \mathbb{R}$ be a function differentiable on an interval $I \subseteq \mathbb{R}$ with the bounded derivative on I; then g is Lipschitz on I. Indeed, let $L = \sup_{x \in I} |g'(x)|$; then g is L-Lipschitz.

Proposition 3.5.3 *The function $g : \mathbb{R} \to \mathbb{R}$ is locally Lipschitz on $\mathbb{R}$ if and only if g is Lipschitz on every compact subset of $\mathbb{R}$.*

Proof

The sufficiency of the condition is obvious.

The necessity. Suppose, reducing to the absurd, that there is a compact $K \subseteq \mathbb{R}$, so that g is not Lipschitz on K. Then, for any $n \in \mathbb{N}$, there exist $x_n, y_n \in K$ such that $|g(x_n) - g(y_n)| > n \cdot |x_n - y_n|$. Since K is compact, possibly passing to a subsequence, we can assume that $x_n \to x$, $y_n \to y$, where $x, y \in K$.

If $x = y$, then there exist an open set $D \subseteq \mathbb{R}$ and $L > 0$ such that $x \in D$ and g is L-Lipschitz on D; since $x_n, y_n \to x$, there exists $n_0 \in \mathbb{N}$ such that $x_n, y_n \in D$, for any $n \geq n_0$. Then

$$L \cdot |x_n - y_n| \geq |g(x_n) - g(y_n)| > n \cdot |x_n - y_n|, \text{ for any } n \geq n_0,$$

which is absurd ($x_n \neq y_n$, for any $n \in \mathbb{N}$).

If $x \neq y$, $|x - y| > 0$. Because $|x_n - y_n| \to |x - y|$, there exists $n_0 \in \mathbb{N}$ such that $|x_n - y_n| > \frac{1}{2} \cdot |x - y|$, for any $n \geq n_0$. Since g is continuous on $\mathbb{R}$ (see Remark 3.5.2) and since K is compact, $g(K)$ is also compact, and so it is bounded. Let $d = \sup_{u,v \in K} |g(u) - g(v)| < +\infty$. Then

$$\frac{n}{2} \cdot |x - y| < n \cdot |x_n - y_n| < |g(x_n) - g(y_n)| \leq d, \text{ for any } n \geq n_0,$$

which is also absurd. ■

Remarks 3.5.4

(i) The function $g : \mathbb{R} \to \mathbb{R}$ is said to be **of class** C^1 on $\mathbb{R}$ if g is differentiable and the derivative g' is continuous on $\mathbb{R}$; it is denoted by $g \in C^1(\mathbb{R})$. If g is of class C^1 on $\mathbb{R}$, then g is locally Lipschitz on $\mathbb{R}$. Indeed, for every compact $K \subseteq \mathbb{R}$, there exists an interval compact I such that $K \subseteq I$. The derivative of g is continuous on I, hence it is bounded. According to (ii) of Remark 3.5.2, g is Lipschitz on I, and then it is Lipschitz on K.

(ii) Every locally Lipschitz function on $\mathbb{R}$ is Lipschitz on every bounded subset of $\mathbb{R}$.

Theorem 3.5.5

Let $g : \mathbb{R} \to \mathbb{R}$ be a locally Lipschitz function.

(a) For every null set $N \subseteq \mathbb{R}$, $g(N)$ is a null set.

(b) For every $A \in \mathscr{L}$, $g(A) \in \mathscr{L}$.

Proof

(a) First, let us suppose that N is a bounded null set, and let $a \in \mathbb{R}_+^*$ such that $N \subseteq]-a, a[\subseteq [-a, a]$. The function g is Lipschitz on $[-a, a]$; let $L > 0$ such that

$$|g(x) - g(y)| \leq L \cdot |x - y|, \text{ for every } x, y \in [-a, a].$$

Since N is a null set, for every $\varepsilon > 0$, there exists a sequence of open intervals $(]a_p, b_p[)_p$ such that $N \subseteq \bigcup_{p=1}^{\infty}]a_p, b_p[\subseteq [-a, a]$ and $\sum_{p=1}^{\infty}(b_p - a_p) < \frac{\varepsilon}{2L}$.

For any $p \geq 1$ and for every $x, y \in]a_p, b_p[$, $|g(x) - g(y)| \leq L \cdot |x - y| \leq L \cdot (b_p - a_p)$; therefore $g(]a_p, b_p[)$ is bounded. Then, for any $p \in \mathbb{N}^*$, there exists an open interval $]c_p, d_p[$ such that $g(]a_p, b_p[) \subseteq]c_p, d_p[$ and $d_p - c_p \leq 2L \cdot (b_p - a_p)$. Consequently, $g(N) \subseteq \bigcup_{p=1}^{\infty}]c_p, d_p[$ and $\sum_{p=1}^{\infty}(d_p - c_p) < \varepsilon$. Therefore $g(N)$ is a null set.

If N is a unbounded null set, then we can represent it as a countable union of bounded null sets; for example, $N = \bigcup_{n=1}^{\infty} N_n$, where $N_n = N \cap [-n, n]$. Then $g(N) = \bigcup_{n=1}^{\infty} g(N_n)$, which shows that $g(N)$ is a countable union of null sets, so it is a null set.

(b) As in the previous point, we first assume that $A \in \mathscr{L}$ is bounded. Using Corollary 1.3.9, for all $n \in \mathbb{N}^*$, we can find a closed set $F_n \subseteq A$ such that $\lambda(A \setminus F_n) < \frac{1}{n}$. Let $B = \bigcup_{n=1}^{\infty} F_n \in \mathscr{L}$. Then $\lambda(A \setminus B) = \lambda(\bigcap_{n=1}^{\infty}(A \setminus F_n)) \leq \lambda(A \setminus F_n) < \frac{1}{n}$, for any $n \in \mathbb{N}^*$. Therefore $N = A \setminus B$ is a null set and, it is obvious that $A = B \cup N$.

According to Remark 3.5.2, g is continuous on $\mathbb{R}$, and, since every F_n is compact, $g(F_n)$ is also compact, for any $n \in \mathbb{N}^*$. Therefore $\bigcup_{n=1}^{\infty} g(F_n) \in \mathcal{B}_u \subseteq \mathscr{L}$ (see Definition 1.3.18). As stated by a), $g(N)$ is a null set, so that $g(N) \in \mathscr{L}$. Then $g(A) = g(B) \cup g(N) = \bigcup_{n=1}^{\infty} g(F_n) \cup g(N) \in \mathscr{L}$.

If $A \in \mathscr{L}$ is unbounded, we can represent it as a countable union of bounded measurable sets: $A = \bigcup_{n=1}^{\infty} A_n$, where, for any $n \in \mathbb{N}^*$, $A_n = A \cap [-n, n]$. Then $g(A) = \bigcup_{n=1}^{\infty} g(A_n)$; every A_n is measurable and bounded so that $g(A_n) \in \mathscr{L}$, for any $n \in \mathbb{N}^*$. Therefore $g(A)$ is a countable union of mesurable sets, so it is measurable.

■

We will now present a change of variable theorem for the Lebesgue integral.

Theorem 3.5.6
Let $g : \mathbb{R} \to \mathbb{R}$ be an injective function of class C^1 (g is differentiable with continuous derivative on $\mathbb{R}$); then, for every $A \in \mathscr{L}$, $g(A) \in \mathscr{L}$ and

$$\lambda(g(A)) = \int_A |g'|d\lambda. \tag{*}$$

For every Borel function $f : \mathbb{R} \to \mathbb{R}$ and for every set $A \in \mathscr{L}$, $f \in \mathfrak{L}^1(g(A))$ if and only if $(f \circ g) \cdot |g'| \in \mathfrak{L}^1(A)$ and then

$$\int_{g(A)} f d\lambda = \int_A (f \circ g) \cdot |g'|d\lambda. \tag{**}$$

Proof
Let $g : \mathbb{R} \to \mathbb{R}$ be an injection of class C^1.

We will first note that any continuous injection on $\mathbb{R}$ is strictly monotonic. Let us suppose that g is strictly increasing (the demonstration is similar if we assume that g is strictly decreasing). Then $g'(x) \geq 0$, for every $x \in \mathbb{R}$. According to (i) of Remark 3.5.4, g is locally Lipschitz.

Let $A \in \mathscr{L}$; point b) of the previous theorem assures us that $g(A) \in \mathscr{L}$. Let's calculate $\lambda(g(A))$.

Because we assumed that g is an increasing function, for every open interval $I =]a, b[\subseteq \mathbb{R}$, $g(I) =]g(a), g(b)[$ is still an open interval and

$$\lambda(g(I)) = g(b) - g(a) = \int_a^b g'(x)dx = \int_I g'd\lambda = \int_I |g'|d\lambda.$$

Let now $D \subseteq \mathbb{R}$ be an open set, and let $D = \bigcup_{n=1}^{\infty} I_n$ be the representation of D (see Theorem 1.1.3); then, since the open intervals $g(I_n)$ are pairwise disjoint (g is injective),

$$\lambda(g(D)) = \lambda\left(\bigcup_{n=1}^{\infty} g(I_n)\right) = \sum_{n=1}^{\infty} \lambda(g(I_n)) = \sum_{n=1}^{\infty} \int_{I_n} |g'|d\lambda = \int_D |g'|d\lambda$$

(at the last above equality, we used Corollary 3.1.11).

Let $A \in \mathscr{L}$ be a bounded measurable set. For any $n \in \mathbb{N}^*$, let $D_n \in \tau_u$ such that $A \subseteq D_n$ and $\lambda(D_n \setminus A) < \frac{1}{n}$. It is obvious that we can assume that all the open sets D_n are bounded and that $D_{n+1} \subseteq D_n$.

Let $B = \bigcap_{n=1}^{\infty} D_n$; then $N = B \setminus A$ is a null set and $A = B \setminus N$. Then $g(A) = g(B) \setminus g(N)$, and, since N is a null set, $\lambda(g(A)) = \lambda(g(B))$ (see a) of previous theorem). The function g carries bounded sets into bounded sets (g is strictly increasing). Then $g(B) = \bigcap_{n=1}^{\infty} g(D_n)$, $g(D_{n+1}) \subseteq g(D_n)$ and $\lambda(g(D_n)) < +\infty$. The measure λ is continuous from above (see 7) of Theorem 1.3.11); therefore $\lambda(g(B)) = \lim_n \lambda(g(D_n))$. Hence,

$$\lambda(g(A)) = \lambda(g(B)) = \lim_n \lambda(g(D_n)) = \lim_n \int_{D_n} |g'| d\lambda = \int_B |g'| d\lambda = \int_A |g'| d\lambda.$$

In the penultimate equality, we took into account the fact that the mapping $A \mapsto \gamma(A) = \int_A |g'| d\lambda$ is a finite measure on $\mathscr{L}(\mathbb{R})$ (see Theorem 3.1.16) and this measure is continuous from above (see 7) of Theorem 1.4.7); the last equality is the consequence of Remark 3.2.12.

If $A \in \mathscr{L}$ is unbounded, then it is expressed as an union of an increasing sequence of measurable bounded sets: $A = \bigcup_{n=1}^{\infty} A_n$, where $A_n = A \cap [-n, n]$, for any $n \in \mathbb{N}$. Then $g(A) = \bigcup_{n=1}^{\infty} g(A_n)$, and, using the continuity property from below of measure (the property 6) of Theorem 1.3.11) and the monotone convergence theorem (Theorem 3.1.7),

$$\lambda(g(A)) = \lim_n \lambda(g(A_n)) = \lim_n \int_{A_n} |g'| d\lambda = \lim_n \int_A \chi_{A_n} \cdot |g'| d\lambda = \int_A |g'| d\lambda.$$

Now consider a Borel function $f : \mathbb{R} \to \mathbb{R}$ and a set $A \in \mathscr{L}$; as demonstrated above, $g(A) \in \mathscr{L}$ and the relation $(*)$ is verified. Moreover, according to Theorem 2.1.15, $f \circ g \in \mathfrak{L}(A)$ and then $(f \circ g) \cdot |g'| \in \mathscr{L}(A)$ (see Theorem 2.1.20). We will prove the theorem for different situations in which the function f can be found.

(1) Suppose that $f = \chi_B$, where $B \in \mathcal{B}_u$. Then, using the relation $(*)$,

$$\int_A (f \circ g) \cdot |g'| d\lambda = \int_A \chi_{g^{-1}(B)} \cdot |g'| d\lambda = \int_{A \cap g^{-1}(B)} |g'| d\lambda =$$

$$= \lambda(g(A \cap g^{-1}(B)) = \lambda(g(A) \cap B) = \int_{g(A)} f d\lambda.$$

We can remark that $f \in \mathfrak{L}^1(g(A))$ if and only if $\lambda(g(A) \cap B) < +\infty$ and that this last inequality occurs if and only if $(f \circ g) \cdot |g'| \in \mathfrak{L}^1(A)$. The above equality is precisely the relation $(**)$ in this case.

(2) Let now $f = \sum_{i=1}^{p} a_i \cdot \chi_{B_i}$ be a positive Borel simple function; then $a_i \geq 0$ and $B_i \in \mathcal{B}_u$, for any $i = 1, \cdots, p$. For any $i = 1, \cdots, p$, $\chi_{B_i} \in \mathfrak{L}^1(g(A))$ if and only if $(\chi_{B_i} \circ g) \cdot |g'| \in \mathfrak{L}^1(A)$ and

$$\int_{g(A)} f d\lambda = \sum_{i=1}^{p} a_i \cdot \int_{g(A)} \chi_{B_i} d\lambda = \sum_{i=1}^{p} a_i \cdot \int_A (\chi_{B_i} \circ g) \cdot |g'| d\lambda =$$

$$= \sum_{i=1}^{p} a_i \cdot \int_A \chi_{g^{-1}(B_i)} \cdot |g'| d\lambda = \int_A \sum_{i=1}^{p} a_i \cdot \chi_{g^{-1}(B_i)} \cdot |g'| d\lambda = \int_A (f \circ g) \cdot |g'| d\lambda.$$

(3) Let f be a positive Borel function; according to Remark 2.3.4, there exists an increasing sequence of positive Borel simple functions $(f_n)_n$ such that $f_n \uparrow f$. Using the monotone convergence theorem (Theorem 3.1.7), we obtain

$$\int_{g(A)} f d\lambda = \lim_n \int_{g(A)} f_n d\lambda = \lim_n \int_A (f_n \circ g) \cdot |g'| d\lambda = \int_A (f \circ g) \cdot |g'| d\lambda.$$

(4) Finally, if f is any Borel function, then $f = f^+ - f^-$, where f^+ and f^- are positive Borel functions. The proof results from the simple observation that $(f \circ g)^+ = (f^+ \circ g)$ and $(f \circ g)^- = (f^- \circ g)$. ■

3.5.2 Integrals Depending on a Parameter

When we integrate a function which depends on a parameter, the problems which arise concern the study of the integral depending on the mentioned parameter.

Let $A \in \mathscr{L}(\mathbb{R})$, let $T \subseteq \mathbb{R}$ and let $f : T \times A \to \mathbb{R}$ such that $f(t, \cdot) \in \mathfrak{L}^1(A)$, for every $t \in T$. Then, the function $F : T \to \mathbb{R}$, $F(t) = \int_A f(t, x) d\lambda(x)$, is well defined; F is said to be an **integral depending on a parameter**.

We want to specify the conditions under which the function F is continuous or differentiable on T.

Theorem 3.5.7
With the above notations, suppose that there is a positive function $g \in \mathfrak{L}^1(A)$ such that $|f(t, x)| \leq g(x)$, for every $t \in T$ and for almost every $x \in A$. Let t_0 be a limit point of T (there exists a sequence $(t_n)_n \subseteq T$, $t_n \neq t_0$, $t_n \to t_0$).

(1) If it exists $\lim_{t \to t_0} f(t, x) = \varphi(x)$, for almost every $x \in A$, then $\varphi \in \mathfrak{L}^1(A)$ and $\lim_{t \to t_0} \int_A f(t, x) d\lambda(x) = \int_A \varphi(x) d\lambda(x)$.

(2) If $f(\cdot, x)$ is continuous at $t_0 \in T$, for almost every $x \in A$, then $F = \int_A f(\cdot, x) d\lambda(x)$ is continuous at t_0.

(3) If $f(\cdot, x)$ is continuous on T, for almost every $x \in A$, then F is continuous on T.

Proof
(1) Let $(t_n)_n \subseteq T, t_n \to t_0, t_n \neq t_0, \forall n \in \mathbb{N}$; then $f(t_n, \cdot) \underset{A}{\overset{\cdot}{\to}} \varphi$. According to (7) of Theorem 2.1.18, $\varphi \in \mathfrak{L}(A)$. Since $|f(t_n, \cdot)| \leq g$ almost everywhere, then $|\varphi| \leq g$ almost everywhere. Theorem 3.2.6 assures us that $\varphi \in \mathfrak{L}^1(A)$. In conformity with the dominated convergence theorem (3.3.9), $\int_A f(t_n, x) d\lambda(x) \to \int_A \varphi d\lambda$. The sequence $(t_n)_n$ being arbitrary, there exists $\lim_{t \to t_0} \int_A f(t, x) d\lambda(x) = \int_A \varphi(x) d\lambda(x)$.

(2) and (3) are consequences of (1). ■

Theorem 3.5.8
Let $A \in \mathscr{L}$, let $T \subseteq \mathbb{R}$ be an open interval, and let $f : T \times A \to \mathbb{R}$, such that $f(t, \cdot) \in \mathfrak{L}^1(A)$, for every $t \in T$ and $g \in \mathfrak{L}^1(A)$. We suppose that
(a) There exists $\frac{\partial f}{\partial t}(t, x)$, for every $t \in T$ and almost for every $x \in A$, and
(b) $\left|\frac{\partial f}{\partial t}(t, x)\right| \leq g(x)$, for every $t \in T$ and for almost every $x \in A$.

Then
(1) $\frac{\partial f}{\partial t}(t, \cdot) \in \mathfrak{L}^1(A)$, for every $t \in T$,
(2) $F : T \to \mathbb{R}$, $F(t) = \int_A f(t, x)d\lambda(x)$, is differentiable on T and

$$F'(t) = \int_A \frac{\partial f}{\partial t}(t, x)d\lambda(x), \text{ for every } t \in T.$$

Proof
Let $t \in T$ be an arbitrary but fixed point, and let $a > 0$ such that $]t - a, t + a[\subseteq T$. The function $r : (]-a, a[\setminus\{0\}) \times A \to \mathbb{R}$, $r(h, x) = \frac{1}{h} \cdot [f(t + h, x) - f(t, x)]$, verifies the hypotheses (1) of the previous theorem. Indeed, for every $h \in]-a, a[$ and every $x \in A$, there exists s between t and $t + h$ such that $f(t + h, x) - f(t, x) = \frac{\partial f}{\partial t}(s, x) \cdot h$ and then, for every $h \in]-a, a[\setminus\{0\}$ and for almost every $x \in A$,

$$|r(h, x)| = \frac{1}{|h|} \cdot |f(t + h, x) - f(t, x)| \leq \left|\frac{\partial f}{\partial t}(s, x)\right| \leq g(x).$$

Moreover, for almost every $x \in A$, there exists $\lim_{h\to 0} r(h, x) = \frac{\partial f}{\partial t}(t, x)$.

Then $\lim_{h\to 0} r(h, \cdot) = \frac{\partial f}{\partial t}(t, \cdot) \in \mathfrak{L}^1(A)$ and

$$\lim_{h\to 0} \int_A r(h, x)d\lambda(x) = \int_A \frac{\partial f}{\partial t}(t, x)d\lambda(x).$$

The proof ends if we notice that

$$\int_A r(h, x)d\lambda(x) = \frac{1}{h}[F(t + h) - F(t)].$$

■

3.5.3 Jensen's Inequality

Let $J \subseteq \mathbb{R}$ be an interval and let $f : J \to \mathbb{R}$; f is called **convex** on J if, for every $x, y \in J$ and every $t \in]0, 1[$ we have $f(tx + (1 - t)y) \leq tf(x) + (1 - t)f(y)$.

We will need some useful lemmas.

Lemma 3.5.9 *$f : J \to \mathbb{R}$ is convex on J if and only if, for every $x \in J$, the function $g_x : J \setminus \{x\} \to \mathbb{R}$, defined by*

$$g_x(y) = \frac{f(y) - f(x)}{y - x}, \text{ for every } y \in J,$$

is increasing on $J \setminus \{x\}$.

Proof

Let us suppose that f is convex on J and let an arbitrary $x \in J$. Every two points $y, z \in J \setminus \{x\}$ with $y < z$ can be found in one of the situations:

If $y < z < x$, then $t = \frac{x-z}{x-y} \in]0, 1[$ and $z = ty + (1 - t)x$. Hence $f(z) \leq \frac{x-z}{x-y} \cdot f(y) + \frac{z-y}{x-y} \cdot f(x)$. In the two members of the inequality, we multiply by $x - y$ and add $xf(x)$; after some calculations, we obtain $g_x(y) \leq g_x(z)$.

In the cases $y < x < z$ and $x < y < z$, the procedure is similar, getting every time $g_x(y) \leq g_x(z)$.

Conversely, for every $x, y \in J$ and $t \in (0, 1)$, let $z = tx + (1 - t)y$; if $x < y$, then $x < z < y$. Since g_z is increasing, $g_z(x) \leq g_z(y)$, from where $\dfrac{f(x) - f(z)}{x - z} \leq \dfrac{f(y) - f(z)}{y - z}$. Then we obtain

$$f(tx + (1 - t)y) = f(z) \leq \frac{y - z}{y - x} \cdot f(x) + \frac{z - x}{y - x} \cdot f(u) = tf(x) + (1 - t)f(y). \quad \blacksquare$$

Lemma 3.5.10 *$f : J \to \mathbb{R}$ is convex on J if and only if, for every interior point x_0 of the interval J, there exist $a, b \in \mathbb{R}$ such that $f(x) \geq ax+b$, for every $x \in J$ and $f(x_0) = ax_0+b$.*

Proof

Let the function $f : J \to \mathbb{R}$ be convex on J, and let x_0 be an interior point of J; according to the previous lemma, the function $x \mapsto \frac{f(x)-f(x_0)}{x-x_0}$ is increasing on $J \setminus \{x_0\}$, and then there exists

$$a = \lim_{x \uparrow x_0} \frac{f(x) - f(x_0)}{x - x_0} = \sup \left\{ \frac{f(x) - f(x_0)}{x - x_0} : x \in J, x < x_0 \right\}.$$

It follows that, for every $x < x_0$, $f(x) \geq ax + (f(x_0) - ax_0)$. For all $x > x_0$, $\frac{f(x)-f(x_0)}{x-x_0} \geq a$, and then $f(x) \geq ax + (f(x_0) - ax_0)$. Necessity results if we take $b = f(x_0) - ax_0$.

Conversely, for every $x, y \in J$ and every $t \in (0, 1)$, $x_0 = tx+(1-t)y$ is an interior point of J. Then there exist $a, b \in \mathbb{R}$ such that $f(z) \geq az + b$, for any $z \in J$ and $f(x_0) = ax_0 + b$. Therefore $tf(x) + (1 - t)f(y) \geq t(ax_0 + b) + (1 - t)(ax_0 + b) = ax_0 + b = f(x_0) = f(tx + (1 - t)y)$. $\blacksquare$

Lemma 3.5.11 *Let a convex function $f : J \to \mathbb{R}$; then f is continuous on the interval I formed by the interior points of the interval J.*

Proof

If $J \in \mathcal{I}$, then $I = J$; generally $\text{card}(J \setminus I) \leq 2$.

Let $x \in I$ and let $y, z \in I$, such that $y < x < z$. For every $(x_n)_n \subseteq J$, $x_n \uparrow x$, we can consider without loss of generality that $y < x_n < x < z$, for all $n \in \mathbb{N}$. According to Lemma 3.5.9, g_x, defined by $g_x(u) = \frac{f(u)-f(x)}{u-x}$, is increasing on $J \setminus \{x\}$, and then $g_x(y) \leq g_x(x_n) \leq g_x(z)$. If we multiply the above inequalities by $x_n - x < 0$, we obtain $\frac{x_n - x}{y - x} \cdot [f(y) - f(x)] \geq f(x_n) - f(x) \geq \frac{x_n - x}{z - x} \cdot [f(z) - f(x)]$, for any $n \in \mathbb{N}$. Passing to the limit in the above inequalities, we obtain $f(x_n) \to f(x)$. Therefore $\lim_{u \uparrow x} f(u) = f(x)$.

Similarly, it is shown that $\lim_{u \downarrow x} f(u) = f(x)$; therefore f is continuous at x, for every $x \in I$. ■

Remark 3.5.12 Let $J \in \mathcal{J}$ and let $f : J \to \mathbb{R}$ be a convex function; then f is a Borel function on J, and so it is measurable on J. Indeed, if I is the interval of all interior points of J and if $a \in \mathbb{R}$, then $f^{-1}(]-\infty, a[) = [f^{-1}(]-\infty, a[) \cap I] \cup [f^{-1}(]-\infty, a[) \cap (J \setminus I)]$. According to the previous lemma, f is continuous on I; hence $f^{-1}(]-\infty, a[) \cap I \in \tau_u \subseteq \mathcal{B}_u$; since $\text{card}(f^{-1}(]-\infty, a[) \cap (J \setminus I)) \leq 2$, $f^{-1}(]-\infty, a[) \cap (J \setminus I) \in \mathcal{B}_u$, and then $f^{-1}(]-\infty, a[) \in \mathcal{B}_u$.

Theorem 3.5.13 (Jensen's Inequality)
Let $J \subseteq \mathbb{R}$ be an interval, let $F : J \to \mathbb{R}$ be a convex function, let $A \in \mathscr{L}$ with $0 < \lambda(A) < +\infty$, and let $f : A \to J$, $f \in \mathfrak{L}^1(A)$.
Then $F \circ f$ has an integral on A and

$$F\left(\frac{1}{\lambda(A)} \cdot \int_A f d\lambda\right) \leq \frac{1}{\lambda(A)} \cdot \int_A (F \circ f) d\lambda.$$

Proof

According to Lemma 3.5.10, for every $y_0 \in J$ there exist $a, b \in \mathbb{R}$ such that $F(y) \geq ay + b$, for every $y \in J$ and $F(y_0) = ay_0 + b$. We will take $y_0 = \frac{1}{\lambda(A)} \cdot \int_A f d\lambda \in I$ (see Exercise 14) of 3.7).

For every $x \in A$, $y = f(x) \in J$ and then

$$F(f(x)) \geq af(x) + b.$$

Since F is a Borel function (see Remark 3.5.12), Theorem 2.1.15 assures us that $F \circ f \in \mathfrak{L}(A)$. Since $\lambda(A) < +\infty$, the function $g : A \to \mathbb{R}$, $g(x) = af(x) + b$, is integrable on A, and, since $F \circ f \geq g$, $(F \circ f)^- \leq g^-$. Then $\int_A (F \circ f)^- d\lambda < +\infty$. It follows that $F \circ f$

has an integral on A (see Definition 3.2.1). According to (2) of Theorem 3.2.11,

$$\int_A (F \circ f) d\lambda \geq a \int_A f d\lambda + b\lambda(A) = a y_0 \lambda(A) + b\lambda(A) = F(y_0) \cdot \lambda(A),$$

which is exactly the inequality to be demonstrated. ■

Corollary 3.5.14 *Under the conditions of the previous theorem, if $\lambda(A) = 1$, then*

$$F\left(\int_A f d\lambda\right) \leq \int_A (F \circ f) d\lambda.$$

3.6 Abstract Setting

Let $(X, \mathcal{A})$ be a measurable space, and let γ be a σ-finite measure on $\mathcal{A}$. We denote with $\mathcal{E}(X)$ the set of all $\mathcal{A}$-simple functions, with $\mathcal{E}_+(X)$ the subset of positive simple functions and with $\mathcal{M}(X)$ the set of all measurable functions on X. For every $f = \sum_{i=1}^{p} a_i \cdot \chi_{A_i} \in \mathcal{E}_+(X)$ and every $A \in \mathcal{A}$, let's define

$$\int_A f d\gamma = \sum_{i=1}^{p} a_i \cdot \gamma(A_i \cap A) \in [0, +\infty].$$

It is said that f is **integrable** on A if $\int_A f d\gamma < +\infty$.

Let $\mathcal{E}_+^1(X)$ be the set of all positive simple functions integrable on X.

If we replace $\mathscr{L}$ with $\mathcal{A}$ and $\mathcal{E}_+(A)$ with $\mathcal{E}_+(X)$ we find, in this abstract framework, the properties of Proposition 3.1.3:

Proposition 3.6.1 *Let $f, g \in \mathcal{E}_+(X)$ and $c \in \mathbb{R}_+$; then*

(1) *$cf \in \mathcal{E}_+(X)$ and $\int_X cf d\gamma = c \int_X f d\gamma$.*
(2) *$f + g \in \mathcal{E}_+(X)$ and $\int_X (f + g) d\gamma = \int_X f d\gamma + \int_X g d\gamma$.*
(3) *$\int_X f d\gamma \leq \int_X g d\gamma$, if $f \leq g$, γ-a.e.*
(4) *$\int_B f d\gamma \leq \int_C f d\gamma$, for every $B, C \in \mathcal{A}$ with $B \subseteq C$.*
(5) *For every $\varepsilon > 0$, there exists $\delta > 0$ such that $\int_B f d\gamma < \varepsilon$, for every $B \in \mathcal{A}$ with $\gamma(B) < \delta$.*

Definition 3.6.2

Let $f \in \mathcal{M}_+(X)$ be a positive measurable function on X and let $A \in \mathcal{A}$; we define

$$\int_A f d\gamma = \sup \left\{ \int_A \varphi d\gamma : \varphi \in \mathcal{E}_+(X), \varphi \leq f \right\} \in [0, +\infty].$$

f is **integrable** if $\int_X f d\gamma < +\infty$. Let $\mathfrak{L}_+^1(X)$ be the set of all positive integrable functions on X.

We find the results of Proposition 3.1.6:

Proposition 3.6.3 *Let $A \in \mathcal{A}$, $f, g \in \mathcal{M}_+(X)$ and $c \geq 0$; then:*

(1) $cf \in \mathcal{M}_+(X)$ *and* $\int_A cf d\gamma = c \int_A f d\gamma$.

(2) $\int_A f d\gamma \leq \int_A g d\gamma$, *if* $f \leq g$.

(3) $\int_B f d\gamma \leq \int_C f d\gamma$, *for every* $B, C \in \mathcal{A}$, $B \subseteq C$.

We can also prove the monotone convergence theorem, Beppo Levi's theorem, and Fatou's lemma.

Theorem 3.6.4 (Convergence Monotone Theorem)
Let $f : X \to \mathbb{R}_+$ and $(f_n) \subseteq \mathcal{M}_+(X)$ such that $f_n \leq f_{n+1}$, for any $n \in \mathbb{N}$ and $f_n \uparrow f$; then $f \in \mathcal{M}_+(X)$ and $\int_X f_n d\gamma \uparrow \int_X f d\gamma$.

Theorem 3.6.5 (Beppo Levi)
Let $(f_n)_n \subseteq \mathcal{M}_+(X)$ be a sequence for which the corresponding series $\sum_{n=1}^{\infty} f_n$ is pointwise convergent on X and let $f = \sum_{n=1}^{\infty} f_n$; then $f \in \mathcal{M}_+(X)$ and

$$\int_X f d\gamma = \sum_{n=1}^{\infty} \int_X f_n d\gamma.$$

Theorem 3.6.6 (Fatou's Lemma)
Let $(f_n) \subseteq \mathcal{M}_+(X)$ such that $f = \liminf_n f_n < +\infty$; then

$$\int_X \liminf_n f_n d\gamma \leq \liminf_n \int_X f_n d\gamma.$$

Let $f \in \mathcal{M}(X)$; then $f^+ = \sup\{f, \underline{0}\} \in \mathcal{M}_+(X)$, $f^- = \sup\{-f, \underline{0}\} \in \mathcal{M}_+(X)$, and $f = f^+ - f^-$, $|f| = f^+ + f^-$.

If one of the functions f^+ or f^- is integrable, then we can define

$$\int_X f d\gamma = \int_X f^+ d\gamma - \int_X f^- d\gamma.$$

If the two functions f^+ and f^- are integrable, then it is said that f is **integrable**. We denote $\mathfrak{L}^1(X)$ the set of all integrable functions on X.

For every $A \in \mathcal{A}$ and every $f \in \mathfrak{L}(X)$, $\int_A f d\gamma = \int_X \chi_A \cdot f d\gamma$. The function f is integrable on A if $\chi_A \cdot f \in \mathfrak{L}^1(X)$; $\mathfrak{L}^1(A)$ is the set of functions integrable on A.

Let $f = \sum_{i=1}^{n} a_i \chi_{A_i}$ be a $\mathcal{A}$-simple function: $\{a_1, \ldots, a_p\} \subseteq \mathbb{R}$ and $\{A_1, \ldots, A_p\}$ is a partition $\mathcal{A}$-measurable of X. We can remark that $f \in \mathfrak{L}^1(X)$ if and only if $a_i = 0$ for any $i \in \{1, \cdots, p\}$ for which $\gamma(A_i) = +\infty$. Let $\mathcal{E}^1(X) \subseteq \mathfrak{L}^1(X)$ be the set of all $\mathcal{A}$-simple integrable functions.

With the appropriate adaptations, the results 3.2.2–3.2.8, 3.2.10, 3.3.1, 3.3.3, and 3.3.6–3.3.10 also work in an abstract framework.

Theorem 3.6.7
Let $f \in \mathcal{M}(X)$; then $f \in \mathfrak{L}^1(X)$ if and only if $|f| \in \mathfrak{L}^1_+(X)$, and, in this case,

$$\left| \int_X f d\gamma \right| \leq \int_X |f| d\gamma.$$

Theorem 3.6.8
Let $f \in \mathcal{M}(X)$.
(1) *If $f = \underline{0}$ γ-a.e. on X, then $f \in \mathfrak{L}^1(X)$ and $\int_X f d\gamma = 0$.*
(2) *If $\gamma(A) = 0$, then $f \in \mathfrak{L}^1(A)$ and $\int_A f d\gamma = 0$.*

Theorem 3.6.9
Let $f \in \mathfrak{L}^1(X)$.
(a) *If $\int_A f d\gamma \geq 0$, for every $A \in \mathcal{A}$, then $f \geq \underline{0}$ γ-a.e. on X.*
(b) *If $\int_A f d\gamma \leq 0$, for every $A \in \mathcal{A}$, then $f \leq \underline{0}$ γ-a.e. on X.*
(c) *If $\int_A f d\gamma = 0$, for every $A \in \mathcal{A}$, then $f = \underline{0}$ γ-a.e. on X.*

Theorem 3.6.10
Let $f \in \mathcal{M}(X)$ and $g \in \mathfrak{L}^1_+(X)$ such that $|f| \leq g$ γ-a.e.; then $f \in \mathfrak{L}^1(X)$.

Theorem 3.6.11
If $f \in \mathfrak{L}^1(X)$ and $A \in \mathcal{A}$, then $f \in \mathfrak{L}^1(A)$.

Theorem 3.6.12
A bounded measurable function is integrable on a finite measure set.

Theorem 3.6.13
$\mathfrak{L}^1(X)$ is a real vector space, and the mapping $I : \mathfrak{L}^1(X) \to \mathbb{R}$, $I(f) = \int_X f d\gamma$, is a linear operator:
(1) $\int_X (f+g)d\gamma = \int_X f d\gamma + \int_X g d\gamma$, *for every* $f, g \in \mathfrak{L}^1(X)$;
(2) $\int_X cf d\gamma = c\int_X f d\gamma$, *for every* $f \in \mathfrak{L}^1(X)$ *and any* $c \in \mathbb{R}$.

Theorem 3.6.14
Let $f, g \in \mathfrak{L}^1(X)$; *then*
(1) $\int_X f d\gamma \le \int_X g d\gamma$, *if* $f \le g$.
(2) *For every* $\varepsilon > 0$, *there exists* $\delta > 0$ *such that* $\int_A |f| d\gamma < \varepsilon$, *for every* $A \in \mathcal{A}$ *with* $\gamma(A) < \delta$.
(3) *For every* $\varepsilon > 0$, *there exists* $A_0 \in \mathcal{A}$ *with* $\gamma(A_0) < +\infty$ *such that* $\int_{X \setminus A_0} |f| d\gamma < \varepsilon$.
(4) $\int_{\bigcup_{n=1}^{\infty} A_n} f d\gamma = \sum_{n=1}^{\infty} \int_{A_n} f d\gamma$, *for every* $(A_n)_n \subseteq \mathcal{A}$ *with* $A_n \cap A_m = \emptyset$, *for all* $n \neq m$.

Theorem 3.6.15
Let $\|\cdot\|_1 : \mathfrak{L}^1(X) \to \mathbb{R}_+$, *defined by* $\|f\|_1 = \int_X |f| d\gamma$, *for every* $f \in \mathfrak{L}^1(X)$; *then*
(1) $\|f\|_1 = 0$ *if and only if* $f = \underline{0}$ γ*-a.e.*
(2) $\|cf\|_1 = |c| \cdot \|f\|_1$, *for every* $f \in \mathfrak{L}^1(X)$ *and any* $c \in \mathbb{R}$.
(3) $\|f+g\|_1 \le \|f\|_1 + \|g\|_1$, *for every* $f, g \in \mathfrak{L}^1(X)$.

Theorem 3.6.16
Let $(f_n) \subseteq \mathfrak{L}^1(X)$ *and let* $f \in \mathfrak{L}^1(X)$.
(1) *If* $f_n \xrightarrow[X]{\|\cdot\|_1} f$, *then* $f_n \xrightarrow[X]{\gamma} f$.
(2) *If* $(f_n)_n$ *is* $\|\cdot\|_1$*-Cauchy, then* $(f_n)_n$ *is Cauchy in measure.*

Theorem 3.6.17 (Dominated Convergence Theorem)
Let $(f_n) \subseteq \mathcal{M}(X)$ *and let* $g \in \mathfrak{L}^1(X)$ *such that*
(1) $f_n \underset{X}{\dot{\rightarrow}} f$ *and*
(2) $|f_n| \le g$, *for any* $n \in \mathbb{N}$.

Then $(f_n) \subseteq \mathfrak{L}^1(X)$, $f \in \mathfrak{L}^1(X)$, $f_n \xrightarrow[X]{\|\cdot\|_1} f$ *and* $\int_X f_n d\gamma \to \int_X f d\gamma$.

Theorem 3.6.18 (Bounded Convergence Theorem)
Let us suppose that $\gamma(X) < +\infty$, $(f_n) \subseteq \mathcal{M}(X)$ *and* $c \in \mathbb{R}_+$ *such that*
(1) $f_n \underset{X}{\dot{\rightarrow}} f$.
(2) $|f_n| \le c$, *for any* $n \in \mathbb{N}$.

Then $(f_n) \subseteq \mathfrak{L}^1(X)$, $f \in \mathfrak{L}^1(X)$, $f_n \xrightarrow[X]{\|\cdot\|_1} f$ *and* $\int_X f_n d\gamma \to \int_X f d\gamma$.

The seminormed space $(\mathfrak{L}^1(X), \|\cdot\|_1)$ is complete, and the set of integrable simple functions, $\mathcal{E}^1(X) = \mathcal{E}(X) \cap \mathfrak{L}^1(X)$, is dense in $\mathfrak{L}^1(X)$ with respect to the topology of $\|\cdot\|_1$- convergence.

3.7 Exercises

(1) If $A \in \mathscr{L}$ and $f \in \mathfrak{L}_+(A)$ with $0 \le f(x) \le a$, then $0 \le \int_A f d\lambda \le a\lambda(A)$.
(2) If $f \in \mathfrak{L}_+(A)$, then $\int_A f d\lambda \ge a\lambda((f \ge a))$, for any $a > 0$.
If $f \in \mathfrak{L}^1_+(A)$, then $\lim_{a \to +\infty} a\lambda((f \ge a)) = 0$.
(3) Compute $\int_{[0,+\infty[} e^{-[x]} d\lambda(x)$ ($[\cdot]$ is the ceiling function—$[x]$ is the least integer greater than or equal to x).
(4) Let $f : [a,b] \to \mathbb{R}$ defined by $f(x) = \alpha x + \beta$, with $\alpha > 0$ and $f(a) > 0$. For any $n \in \mathbb{N}^*$, we denote $c_k^n = a + k \cdot \frac{b-a}{n}$, $k = 0, 1, \ldots, n$ and le $f_n : [a,b] \to \mathbb{R}$, $f_n(x) = \sum_{k=0}^{n-1} f(c_k^n) \cdot (c_{k+1}^n - c_k^n)$. Show that $(f_n)_n \subseteq \mathcal{E}_+([a,b])$ and that $f_n \uparrow f$. Compute the integral $\int_{[a,b]} f d\lambda$, and compare with the Riemann integral $\int_a^b f(x)dx$.
(5) Let $f : [a,b] \to [m,M]$ be a bounded measurable function (then f is a Lebesgue integrable function; see Corollary 3.2.8); let $I = \int_{[a,b]} f d\lambda \in \mathbb{R}$. For every partition $\Delta = \{y_0, y_1, \cdots, y_n\}$ of the interval $[m,M]$, let $\sigma_\Delta = \sum_{k=1}^n y_{k-1} \cdot \lambda(E_k)$, $\Sigma_\Delta = \sum_{k=1}^n y_k \cdot \lambda(E_k)$, where, for any $k = 1, \cdots n$, $E_k = \{x \in [a,b] : y_{k-1} \le f(x) < y_k\} \in \mathfrak{L}([a,b])$. Show that $\sup_{\Delta \in \mathcal{D}([a,b])} \sigma_\Delta = \inf_{\Delta \in \mathcal{D}([a,b])} \Sigma_\Delta = I$. Show that, for every $\varepsilon > 0$, there exists $\delta > 0$ such that, for every $\Delta \in \mathcal{D}([a,b])$ with $\|\Delta\| < \delta$, $\Sigma_\Delta - \sigma_\Delta < \varepsilon$.

(6) Let $[c,d] \subseteq [a,b]$ and $f : [a,b] \to \mathbb{R}$, $f(x) = \begin{cases} 1, & x \in [c,d], \\ 0, & x \in [a,b] \setminus [c,d]. \end{cases}$

Show that, for every $\varepsilon > 0$, there exists a continuous function $g : [a,b] \to \mathbb{R}$ such that

$$\int_{[a,b]} |f - g| d\lambda < \varepsilon.$$

Indication: We define $g(x) = f(x)$ if $x \in [a, c - \varepsilon] \cup [c,d] \cup [d + \varepsilon, b]$ and piecewise linear on $[c - \varepsilon, c]$ and $[d, d + \varepsilon]$.

(7) Let $f \in \mathfrak{L}^1(A)$ and $B, C \in \mathscr{L}(A)$; show that

$$\int_{B \cup C} f d\lambda = \int_B f d\lambda + \int_C f d\lambda - \int_{B \cap C} f d\lambda.$$

(8) Let $f, g \in \mathfrak{L}^1(A)$ two bounded functions on A; show that $fg,\ f^2,\ g^2 \in \mathfrak{L}^1(A)$ and that

$$\int_A |fg| d\lambda \le \frac{1}{2}\left[\int_A f^2 d\lambda + \int_A g^2 d\lambda\right].$$

(9) For every function $f \in \mathfrak{L}^1_+(A)$ and for any $p \in \mathbb{N}$, we define $f_p : A \to \mathbb{R}_+$ by $f_p(x) = \begin{cases} f(x), & f(x) \le p \\ p, & f(x) > p \end{cases}$. Show that $(f_p) \subseteq \mathfrak{L}^1_+(A)$ and $\int_A f_p d\lambda \uparrow \int_A f d\lambda$.

Calculate this way $\int_{]0,1]} f d\lambda$, where $f(x) = \dfrac{1}{\sqrt[3]{x}}$, for every $x \in]0,1]$.

(10) We consider the function $f : \mathbb{R} \to \mathbb{R}$, $f = \sum_{n=1}^{+\infty} \frac{(-1)^n}{n} \chi_{[n,\, n+1[}$; show that it is measurable and bounded, but not integrable on $\mathbb{R}$ (see Corollary 3.2.8).

(11) Show that the function $f = \sum_{n=1}^{+\infty} \frac{(-1)^n}{n^2} \chi_{[n,\, n+1[}$ is integrable on $\mathbb{R}$.

(12) Let $f = \sum_{i=1}^{p} a_i \cdot \chi_{A_i} \in \mathcal{E}(A)$; show that $f \in \mathfrak{L}^1(A)$ if and only if, for any $i \in \{1, \cdots, p\}$ for which $a_i \neq 0$ it follows that $\lambda(A_i) < +\infty$. Deduce by that $\mathcal{E}(A) \subseteq \mathfrak{L}^1(A)$, if $\lambda(A) < +\infty$.

(13) Let $A \in \mathscr{L}$, $a, b \in \mathbb{R}$, $a \neq 0$, $f : aA + b \to \mathbb{R}$ and $g : A \to \mathbb{R}$, $g(x) = f(ax + b)$, for every $x \in A$. Show that $f \in \mathfrak{L}^1(aA + b)$ if and only if $g \in \mathfrak{L}^1(A)$ and that

$$\int_{aA+b} f(y) d\lambda(y) = |a| \cdot \int_A g(x) d\lambda(x) = |a| \cdot \int_A f(ax + b) d\lambda(x).$$

Note that in this case, unlike that of Theorem 3.5.6 *(where f is assumed to be Borel function), the function f is an arbitrary measurable function.*

Indication. According to Exercise (5) of 2.5, $f \in \mathfrak{L}(aA + b)$ if and only if $g \in \mathfrak{L}(A)$. The demonstration is made in the following cases: (1) $f = \chi_E \in \mathfrak{L}(aA + b)$, (2) $f = \sum_{k=1}^{p} a_i \cdot \chi_{E_i} \in \mathcal{E}(aA + b)$, (3) $f \in \mathfrak{L}_+(aA + b)$ and (4) $f \in \mathfrak{L}(aA + b)$.

(14) Let $A \in \mathscr{L}$ with $0 < \lambda(A) < +\infty$ and let $f : A \to I$, where I is an interval and $f \in \mathfrak{L}^1(A)$. Show that $\frac{1}{\lambda(A)} \cdot \int_A f d\lambda \in I$.

(15) Show that $\sum_{n=1}^{\infty} f_n$ converges a.e., in the case where $(f_n) \subseteq \mathfrak{L}_+^1(A)$ and $\sum_{n=1}^{\infty} \int_A f_n d\lambda < +\infty$.

(16) Show that $f : [0, +\infty[\to \mathbb{R}, f(x) = \dfrac{\sin^2 x}{x^2}$, is Riemann and Lebesgue integrable on $[0, +\infty[$.

(17) Let $f_n :]0, +\infty[\to \mathbb{R}, f_n(x) = \begin{cases} \dfrac{x}{n^2}, & 0 < x < n, \\ 0, & x \geq n. \end{cases}$

Check if $\lim_n \left(\int_{]0,\infty[} f_n d\lambda \right) = \int_{]0,\infty[} \left(\lim_n f_n \right) d\lambda$ and explain the result.

(18) Let $f \in \mathfrak{L}^1(\mathbb{R})$; show that

$$\lim_{h \downarrow 0} \int_{\mathbb{R}} |f(x+h) - f(x)| d\lambda(x) = 0.$$

(19) Let $f \in \mathfrak{L}^1(\mathbb{R})$; compute

$$\lim_{n \to \infty} \int_{\mathbb{R}} f(x) \sin^n x d\lambda(x).$$

(20) Let $f \in \mathfrak{L}^1(\mathbb{R})$; show that

$$\lim_{n \to +\infty} \int_{(|x|>n)} f(x) d\lambda(x) = 0.$$

(21) Let $(X, \mathcal{A}, \gamma)$ be an abstract measure space where γ is a positive σ-finite and complete measure and let $A \in \mathcal{A}$; let's define $\gamma_A : \mathcal{A} \to \bar{\mathbb{R}}_+$ by $\gamma_A(B) = \gamma(A \cap B)$. Show that every γ_A-integrable function is γ-integrable.

(22) Let $A \in \mathscr{L}$ with $\lambda(A) = 1$. Show that

$$e^{\int_A f(x) d\lambda(x)} \leq \int_A e^{f(x)} d\lambda(x), \text{ for every } f \in \mathfrak{L}^1(A),$$

$$\int_A \ln(g(x)) d\lambda(x) \leq \ln \left(\int_A g(x) d\lambda(x) \right), \text{ for every } g \in \mathfrak{L}^1(A), g > 0.$$

(23) Let $A \in \mathscr{L}$ with $0 < \lambda(A) < +\infty$; show that, for every $f \in \mathfrak{L}^1(A)$ and any $p \geq 1$,

$$\left| \int_A f d\lambda \right| \leq (\lambda(A))^{\frac{p-1}{p}} \cdot \left(\int_A |f|^p d\lambda \right)^{\frac{1}{p}}.$$

Indication. Jensen's inequality will be applied to the function $F(y) = |y|^p$.

The L^p Spaces

This chapter is devoted to a class of Banach spaces constructed using the notion of an integrable function—Lebesgue spaces or classical Banach spaces.

In the first paragraph, we will present the algebraic and topological structure of these spaces. Paragraph 2 is reserved for the study of density properties in L^p.

A limit case of Lebesgue spaces, the space L^∞, is studied in the third paragraph, and in the last paragraph, an introduction in the theory of Fourier series on L^2 is made.

4.1 Algebraic and Topological Structure

Definition 4.1.1

Let $A \in \mathscr{L}$, $p \in \mathbb{R}$, $p \geq 1$; a function $f : A \to \mathbb{R}$ is said to be ***p*-integrable** on A if $f \in \mathfrak{L}(A)$ and $|f|^p \in \mathfrak{L}^1(A)$.

We denote with $\mathfrak{L}^p(A)$ the set of all p-integrable functions on A.

In the particular case $p = 1$, we find the space $\mathfrak{L}^1(A)$ studied in the previous chapter; indeed, $f \in \mathfrak{L}^1(A)$ if and only if $|f| \in \mathfrak{L}^1(A)$ (see Theorem 3.2.2).

We define the application $\|\cdot\|_p : \mathfrak{L}^p(A) \to \mathbb{R}_+$ by

$$\|f\|_p = \left(\int_A |f|^p d\lambda\right)^{\frac{1}{p}}, \text{ for every } f \in \mathfrak{L}^p(A).$$

Proposition 4.1.2 *$\mathfrak{L}^p(A)$ is a real vector space.*

Proof

The set $F(A, \mathbb{R})$ of all real functions defined on A is a vector space with respect to the usual operations of addition and multiplication with scalars (operations punctually defined). To show that $\mathfrak{L}^p(A) \subseteq F(A, \mathbb{R})$ is a vector subspace, it is enough to show that the sum of two functions of $\mathfrak{L}^p(A)$ remains in $\mathfrak{L}^p(A)$ and that the product of a real scalar with a function of $\mathfrak{L}^p(A)$ remains in $\mathfrak{L}^p(A)$.

L. C. Florescu, *Lebesgue Integral*, Compact Textbooks in Mathematics,
https://doi.org/10.1007/978-3-030-60163-8_4

Let $f, g \in \mathfrak{L}^p(A)$; then $f, g \in \mathfrak{L}(A)$, and so $f + g \in \mathfrak{L}(A)$. On the other hand, the function $h : \mathbb{R} \to \mathbb{R}$, $h(y) = |y|^p$, is continuous, and therefore $h \circ (f+g) = |f+g|^p \in \mathfrak{L}(A)$ (see Theorem 2.1.15).

$$|f + g|^p \leq (|f| + |g|)^p \leq 2^p \cdot \max\{|f|^p, |g|^p\} \leq 2^p \cdot (|f|^p + |g|^p). \qquad (*)$$

The function $2^p \cdot (|f|^p + |g|^p)$ is integrable, and, from $(*)$, it dominates the measurable function $|f + g|^p$; Theorem 3.2.6 assures us that $|f + g|^p \in \mathfrak{L}^1(A)$, and so $f + g \in \mathfrak{L}^p(A)$.

For any $c \in \mathbb{R}$ and every $f \in \mathfrak{L}^p(A)$, $c \cdot f \in \mathfrak{L}(A)$ and $|c \cdot f|^p = |c|^p \cdot |f|^p \in \mathfrak{L}^1(A)$; hence $c \cdot f \in \mathfrak{L}^p(A)$. ∎

Lemma 4.1.3 *Let $p, q > 1$ such that $\frac{1}{p} + \frac{1}{q} = 1$; for all $a, b \geq 0$,*

$$a \cdot b \leq \frac{a^p}{p} + \frac{b^q}{q}.$$

Proof

The function defined by $x \mapsto f(x) = ax - \frac{a^p}{p} - \frac{x^q}{q}$ is differentiable over $[0, +\infty[$; its derivative, $f'(x) = a - x^{q-1}$, takes the value zero for $x_0 = a^{\frac{1}{q-1}}$. We immediately notice that f is increasing on $[0, x_0]$ and decreasing on $[x_0, +\infty[$. Hence $f(x) \leq f(x_0) = f(a^{\frac{1}{q-1}}) = a \cdot a^{\frac{1}{q-1}} - \frac{a^p}{p} - \frac{a^{\frac{q}{q-1}}}{q} = a^p \cdot (1 - \frac{1}{p} - \frac{1}{q}) = 0$, for every $x \geq 0$, which proves the lemma. ∎

Theorem 4.1.4 (Hölder Inequality)
Let $p, q > 1$ such that $\frac{1}{p} + \frac{1}{q} = 1$.
For every $f \in \mathfrak{L}^p(A)$ and every $g \in \mathfrak{L}^q(A)$, $f \cdot g \in \mathfrak{L}^1(A)$ and

$$\int_A |f \cdot g| d\lambda \leq \|f\|_p \cdot \|g\|_q = \left(\int_A |f|^p d\lambda\right)^{\frac{1}{p}} \cdot \left(\int_A |g|^q d\lambda\right)^{\frac{1}{q}}.$$

Proof

If $\|f\|_p = 0$, then $\int_A |f|^p d\lambda = 0$, from where $f = \underline{0}$ a.e. (Theorem 3.3.3). Then $f \cdot g = \underline{0}$ a.e., and, applying Theorem 3.3.3 again, $\|f \cdot g\|_1 = \int_A |fg| d\lambda = 0 \leq \|f\|_p \cdot \|g\|_q$. We reason in the same way if $\|g\|_q = 0$.

Let us suppose that $\|f\|_p \neq 0 \neq \|g\|_q$. In the inequality of the previous lemma, let us replace $a = \frac{|f(x)|}{\|f\|_p}$ and $b = \frac{|g(x)|}{\|g\|_q}$; we obtain then

$$\frac{|f(x)g(x)|}{\|f\|_p \cdot \|g\|_q} \leq \frac{|f(x)|^p}{p\|f\|_p^p} + \frac{|g(x)|^q}{q\|g\|_q^q}, \text{ for every } x \in A.$$

The same inequality can be written:

$$|fg| \leq ||f||_p \cdot ||g||_q \left(\frac{1}{p||f||_p^p} \cdot |f|^p + \frac{1}{q||g||_q^q} \cdot |g|^q \right). \qquad (*)$$

As $f \in \mathfrak{L}^p(A)$ and $g \in \mathfrak{L}^q(A)$, it follows that $fg \in \mathfrak{L}(A)$ and $|f|^p, |g|^q \in \mathfrak{L}^1(A)$. Then

$$h = \|f\|_p \cdot \|g\|_q \left(\frac{1}{p\|f\|_p^p} |f|^p + \frac{1}{q\|g\|_q^q} |g|^q \right) \in \mathfrak{L}^1(A).$$

From $(*)$, $|fg| \leq h$; Theorem 3.2.6 leads to $fg \in \mathfrak{L}^1(A)$.

The integral being monotonic (see 1) of Theorem 3.3.1) and linear (Theorem 3.2.10), we can now integrate the inequality $(*)$ and then

$$\int_A |fg| d\lambda \leq \|f\|_p \cdot \|g\|_q \left(\frac{1}{p\|f\|_p^p} \int_A |f|^p d\lambda + \frac{1}{q\|g\|_q^q} \int_A |g|^q d\lambda \right) =$$

$$= \|f\|_p \cdot \|g\|_q \left(\frac{1}{p||f||_p^p} \cdot ||f||_p^p + \frac{1}{q\|g\|_q^q} \cdot \|g\|_q^q \right) = \|f\|_p \cdot \|g\|_q.$$

■

Theorem 4.1.5 (Minkowski Inequality)
For any $p \geq 1$ and every $f, g \in \mathfrak{L}^p(A)$,

$$\|f + g\|_p \leq \|f\|_p + \|g\|_p.$$

Proof
If $p = 1$, then the inequality is obvious (see 3) of Theorem 3.3.3).

Let us suppose that $p > 1$ and $f, g \in \mathfrak{L}^p(A)$. From Proposition 4.1.2, $f + g \in \mathfrak{L}^p(A)$.

If $\|f + g\|_p = 0$, then the Minkowski inequality is obvious.

Let's then suppose, more, that $\|f + g\|_p > 0$. Let $q = \frac{p}{p-1} > 1$; then $\frac{1}{p} + \frac{1}{q} = 1$, and the function $h = |f + g|^{p-1} \in \mathfrak{L}^q(A)$. Indeed, h is the composition between the measurable function $f + g$ and the continuous function $l : \overline{\mathbb{R}} \to \mathbb{R}, l(y) = |y|^{p-1}$; from Proposition 2.1.15, $h = l \circ (f + g) \in \mathfrak{L}(A)$. Moreover, $|h|^q = |f + g|^p \in \mathfrak{L}^1(A)$ ($f + g \in \mathfrak{L}^p(A)$), and so $h \in \mathfrak{L}^q(A)$.

From the inequality of Hölder, it follows that $|f + g|^{p-1} \cdot |f| = h|f| \in \mathfrak{L}^1(A)$ and $|f+g|^{p-1} \cdot |g| = h|g| \in \mathfrak{L}^1(A)$. Then, using the properties of monotonicity and linearity of the integral,

$$(\|f + g\|_p)^p = \int_A |f + g|^p d\lambda = \int_A |f + g|^{p-1} \cdot |f + g| d\lambda \leq$$

$$\leq \int_A |f+g|^{p-1}(|f|+|g|)d\lambda = \int_A |f+g|^{p-1}|f|d\lambda + \int_A |f+g|^{p-1}|g|d\lambda.$$

But, according to the Hölder inequality,

$$\int_A |f+g|^{p-1} \cdot |f|d\lambda = \|fh\|_1 \leq \|f\|_p \cdot \|h\|_q,$$

$$\int_A |f+g|^{p-1} \cdot |g|d\lambda = \|gh\|_1 \leq \|g\|_p \cdot \|h\|_q.$$

Then

$$\left(||f+g||_p\right)^p \leq \left(||f||_p + ||g||_p\right) \cdot ||h||_q =$$

$$= \left(||f||_p + ||g||_p\right) \cdot \left(\int_A |f+g|^{(p-1)q} d\lambda\right)^{\frac{1}{q}} =$$

$$= \left(||f||_p + ||g||_p\right) \cdot \left(\int_A |f+g|^p d\lambda\right)^{\frac{p-1}{p}} = \left(||f||_p + ||g||_p\right) \cdot \left(||f+g||_p\right)^{p-1}.$$

Simplifying the previous inequality with $\left(||f+g||_p\right)^{p-1}$, we obtain

$$||f+g||_p \leq ||f||_p + ||g||_p.$$ ■

Theorem 4.1.6
$(\mathfrak{L}^p(A), \|\cdot\|_p)$ is a seminormed space.

Proof
$\|f\|_p = 0$ if and only if $f = \underline{0}$, a.e. (Theorem 3.3.3).

For any $c \in \mathbb{R}$ and for every $f \in \mathfrak{L}^p(A)$, $\|c \cdot f\|_p = \left(\int_A |c \cdot f|^p d\lambda\right)^{\frac{1}{p}} = |c| \cdot \|f\|_p$.

Triangle inequality is actually Minkowski inequality. ■

Definition 4.1.7
If $\|f_n - f\|_p \xrightarrow[n\to+\infty]{} 0$, then we say that the sequence $(f_n) \subseteq \mathfrak{L}^p(A)$ is **L^p-convergent** to $f \in \mathfrak{L}^p(A)$ and we write $f_n \xrightarrow[A]{\|\cdot\|_p} f$. The mapping $\|\cdot\|_p$ is called the **seminorm of L^p-convergence**.

(Continued)

Definition 4.1.7 (continued)

The sequence (f_n) is L^p**-Cauchy** if $\lim_{m,n\to+\infty}\|f_m - f_n\|_p = 0$.
$f \in \mathfrak{L}^p(A)$ is L^p**-adherent point** for $F \subseteq \mathfrak{L}^p(A)$ if there exists a sequence $(g_n) \subseteq F$ such that $g_n \xrightarrow[A]{\|\cdot\|_p} f$. We denote with $\overline{F}^p$ the set of all L^p-adherent points of F. It can be easily shown that $f \in \overline{F}^p$ if and only if, for every $\varepsilon > 0$, there exists $g \in F$ such that $\|f - g\|_p < \varepsilon$. The set $F \subseteq \mathfrak{L}^p(A)$ is **dense** if $\overline{F}^p = \mathfrak{L}^p(A)$.

Definition 4.1.8

Let $p \geq 1$; the la relation $f \sim g$ if and only if $f = g$ a.e. is an equivalence relation on $\mathfrak{L}^p(A)$ ($\sim$ is reflexive, symmetric, and transitive). We remark that if $f \in \mathfrak{L}^p(A)$ and $f = g$ a.e., then $g \in \mathfrak{L}^p(A)$ and $\|f\|_p = \|g\|_p$. Indeed, if $f \in \mathfrak{L}^p$, then $g \in \mathfrak{L}(A)$ (see 1) of Theorem 2.1.13); since $|f|^p = |g|^p$ a.e., $\int_A |f|^p d\lambda = \int_A |g|^p d\lambda$ (Theorem 3.2.11), and so $g \in \mathfrak{L}^p(A)$ and $\|f\|_p = \|g\|_p$.
Let $\mathfrak{L}^p(A)|_\sim = L^p(A)$ be the quotient set. The elements of this set are of the form: $[f] = \{g \in \mathfrak{L}^p(A) : f \sim g\}$.

We remark that $L^p(A)$ is a real vector space: $[f] + [g] = [f + g], c \cdot [f] = [c \cdot f] \in L^p(A)$, for every $f, g \in \mathfrak{L}^p(A)$ and for any $c \in \mathbb{R}$. Moreover, for every $g \in [f]$, $\int_A |f|^p d\lambda = \int_A |g|^p d\lambda$. Therefore, we can correctly define the application:

$$\|\cdot\|_p : L^p(A) \to \mathbb{R}_+,\ \|[f]\|_p = \|f\|_p, \text{ for every } [f] \in L^p(A).$$

Theorem 4.1.9

The space $(L^p(A), \|\cdot\|_p)$ is a real normed space.

Theorem 4.1.10

Let $p \geq 1$, $(f_n)_n \subseteq \mathfrak{L}^p(A)$ and $f \in \mathfrak{L}^p(A)$.

(1) *If $f_n \xrightarrow[A]{\|\cdot\|_p} f$, then $f_n \xrightarrow[A]{\lambda} f$.*
(2) *If $(f_n)_n$ is a L^p-Cauchy sequence, then $(f_n)_n$ is Cauchy in measure.*

Proof

(1) For every $\varepsilon > 0$ and any $n \in \mathbb{N}$, let $A_n(\varepsilon) = (|f_n - f| > \varepsilon) \in \mathscr{L}$. Then

$$|f_n - f|^p \geq |f_n - f|^p \chi_{A_n(\varepsilon)} \geq \varepsilon^p \chi_{A_n(\varepsilon)}.$$

Now using the monotonicity of the integral, we get

$$\|f_n - f\|_p^p = \int_A |f_n - f|^p d\lambda \geq \varepsilon^p \lambda(A_n(\varepsilon)), \text{ for every } \varepsilon > 0, \text{ and any } n \in \mathbb{N}.$$

It follows that, for every $\varepsilon > 0$ and any $n \in \mathbb{N}$, $\lambda(A_n(\varepsilon)) \leq \frac{1}{\varepsilon^p} \cdot \|f_n - f\|_p^p$.

Since $f_n \xrightarrow[A]{\|\cdot\|_p} f$, it follows that $\lim_n \lambda(A_n(\varepsilon)) = 0$, for every $\varepsilon > 0$, that implies $f_n \xrightarrow[A]{\lambda} f$. We can show (2) similarly, by replacing $A_n(\varepsilon)$ with $A_{m,n}(\varepsilon) = (|f_m - f_n| > \varepsilon)$. ∎

Using Theorem 4.1.10, we will show that the space $(\mathfrak{L}^p(A), \|\cdot\|_p)$ is a complete seminormed space; the proof is an adapted version of the proof of Theorem 3.3.8.

Theorem 4.1.11
The seminormed space $(\mathfrak{L}^p(A), \|\cdot\|_p)$ is complete (every L^p-Cauchy sequence is L^p-convergent). The normed space $(L^p(A), \|\cdot\|_p)$ is a Banach space (complete normed space).

Proof
From point (2) of Theorem 4.1.10, it follows that any L^p-Cauchy sequence, $(f_n)_n \subseteq \mathfrak{L}^p(A)$, is Cauchy in measure. From (1) of Riesz theorem (Theorem 2.2.5), there is a subsequence $(f_{k_n})_n$ of $(f_n)_n$ almost uniform convergent to a function $f : A \to \mathbb{R}$. Point (2) of Theorem 2.2.2 tells us that $f_{k_n} \xrightarrow[A]{\cdot} f$ and then $f \in \mathfrak{L}(A)$ (see 3) of Theorem 2.1.13).

Let's fix $m \in \mathbb{N}$; for any $n \in \mathbb{N}$, let $g_n = |f_m - f_{k_n}|^p \in \mathfrak{L}_+(A)$. We apply to the sequence $(g_n)_n$ the Fatou's lemma (Corollary 3.1.12):

$$\int_A \liminf_n g_n d\lambda \leq \liminf_n \int_A g_n d\lambda.$$

As $\liminf_n g_n = |f_m - f|^p$ a.e., it follows that

$$\int_A |f_m - f|^p d\lambda \leq \liminf_n \|f_m - f_{k_n}\|_p^p.$$

From $\lim_{m,n\to+\infty} \|f_m - f_{k_n}\|_1 = 0$, it follows, on one hand, that $f_m - f \in \mathfrak{L}^p(A)$ and then $f = f_m - (f_m - f) \in \mathfrak{L}^p(A)$ and, on the other hand, that $\|f_m - f\|_p \to 0$. ∎

If $\lambda(A) < +\infty$, then we can compare the spaces $\mathfrak{L}^p(A)$ and the topologies generated by the seminorms $\|\cdot\|_p$ on these spaces for $p \geq 1$.

Theorem 4.1.12

If $\lambda(A) < +\infty$ and $1 \leq p < r$, then $\mathfrak{L}^r(A) \subseteq \mathfrak{L}^p(A)$. If $f_n \xrightarrow[A]{\|\cdot\|_r} f$, then $f_n \xrightarrow[A]{\|\cdot\|_p} f$, for every $(f_n) \subseteq \mathfrak{L}^r(A)$, $f \in \mathfrak{L}^r(A)$.

Proof

Let $p_1 = \frac{r}{p} > 1$ and $q_1 = \frac{r}{r-p} > 1$; then $\frac{1}{p_1} + \frac{1}{q_1} = 1$. For every $f \in \mathfrak{L}^r(A)$, $f \in \mathfrak{L}(A)$ and $|f|^r \in \mathfrak{L}^1(A)$. It follows that $|f|^p \in \mathfrak{L}^{\frac{r}{p}}(A) = \mathfrak{L}^{p_1}(A)$.

Since $\lambda(A) < +\infty$, $\underline{1} \in \mathfrak{L}^{q_1}(A)$.

We can now apply the Hölder inequality (see Theorem 4.1.4) to the functions $|f|^p$ and $\underline{1}$; therefore $|f|^p \cdot \underline{1} = |f|^p \in \mathfrak{L}^1(A)$.

It follows, on the one hand, that $f \in \mathfrak{L}^p(A)$, and then $\mathfrak{L}^r(A) \subseteq \mathfrak{L}^p(A)$. On the other hand, the inequality of Hölder leads to

$$\int_A |f|^p d\lambda \leq \left(\int_A |f|^{p \cdot p_1} d\lambda\right)^{\frac{1}{p_1}} \cdot \left(\int_A \underline{1} d\lambda\right)^{\frac{1}{q_1}},$$

or equivalent to

$$\|f\|_p^p \leq \left(\int_A |f|^r d\lambda\right)^{\frac{p}{r}} \cdot \lambda(A)^{\frac{r-p}{r}} = \|f\|_r^p \cdot \lambda(A)^{\frac{r-p}{r}}.$$

The last inequality implies

$$\|f\|_p \leq [\lambda(A)]^{\frac{r-p}{pr}} \cdot \|f\|_r, \text{ for every } f \in \mathfrak{L}^r(A).$$

Hence, for every $(f_n)_n \subseteq \mathfrak{L}^r(A)$ and $f \in \mathfrak{L}^r(A)$,

$$\|f_n - f\|_p \leq [\lambda(A)]^{\frac{r-p}{pr}} \cdot \|f_n - f\|_r,$$

which shows that if the sequence $(f_n)_n$ L^r-converges to f, then it L^p-converges to f. ■

Remarks 4.1.13

(i) In the previous theorem, the condition $\lambda(A) < +\infty$ is essential. Indeed, let the function $f : [1, +\infty[\to \mathbb{R}$, $f(x) = \dfrac{1}{x}$. Then $f \in C([1, +\infty[) \subseteq \mathfrak{L}([1, +\infty[)$ and, for any $p > 1$,

$$\int_{[1,+\infty[} |f|^p d\lambda = \int_1^{+\infty} \frac{1}{x^p} dx = \frac{1}{1-p} \cdot x^{1-p}\Big|_1^{+\infty} = \frac{1}{p-1}.$$

It follows that $f \in \mathfrak{L}^p([1, +\infty[)$, for any $p > 1$, but $f \notin \mathfrak{L}^1([1, +\infty[)$.

(ii) If A is a set of finite measure and if $p < r$, the trace of the topology induced by the seminorm $\|\cdot\|_p$ on the subset $\mathfrak{L}^r(A) \subseteq \mathfrak{L}^p(A)$ is less fine than the topology induced by the seminorm $\|\cdot\|_r$ on the same subset.

4.2 Density Properties in $\mathfrak{L}^p$

In this paragraph we will highlight various dense sets in $\mathfrak{L}^p(A)$ (see Definition 4.1.7).

Theorem 4.2.1
Let $A \in \mathscr{L}$ and $p \geq 1$; then
(1) $\mathcal{E}^1(A) = \mathcal{E}(A) \cap \mathfrak{L}^p(A)$.
(2) $\overline{\mathcal{E}^1(A)}^p = \mathfrak{L}^p(A)$.

Proof

(1) Let $f \in \mathcal{E}(A)$; then $f = \sum_{i=1}^n a_i \cdot \chi_{A_i}$, where $A_i \in \mathscr{L}(A)$ are pairwise disjoints. Then $f \in \mathfrak{L}^p(A)$ if and only if $\sum_{i=1}^n |a_i|^p \lambda(A_i) < +\infty$ which says that, for any i for which $\lambda(A_i) = +\infty$, $a_i = 0$ and this one is equivalent to $f \in \mathcal{E}^1(A)$.
(2) The inclusion $\overline{\mathcal{E}^1(A)}^p \subseteq \mathfrak{L}^p(A)$ is obvious. Let's prove reverse inclusion.

For every $f \in \mathfrak{L}^p(A)$, $f = f^+ - f^-$, where $f^+, f^- \in \mathfrak{L}_+(A)$. There exist two sequences $(u_n)_n, (v_n)_n \subseteq \mathcal{E}_+(A)$ such that $u_n \uparrow f^+$ and $v_n \uparrow f^-$ (see 1) of Theorem 2.3.3). The sequence $(f_n)_n \subseteq \mathcal{E}(A)$, $f_n = u_n - v_n$, is pointwise convergent on A to f; moreover, for any $n \in \mathbb{N}$,

$$|f_n|^p \leq (u_n + v_n)^p \leq (f^+ + f^-)^p = |f|^p \in \mathfrak{L}^1(A),$$

from where, using Theorem 3.2.6,

$$(f_n)_n \subseteq \mathcal{E}(A) \cap \mathfrak{L}^p(A) = \mathcal{E}^1(A).$$

On the other hand,

$$|f_n - f|^p \leq (|f_n| + |f|)^p \leq 2^p(|f_n|^p + |f|^p) \leq 2^{p+1} \cdot |f|^p,$$

and, since $f_n \to f$, it follows from the dominated convergence theorem (Theorem 3.3.9),

$$\|f_n - f\|_p^p = \int_A |f_n - f|^p d\lambda \to 0,$$

which is equivalent to $f_n \xrightarrow[A]{\|\cdot\|_p} f$. Therefore $f \in \overline{\mathcal{E}^1(A)}^p$. ■

Remark 4.2.2 $\mathfrak{L}^p(A)$ is the completion of $(\mathcal{E}^1(A), \|\cdot\|_p)$.

Theorem 4.2.3
Let $p \geq 1$ and let $C_p(A) = C(A) \cap \mathfrak{L}^p(A)$ be the set of all continuous p-integrable functions; then

$$\overline{C_p(A)}^p = \mathfrak{L}^p(A).$$

The proof is an immediate adaptation of the case $p = 1$ (see Theorem 3.3.11), and we leave it to the reader.

Remark 4.2.4 If A is a compact set, then $C_p(A) = C(A)$, and so, in this case, $C(A)$ is dense in $\mathfrak{L}^p(A)$.

We conclude this section with an important property of spaces $\mathfrak{L}^p([a, b])$—the separability.

Theorem 4.2.5
For every compact interval $[a, b] \subseteq \mathbb{R}$ and any $p \geq 1$, the space $(\mathfrak{L}^p([a, b]), \|\cdot\|_p)$ is separable (contains a countable, dense subset).

Proof
Let P be the set of restrictions of polynomial functions with rational coefficients at the interval $[a, b]$; P is a countable set.

Let us show that P is a dense subset of $\mathfrak{L}^p(A)$, therefore that $\overline{P}^p = \mathfrak{L}^p([a, b])$. On the one hand, $P \subseteq C([a, b])$, and, from the previous remark, $C([a, b]) = C_p([a, b]) \subseteq \mathfrak{L}^p([a, b])$; hence $P \subseteq \mathfrak{L}^p([a, b])$. On the other hand, according to Theorem 4.2.3, for every $f \in \mathfrak{L}^p([a, b])$, and every $\varepsilon > 0$, there exists $g \in C([a, b])$ such that $\|f - g\|_p < \frac{\varepsilon}{3}$.

Now let us remember Weierstrass theorem of uniform approximation of continuous functions with polynomials. From this theorem, there is a polynomial h on $[a, b]$ such that

$$\|g - h\|_\infty = \sup_{x \in [a,b]} |g(x) - h(x)| < \frac{\varepsilon}{3(b-a)^{\frac{1}{p}}}.$$

Obviously, due to the density of rationals in $\mathbb{R}$, we can uniformly approximate h with polynomials of P. Therefore there exists $l \in P$ such that

$$\|h - l\|_\infty < \frac{\varepsilon}{3(b-a)^{\frac{1}{p}}}.$$

It follows that

$$\|f-l\|_p \le \|f-g\|_p + \|g-h\|_p + \|h-l\|_p < \frac{\varepsilon}{3} + (b-a)^{\frac{1}{p}}\|g-h\|_\infty +$$

$$+(b-a)^{\frac{1}{p}}\|h-l\|_\infty < \frac{\varepsilon}{3} + 2(b-a)^{\frac{1}{p}}\frac{\varepsilon}{3(b-a)^{\frac{1}{p}}} = \varepsilon.$$

So, for every $\varepsilon > 0$, there exists $l \in P$ such that $\|f-l\|_p < \varepsilon$; it follows that $f \in \overline{P}^p$, which shows that P is dense in $\mathfrak{L}^p([a,b]), \|\cdot\|_p)$. ■

Corollary 4.2.6 *For any $p \ge 1$, $(\mathfrak{L}^p(\mathbb{R}), \|\cdot\|_p)$ is separable.*

Proof

For any $n \in \mathbb{N}^*$, we define a mapping $\varphi_n : \mathfrak{L}^p([-n,n]) \to \mathfrak{L}^p(\mathbb{R})$, letting $\varphi_n(f)(x) = \begin{cases} f(x), & x \in [-n,n] \\ 0, & x \in \mathbb{R}\setminus[-n,n] \end{cases}$, for every $f \in \mathfrak{L}^p([-n,n])$ and every $x \in \mathbb{R}$. The functions φ_n are well defined. Indeed, for any $n \in \mathbb{N}$ and for every $f \in \mathfrak{L}^p([-n,n])$, $\varphi_n(f) \in \mathfrak{L}(\mathbb{R})$ (see (iii) of Remark 2.1.3), and, from Corollary 3.1.11, $\int_{\mathbb{R}} |\varphi_n(f)|^p d\lambda = \int_{[-n,n]} |f|^p d\lambda < +\infty$. From the preceding equality, it follows that

$$\|\varphi_n(f)\|_p = \|f\|_p, \text{ for every } f \in \mathfrak{L}^p([-n,n]) \text{ and any } n \in \mathbb{N}. \tag{1}$$

In addition, one can easily verify that φ_n is linear:

$$\varphi_n(f + c\cdot g) = \varphi_n(f) + c\cdot\varphi_n(g), \forall f, g \in \mathfrak{L}^p([-n,n]), \forall c \in \mathbb{R}. \tag{2}$$

The preceding theorem states that, for any $n \in \mathbb{N}$, $\mathfrak{L}^p([-n,n])$ is separable; we can thus find F_n a countable and dense set in $\mathfrak{L}^p([-n,n])$.

Let's show that the countable set $F = \bigcup_{n=1}^{\infty} \varphi_n(F_n)$ is dense in $\mathfrak{L}^p(\mathbb{R})$; let $f \in \mathfrak{L}^p(\mathbb{R})$ and let an arbitrary $\varepsilon > 0$.

Since $|f|^p \in \mathfrak{L}^1_+(\mathbb{R})$, we can use point (2) of Theorem 3.1.16; so there is $k \in \mathbb{N}$ such that

$$\int_{\mathbb{R}\setminus[-k,k]} |f|^p d\lambda < \left(\frac{\varepsilon}{2}\right)^p. \tag{3}$$

The function $f_k = f|_{[-k,k]}$ belongs to the space $\mathfrak{L}^p([-k,k])$. Indeed, from (ii) of Remark 2.1.3, $f_k \in \mathfrak{L}([-k,k])$ and $\int_{[-k.k]} |f_k|^p d\lambda = \int_{[-k,k]} |f|^p d\lambda \le \int_{\mathbb{R}} |f|^p d\lambda < +\infty$. Therefore, from (3),

$$\|\varphi_k(f_k) - f\|_p = \left(\int_{\mathbb{R}\setminus[-k,k]} |f|^p d\lambda\right)^{\frac{1}{p}} < \frac{\varepsilon}{2}. \tag{4}$$

As F_k is dense in $\mathfrak{L}^p([-k, k])$, there exists $g \in F_k$ such that

$$\|f_k - g\|_p < \frac{\varepsilon}{2}. \tag{5}$$

Using the relations (2), (1), (5), and (4), we obtain

$$\|\varphi_k(g) - f\|_p \le \|\varphi_k(g) - \varphi_k(f_k)\|_p + \|\varphi_k(f_k) - f\|_p = \|\varphi_k(g - f_k)\|_p +$$

$$+ \|\varphi_k(f_k) - f\|_p = \|g - f_k\|_p + \|\varphi_k(f_k) - f\|_p < \varepsilon.$$

But $\varphi_k(g) \in F$, so that F is countable and dense in $\mathfrak{L}^p(\mathbb{R})$. ■

4.3 The L^∞ Space

Definition 4.3.1

Let $A \in \mathscr{L}$ and $f \in \mathfrak{L}(A)$; we say that

$$\|f\|_\infty = \inf\{\alpha \in \overline{\mathbb{R}}_+ : |f| \le \alpha \text{ a.e.}\} \in [0, +\infty]$$

is the **essential supremum** of $|f|$.

Remarks 4.3.2

(i) $|f| \le \alpha$ a.e. means that $\lambda(|f| > \alpha) = 0$.
If, for any $\alpha \in \mathbb{R}_+$, $\lambda(|f| > \alpha) > 0$, then the only element $\alpha \in \overline{\mathbb{R}}_+$ for which $|f| \le \alpha$ a.e. is $\alpha = +\infty$ and therefore $\|f\|_\infty = +\infty$.
If it exists $\alpha \in \mathbb{R}_+$ such that $\lambda(|f| > \alpha) = 0$, then $\|f\|_\infty < +\infty$.

(ii) In general, the essential supremum of a function $|f|$ is smaller than its supremum:

$$\|f\|_\infty \le \sup_{x \in A} |f(x)|.$$

Indeed, if $f = \chi_{\mathbb{Q}}$, then $\sup_{x\in\mathbb{R}} |f(x)| = 1$, and $\|f\|_\infty = 0$ ($\lambda(|f| > 0) = \lambda(\mathbb{Q}) = 0$; hence $|f| = \underline{0}$ a.e.).

(iii) Let $J \subseteq \mathbb{R}$ be an interval, and let $f : J \to \mathbb{R}$ be a continuous function on J; then $\|f\|_\infty = \sup_{x\in J} |f(x)|$. Indeed, as we saw in the previous point, $\|f\|_\infty \le \sup_{x\in J} |f(x)|$. For any $c \in \mathbb{R}$ with $c < \sup_{x\in J} |f(x)|$, there exists $x_0 \in J$ such that $c < |f(x_0)|$. From the continuity of $|f|$ at x_0, there exists $\delta > 0$ such that $(x_0 - \delta, x_0 + \delta) \cap J \subseteq (|f| > c)$ from where $0 < \lambda((x_0 - \delta, x_0 + \delta) \cap J) \le \lambda(|f| > c)$. Then $c \le \|f\|_\infty$, which implies that $\sup_{x\in J} |f(x)| \le \|f\|_\infty$.

Lemma 4.3.3 *Let $f \in \mathfrak{L}(A)$; then*

(1) $\|f\|_\infty = \inf\{\sup_{x\in A\setminus N} |f(x)| : N \in \mathscr{L}, \lambda(N) = 0\}$.
(2) $|f| \le \|f\|_\infty$ *almost everywhere.*

Proof

(1) Let $\alpha_0 = \inf\{\sup_{x\in A\setminus N} |f(x)| : N \in \mathscr{L}, \lambda(N) = 0\} \in \overline{\mathbb{R}}_+$.

For any $\alpha \in \overline{\mathbb{R}}$ with $|f| \leq \alpha$ a.e., we denote $N = (|f| > \alpha) \in \mathscr{L}$; then $\lambda(N) = 0$. By definition, $\alpha_0 \leq \sup_{x\in A\setminus N} |f(x)| \leq \alpha$. It follows that $\alpha_0 \leq \|f\|_\infty$.

If $\alpha_0 = +\infty$, then the equality is obvious.

Let us suppose that $\alpha_0 < +\infty$; by the definition of α_0, for every $\varepsilon > 0$, there exists $N \in \mathscr{L}$ with $\lambda(N) = 0$ such that $\alpha_0 + \varepsilon > \sup_{x\in A\setminus N} |f(x)|$. It follows that $|f(x)| \leq \alpha_0 + \varepsilon$, for every $x \in A \setminus N$ or $(|f| > \alpha_0 + \varepsilon) \subseteq N$, and then $|f| \leq \alpha_0 + \varepsilon$ a.e. Considering the meaning of the essential supremum of f, $\|f\|_\infty \leq \alpha_0 + \varepsilon$, for every $\varepsilon > 0$, and so $\|f\|_\infty \leq \alpha_0$.

The two inequalities show that $\|f\|_\infty = \alpha_0$.

(2) If $\|f\|_\infty = +\infty$, the inequality is obvious.

Let us suppose that $\|f\|_\infty < +\infty$; from the previous point, for any $n \in \mathbb{N}^*$, there exists $N_n \in \mathscr{L}$ with $\lambda(N_n) = 0$ such that

$$\sup_{x\in A\setminus N_n} |f(x)| < \|f\|_\infty + \frac{1}{n}.$$

Let $N = \bigcup_{n=1}^{\infty} N_n \in \mathscr{L}$; then $\lambda(N) = 0$ and

$$|f(x)| < \|f\|_\infty + \frac{1}{n}, \quad \text{for every } x \in A \setminus N = \bigcap_{n=1}^{\infty}(A \setminus N_n), \quad \text{for any } n \in \mathbb{N}^*.$$

Passing to the limit in the last inequality, we obtain $|f(x)| \leq \|f\|_\infty$, for every $x \in A \setminus N$ or $|f| \leq \|f\|_\infty$ a.e. ■

Remark 4.3.4 From point (2) of the previous lemma, it follows that $\|f\|_\infty$ is the smallest element of the set $\{\alpha \in \overline{\mathbb{R}}_+ : |f| \leq \alpha \text{a.e.}\}$; because this last set is an interval, $\{\alpha \in \overline{\mathbb{R}}_+ : |f| \leq \alpha \text{ a.e.}\} = [\|f\|_\infty, +\infty]$.

Definition 4.3.5

For every $A \in \mathscr{L}$, $\mathfrak{L}^\infty(A) = \{f \in \mathfrak{L}(A) : \|f\|_\infty < +\infty\}$.

Theorem 4.3.6

Equipped with the usual operations of addition and multiplication with scalars, $\mathfrak{L}^\infty(A)$ is a real vector space and $\|\cdot\|_\infty$ is a seminorm on this space.

Proof

Let $f, g \in \mathfrak{L}^\infty(A)$; then $|f+g| \leq |f| + |g| \leq \|f\|_\infty + \|g\|_\infty$ a.e. Therefore $\|f+g\|_\infty \leq \|f\|_\infty + \|g\|_\infty < +\infty$; it follows that $f+g \in \mathfrak{L}^\infty(A)$ and that $\|\cdot\|_\infty$ satisfies the triangle inequality.

For every $f \in \mathfrak{L}^\infty(A)$ and for any $c \in \mathbb{R}$, $|c \cdot f| = |c| \cdot |f| \leq |c| \cdot \|f\|_\infty$ a.e. from where

$$\|c \cdot f\|_\infty \leq |c| \cdot \|f\|_\infty < +\infty. \tag{*}$$

Therefore $c \cdot f \in \mathfrak{L}^\infty(A)$.

If $c = 0$, then it is obvious that $\|c \cdot f\|_\infty = |c| \cdot \|f\|_\infty$.

If $c \neq 0$, then we apply the inequality of $(*)$ for $\frac{1}{c} \in \mathbb{R}$ and $c \cdot f \in \mathfrak{L}^\infty(A)$:

$$\left\| \frac{1}{c} \cdot (c \cdot f) \right\|_\infty \leq \frac{1}{|c|} \cdot \|c \cdot f\|_\infty.$$

Therefore $|c| \cdot \|f\|_\infty \leq \|c \cdot f\|_\infty$ which, with the inequality $(*)$, leads us to equality $\|c \cdot f\|_\infty = |c| \cdot \|f\|_\infty$.

It follows that $\|\cdot\|_\infty$ is a seminorm on $\mathfrak{L}^\infty(A)$. ∎

Remark 4.3.7 The relation $\sim$, defined by $f \sim g$ if and only if $f = g$ a.e., is an equivalence relation on $\mathfrak{L}^\infty(A)$. We denote the quotient set $\mathfrak{L}^\infty(A)|_\sim$ with $L^\infty(A)$; for every $[f] \in L^\infty(A)$, $\|[f]\|_\infty = \|f\|_\infty$. According to Lemma 4.3.3, $\|f\|_\infty = 0$ if and only if $f = \underline{0}$ a.e.; it follows that the previous definition does not depend on the representative f of the equivalence class $[f]$. We can easily deduce that $(L^\infty(A), \|\cdot\|_\infty)$ is a normed space.

Proposition 4.3.8 *Let $(f_n)_n \subseteq \mathfrak{L}^\infty(A)$ and $f \in \mathfrak{L}^\infty(A)$.*

(1) $f_n \xrightarrow[A]{\|\cdot\|_\infty} f \Leftrightarrow$ *there exists $B \in \mathscr{L}(A)$ with $\lambda(B) = 0$ such that* $f_n \xrightarrow[A \setminus B]{\text{u}} f$.

(2) *$(f_n)_n$ is Cauchy sequence in $(\mathfrak{L}^\infty(A), \|\cdot\|_\infty)$ if and only if there exists $B \in \mathscr{L}(A)$ with $\lambda(B) = 0$ such that $(f_n)_n$ is uniformly Cauchy on $A \setminus B$.*

Proof

(1) We suppose that $f_n \xrightarrow[A]{\|\cdot\|_\infty} f$; for any $p \in \mathbb{N}^*$, there exists $n_p \in \mathbb{N}$ such that $\|f_n - f\|_\infty < \frac{1}{p}$, for any $n \geq n_p$. According to Lemma 4.3.3, it follows that $|f_n - f| < \frac{1}{p}$ a.e., or that $A_{n,p} = \left(|f_n - f| \geq \dfrac{1}{p}\right)$ are null sets, for any $p \in \mathbb{N}^*$ and for any $n \geq n_p$. Then the set $B = \bigcup_{p=1}^\infty \bigcup_{n=n_p}^\infty A_{n,p}$ is also a null set.

For every $x \in A \setminus B = \bigcap_{p=1}^\infty \bigcap_{n=n_p}^\infty (A \setminus A_{n,p})$, we have that $|f_n(x) - f(x)| < \frac{1}{p}$, for any $p \in \mathbb{N}^*$ and for any $n \geq n_p$ from where $f_n \xrightarrow[A \setminus B]{\text{u}} f$.

Conversely, if we assume that there is $B \in \mathscr{L}(A)$ with $\lambda(B) = 0$ such that $f_n \xrightarrow[A \setminus B]{\text{u}} f$, then, for every $\varepsilon > 0$, there exists $n_\varepsilon \in \mathbb{N}$ such that $|f_n(x) - f(x)| < \varepsilon$, for any $n \geq n_\varepsilon$ and every $x \in A \setminus B$. It follows that $A \setminus B \subseteq (|f_n - f| < \varepsilon)$, which leads, by passing to the complementary, to $(|f_n - f| \geq \varepsilon) \subseteq B$. Therefore $|f_n - f| < \varepsilon$ a.e. and then $\|f_n - f\|_\infty \leq \varepsilon$, for any $n \geq n_\varepsilon$ or $f_n \xrightarrow[A]{\|\cdot\|_\infty} f$.

(2) is demonstrated in the same way. ■

Theorem 4.3.9
$(\mathfrak{L}^\infty(A), \|\cdot\|_\infty)$ *is a complete seminormed space, and then* $(L^\infty(A), \|\cdot\|_\infty)$ *is a Banach space.*

Proof
Let $(f_n)_n \subseteq (\mathfrak{L}^\infty(A), \|\cdot\|_\infty)$ be a Cauchy sequence. According to (2) of previous proposition, there exists $B \in \mathscr{L}(A)$ with $\lambda(B) = 0$ such that $(f_n)_n$ is uniformly Cauchy on $A \setminus B$; it means that, for every $\varepsilon > 0$, there exists $n_\varepsilon \in \mathbb{N}$ such that, for all $m, n \geq n_\varepsilon$,

$$|f_m(x) - f_n(x)| < \varepsilon, \text{ for every } x \in A \setminus B. \qquad (*)$$

From $(*)$, for every $x \in A \setminus B$, $(f_n(x))_n$ is a Cauchy sequence in $\mathbb{R}$, and then there exists $\lim_n f_n(x) \in \mathbb{R}$.

Let $f : A \to \mathbb{R}$ defined by $f(x) = \begin{cases} \lim_n f_n(x), & x \in A \setminus B, \\ 0, & x \in B. \end{cases}$

The sequence $(f_n)_n$ converges a.e. to f and then, from (3) of Theorem 2.1.13, $f \in \mathfrak{L}(A)$.

From $(*)$, with $n = n_\varepsilon$ and $m \to \infty$, we obtain $|f(x) - f_{n_\varepsilon}(x)| \leq \varepsilon$, for every $x \in A \setminus B$; it follows that $|f(x)| \leq |f(x) - f_{n_\varepsilon}(x)| + |f_{n_\varepsilon}(x)| < \varepsilon + |f_{n_\varepsilon}(x)|$, for every $x \in A \setminus B$, from where $|f| \leq \varepsilon + |f_{n_\varepsilon}| \leq \varepsilon + \|f_{n_\varepsilon}\|_\infty$ a.e. We deduce that $\|f\|_\infty < +\infty$ and then $f \in \mathfrak{L}^\infty(A)$.

We again use the relation $(*)$, in which $n \to \infty$, to obtain $|f_m(x) - f(x)| \leq \varepsilon$, for any $m \geq n_\varepsilon$ and every $x \in A \setminus B$. Since $\lambda(B) = 0$, $|f_m - f| \leq \varepsilon$ a.e. or $\|f_m - f\|_\infty \leq \varepsilon$, for any $m \geq n_\varepsilon$; then $f_n \xrightarrow[A]{\|\cdot\|_\infty} f$. ■

The following proposition is an extension of the Hölder inequality (Theorem 4.1.4).

Proposition 4.3.10 *For every* $f \in \mathfrak{L}^1(A)$ *and every* $g \in \mathfrak{L}^\infty(A)$, *the function* $f \cdot g$ *is integrable* ($f \cdot g \in \mathfrak{L}^1(A)$) *and*

$$\|f \cdot g\|_1 = \int_A |f \cdot g| d\lambda \leq \|f\|_1 \cdot \|g\|_\infty.$$

Proof
From Lemma 4.3.3, $|g| \leq \|g\|_\infty$ almost everywhere, and then $|f \cdot g| \leq \|g\|_\infty \cdot |f|$, a.e. It follows from Theorem 3.2.6 that $f \cdot g \in \mathfrak{L}^1(A)$, and, by integrating the inequality above, we obtain the extension of the inequality of Hölder presented in the statement. ■

If A has a finite measure, we can compare the spaces $\mathfrak{L}^p$, $1 \leq p < +\infty$ with $\mathfrak{L}^\infty$ (see Theorem 4.1.12).

Theorem 4.3.11 (Riesz)
Let $\lambda(A) < +\infty$; then $\mathfrak{L}^\infty(A) \subsetneq \bigcap_{p=1}^\infty \mathfrak{L}^p(A)$ and, for every $f \in \mathfrak{L}^\infty(A)$,

$$\|f\|_\infty = \lim_{p\to+\infty} \|f\|_p.$$

Proof
Let an arbitrary $1 \leq p < +\infty$; if $f \in \mathfrak{L}^\infty(A)$, then $f \in \mathfrak{L}(A)$, and, from Lemma 4.3.3, $|f| \leq \|f\|_\infty$ a.e., from where

$$|f|^p \leq \|f\|_\infty^p \text{ a.e.} \tag{1}$$

But, in the finite measure spaces, the constant functions are integrable. Hence $\|f\|_\infty^p \in \mathfrak{L}^1(A)$ ($\|f\|_\infty \in \mathbb{R}_+$), and then, according to Theorem 3.2.6, $|f|^p \in \mathfrak{L}^1(A)$. It follows that $f \in \mathfrak{L}^p(A)$, which proofs the inclusion $\mathfrak{L}^\infty(A) \subset \bigcap_{p=1}^\infty \mathfrak{L}^p(A)$.

Now, by integrating the inequalities, (1), we get

$$\int_A |f|^p d\lambda \leq \|f\|_\infty^p \cdot \lambda(A),$$

form where

$$\|f\|_p \leq [\lambda(A)]^{\frac{1}{p}} \cdot \|f\|_\infty, \text{ for every } f \in \mathfrak{L}^\infty(A). \tag{2}$$

From (2) it follows that the trace of the topology generated by the seminorm $\|\cdot\|_p$ on $\mathfrak{L}^\infty(A)$ is less fine than the topology generated by $\|\cdot\|_\infty$.

We have observed that inequality (2) occurs for all $f \in \mathfrak{L}^\infty(A)$ and for any $p \geq 1$; passing to the upper limit in this inequality, we get

$$\limsup_{p\to+\infty} \|f\|_p \leq \|f\|_\infty, \text{ for every } f \in \mathfrak{L}^\infty(A). \tag{3}$$

In the above we have assumed that $\lambda(A) > 0$; in the particular case where $\lambda(A) = 0$, it follows that, for any $p \geq 1$, $\mathfrak{L}^1(A) = \mathfrak{L}^p(A) = \mathfrak{L}^\infty(A) = \mathfrak{L}(A)$ (λ is a complete measure) and $\|f\|_p = 0 = \|f\|_\infty$, for any $p \geq 1$. In this situation it is obvious that $\lim_p \|f\|_p = \|f\|_\infty$.

Returning to inequality (3), if $\|f\|_\infty = 0$, then, from (3), it follows that there exists $\lim_p \|f\|_p = 0 = \|f\|_\infty$.

We suppose that $\|f\|_\infty > 0$. For any $0 < \alpha < \|f\|_\infty$, from the definition of $\|f\|_\infty$, it follows that $\lambda(|f| > \alpha) > 0$; let $A_\alpha = (|f| > \alpha) \in \mathscr{L}(A)$. For any $p \geq 1$, we obtain

$$\|f\|_p \geq \left(\int_{A_\alpha} |f|^p d\lambda\right)^{\frac{1}{p}} \geq \left[\alpha^p \cdot \lambda(A_\alpha)\right]^{\frac{1}{p}} = \alpha \cdot [\lambda(A_\alpha)]^{\frac{1}{p}}. \tag{4}$$

Since $\lambda(A_\alpha) \in]0, +\infty[$, there exists $\lim_{p\to+\infty} [\lambda(A_\alpha)]^{\frac{1}{p}} = 1$. If we go to the lower limit for $p \to \infty$ in (4), we obtain

$$\liminf_{p\to+\infty} \|f\|_p \geq \alpha, \text{ for any } \alpha < \|f\|_\infty. \tag{5}$$

But (5) implies that

$$\|f\|_\infty \leq \liminf_{p\to+\infty} \|f\|_p. \tag{6}$$

From (3) and (6) we deduce that there exists $\lim_p \|f\|_p = \|f\|_\infty$. ■

Remark 4.3.12 In the statement of Riesz theorem, we specified that the inclusion $\mathfrak{L}^\infty(A) \subsetneq \bigcap_{p=1}^\infty \mathfrak{L}^p(A)$ is strict. Indeed, if the function $f :]0, 1] \to \mathbb{R}$ is defined by $f(x) = \ln x$, then, according to (iii) of Remark 4.3.2, $\|f\|_\infty = \sup_{x\in]0,1]} |f(x)| = +\infty$. Therefore $f \notin \mathfrak{L}^\infty(]0, 1])$. On the other hand, for any $p \geq 1$, $\|f\|_p^p = \int_{]0,1]} |f|^p d\lambda = \int_{0+0}^1 |\ln x|^p dx < +\infty$ (for $\beta = \frac{1}{2}$ there exists $\lim_{x\downarrow 0} x^\beta \cdot |\ln x|^p = 0 < +\infty$; then the generalized Riemann integral is convergent). It follows that $f \in \mathfrak{L}^p(]0, 1])$, for any $p \geq 1$.

4.4 Fourier Series in $L^2([-\pi, \pi])$

From the start, we observe that a good number of the results of this section remain true if we replace the closed interval $[-\pi, \pi]$ by some other measurable set. Let us remember that the space $L^2([-\pi, \pi])$ is a Banach space with respect to the norm $\|\cdot\|_2 :$ $L^2[-\pi, \pi] \to \mathbb{R}_+$, $\|f\|_2 = \left(\int_{[-\pi,\pi]} f^2 d\lambda\right)^{\frac{1}{2}}$, for every $f \in L^2([-\pi, \pi])$ (we will use the current, instead of equivalence classes $L^2([-\pi, \pi]) = \mathfrak{L}^2([-\pi, \pi])|_\sim$, their representatives).

Moreover, Theorem 4.2.5 assures us that the space $(L^2([-\pi, \pi]), \|\cdot\|_2)$ is a separable Banach space.

Note that if $p = q = \frac{1}{2}$, then we can write the inequality of Hölder: for every $f, g \in L^2([-\pi, \pi])$, $f \cdot g \in L^1([-\pi, \pi])$ and $\|fg\|_1 \leq \|f\|_2 \cdot \|g\|_2$ (see Theorem 4.1.4). So we can define the application $(\cdot, \cdot) : L^2([-\pi, \pi]) \times L^2([-\pi, \pi]) \to \mathbb{R}$ by

$$(f, g) = \int_{[-\pi,\pi]} fg d\lambda.$$

The demonstration of the following proposition is a simple application of the above definition.

Proposition 4.4.1 *The application $(\cdot, \cdot)$ defined above is an inner product on $L^2([-\pi, \pi])$, that is to say it checks the conditions:*

(1) *$(f, f) \geq 0, \forall f \in L^2([-\pi, \pi])$ and $(f, f) = 0$ if and only if $f = \underline{0}$ a.e.*
(2) *$(f, g) = (g, f)$, for every $f, g \in L^2([-\pi, \pi])$.*

(3) $(f+g,h)=(f,h)+(g,h)$, *for every* $f,g,h\in L^2([-\pi,\pi])$.
(4) $(c\cdot f,g)=c\cdot(f,g)$, *for every* $f,g\in L^2([-\pi,\pi])$ *and any* $c\in\mathbb{R}$.

The norm associated with the inner product is defined in a standard way through $\|f\|=\sqrt{(f,f)}$; we immediately note that $\|f\|=\|f\|_2$, for every $f\in L^2([-\pi,\pi])$. The space $L^2([-\pi,\pi])$ is then a separable Hilbert space (a separable Banach space whose norm is induced by an inner product).

In these spaces we can define the notion of orthogonality: two vectors $f,g\in L^2([-\pi,\pi])$ are **orthogonal** when $(f,g)=0$, and this is denoted by $f\perp g$.

A sequence $(f_n)_n\subseteq L^2([-\pi,\pi])$ is said to be **orthogonal** if $f_n\perp f_m$, for any $m\neq n$; moreover, if $\|f_n\|_2=1$, for any $n\in\mathbb{N}$, then the sequence is **orthonormal**.

A general result of the functional analysis assures us that in any separable Hilbert space, there are orthonormal sequences and that each element of the space is expressed by the sum of a series constructed with the elements of a such sequence (see, e.g., Theorems V.9 and V.10 of [3]).

In what follows, we will emphasize an important orthonormal sequence in $L^2([-\pi,\pi])$—the trigonometric system.

Definition 4.4.2

For any $n\in\mathbb{N}$, let $f_n:[-\pi,\pi]\to\mathbb{R}$ defined by:

$$\begin{aligned}
f_0(x) &= 1 &&, \text{ for every } x\in[-\pi,\pi]\,,\\
f_{2n-1}(x) &= \cos nx &&, \text{ for every } x\in[-\pi,\pi]\,, \text{ and any } n\geq 1,\\
f_{2n}(x) &= \sin nx &&, \text{ for every } x\in[-\pi,\pi]\,, \text{ and any } n\geq 1.
\end{aligned}$$

The sequence $(f_n)_n\subseteq C([-\pi,\pi])\subseteq \mathcal{R}([-\pi,\pi])\subseteq L^2([-\pi,\pi])$ is the **trigonometric system**.

We remark that, for every $x\in[-\pi,\pi]$,

$$(f_n(x))_{n\in\mathbb{N}}=(1,\cos x,\sin x,\cos 2x,\sin 2x,\ldots,\cos nx,\sin nx,\ldots)\,.$$

Proposition 4.4.3 *The trigonometric system is an orthogonal sequence in* $L^2([-\pi,\pi])$*:*

$$f_n\perp f_m,\ \textit{for all } n\neq m \textit{ and } \|f_0\|_2=\sqrt{2\pi},\ \|f_n\|_2=\sqrt{\pi},\ \textit{for any } n\geq 1.$$

Proof

Let's show that $(f_n,f_m)=0$, for all $n\neq m$.

Since $(f_n)_n\subseteq\mathcal{R}([-\pi,\pi])$, it follows that, for all $n,m\in\mathbb{N}$,

$$(f_n,f_m)=\int_{[-\pi,\pi]}f_nf_m d\lambda=\int_{-\pi}^{\pi}f_n(x)f_m(x)dx.$$

$$(f_0,f_{2m-1})=\int_{-\pi}^{\pi}\cos mx dx=\frac{1}{m}\sin mx|_{-\pi}^{\pi}=0,\ \text{ for any } m\geq 1.$$

$$(f_0, f_{2m}) = \int_{-\pi}^{\pi} \sin mx dx = -\frac{1}{m} \cos mx|_{-\pi}^{\pi} = 0, \text{ for any } m \geq 1.$$

$$(f_{2n-1}, f_{2m-1}) = \int_{-\pi}^{\pi} \cos nx \cos mx dx =$$

$$= \frac{1}{2} \cdot \int_{-\pi}^{\pi} [\cos(n+m)x + \cos(n-m)x] dx = 0, \text{ for all } n, m \geq 1, n \neq m.$$

$$(f_{2n-1}, f_{2m}) = \int_{-\pi}^{\pi} \cos nx \sin mx dx =$$

$$= \frac{1}{2} \int_{-\pi}^{\pi} [\sin(n+m)x - \sin(n-m)x]\, dx = 0, \text{ for all } n, m \geq 1.$$

$$(f_{2n}, f_{2m}) = \int_{-\pi}^{\pi} \sin nx \sin mx dx =$$

$$= \frac{1}{2} \int_{\pi}^{\pi} [-\cos(n+m)x + \cos(n-m)x]\, dx = 0, \text{ for all } n, m \geq 1, n \neq m.$$

It follows that $f_n \perp f_m$, for all $n, m \in \mathbb{N}, n \neq m$.
$||f_n||_2 = \sqrt{\int_{-\pi}^{\pi} f_n^2(x) dx}$, for any $n \in \mathbb{N}$ and then

$$||f_0||_2 = \sqrt{2\pi},$$

$$||f_{2n-1}||_2 = \sqrt{\int_{-\pi}^{\pi} \cos^2 nx dx} = \sqrt{\int_{-\pi}^{\pi} \frac{1+\cos 2nx}{2} dx} = \sqrt{\pi},$$

$$||f_{2n}||_2 = \sqrt{\int_{-\pi}^{\pi} \sin^2 nx dx} = \sqrt{\int_{-\pi}^{\pi} \frac{1-\cos 2nx}{2} dx} = \sqrt{\pi}.$$ ■

Definition 4.4.4

The sequence $(e_n)_n$, where, for any $n \in \mathbb{N}, e_n = \frac{1}{\|f_n\|_2} \cdot f_n$, is an orthonormal sequence. For every $x \in [-\pi, \pi]$,

$$(e_n(x))_{n\in\mathbb{N}} = \left(\frac{1}{\sqrt{2\pi}}, \frac{1}{\sqrt{\pi}} \cos x, \frac{1}{\sqrt{\pi}} \sin x, \ldots, \frac{1}{\sqrt{\pi}} \cos nx, \frac{1}{\sqrt{\pi}} \sin nx, \ldots \right).$$

For every $f \in L^2([-\pi, \pi])$ the **Fourier coefficients** associated with f are defined by $c_n = (f, e_n) = \frac{1}{\|f_n\|_2} \cdot (f, f_n)$, for any $n \in \mathbb{N}$, and the **Fourier series** associated with f and with the trigonometric system is

(Continued)

Definition 4.4.4 (continued)

$$x \mapsto \sum_{n=0}^{\infty} c_n \cdot e_n(x) = \frac{1}{2\pi}\cdot(f, f_0)+\frac{1}{\pi}\cdot\sum_{n=1}^{\infty}[(f, f_{2n-1}) \cdot \cos nx + (f, f_{2n}) \cdot \sin nx].$$

To simplify the writing, let us denote, for any $n \in \mathbb{N}$,

$$a_n = \frac{1}{\pi} \cdot \int_{[-\pi,\pi]} f(x) \cos nx d\lambda(x),\ b_n = \frac{1}{\pi} \cdot \int_{[-\pi,\pi]} f(x) \sin nx d\lambda(x).$$

Then the Fourier series associated with f is

$$\frac{a_0}{2} + \sum_{n=1}^{\infty}(a_n \cos nx + b_n \sin nx).$$

Theorem 4.4.5

Let $f \in L^2([-\pi,\pi])$, and let $(a_n)_n, (b_n)_n$ the sequences defined above by $a_n = \frac{1}{\pi}(f, f_{2n-1}), b_n = \frac{1}{\pi}(f, f_{2n})$. We denote

$$S_n(x) = \frac{a_0}{2} + \sum_{k=1}^{n}(a_k \cos kx + b_k \sin kx),\ \forall n \in \mathbb{N}^*, \forall x \in [-\pi,\pi].$$

$(S_n)_n$ is the sequence of partial sums of the Fourier series associated with f. For every sequences $(\alpha_n)_n, (\beta_n)_n \subseteq \mathbb{R}$, let

$$T_n(x) = \frac{\alpha_0}{2} + \sum_{k=1}^{n}(\alpha_k \cos kx + \beta_k \sin kx),\ \forall n \in \mathbb{N}^*, \forall x \in [-\pi,\pi]$$

be a trigonometric polynomial. Then

(1) $\|f - S_n\|_2 \leq \|f - T_n\|_2$, *for any $n \in \mathbb{N}$,*

(2) $\frac{a_0^2}{2} + \sum_{n=1}^{\infty}(a_n^2 + b_n^2) \leq \frac{1}{\pi} \cdot \|f\|_2^2 = \frac{1}{\pi} \cdot \int_{[-\pi,\pi]} f^2 d\lambda$,

(3) $\lim_{n\to+\infty} a_n = 0 = \lim_{n\to+\infty} b_n$.

Proof

T_n being an arbitrary trigonometric polynomial of indicated form,

$$(T_n, T_n) = \frac{\alpha_0^2}{4} \cdot (\underline{1}, \underline{1}) + \sum_{k=1}^{n}[\alpha_k^2 \cdot (\cos k\cdot, \cos k\cdot) + \beta_k^2 \cdot (\sin k\cdot, \sin k\cdot)] =$$

$$= \frac{\alpha_0^2}{4} \cdot 2\pi + \sum_{k=1}^{n} (\alpha_k^2 + \beta_k^2) \cdot \pi = \left(\frac{\alpha_0^2}{2} + \sum_{k=1}^{n} (\alpha_k^2 + \beta_k^2) \right) \cdot \pi .$$

Then, for any $n \in \mathbb{N}$,

$$\|f - T_n\|_2^2 = (f - T_n, f - T_n) = (f, f) - 2(f, T_n) + (T_n, T_n) =$$

$$= \|f\|_2^2 - 2 \sum_{k=1}^{n} [\alpha_k (f, f_{2k-1}) + \beta_k (f, f_{2k})] - \alpha_0 (f, f_0) + (T_n, T_n) =$$

$$= \|f\|_2^2 - \pi \alpha_0 a_0 - 2\pi \sum_{k=1}^{n} (\alpha_k a_k + \beta_k b_k) + \frac{\pi}{2} \alpha_0^2 + \pi \sum_{k=1}^{n} (\alpha_k^2 + \beta_k^2) =$$

$$= \|f\|_2^2 + \frac{\pi}{2} (\alpha_0 - a_0)^2 + \pi \sum_{k=1}^{n} [(\alpha_k - a_k)^2 + (\beta_k - b_k)^2] -$$

$$-\pi \left(\frac{a_0^2}{2} + \sum_{k=1}^{n} (a_k^2 + b_k^2) \right) .$$

In particular, if instead of T_n we put S_n, we obtain

$$\|f - S_n\|_2^2 = \|f\|_2^2 - \pi \left(\frac{a_0^2}{2} + \sum_{k=1}^{n} (a_k^2 + b_k^2) \right) . \qquad (*)$$

By comparing the two members of the relation $(*)$, it immediately follows the inequality of point (1) From the last equality, it turns out that, for any $n \in \mathbb{N}$,

$$\frac{a_0^2}{2} + \sum_{k=1}^{n} (a_k^2 + b_k^2) = \frac{1}{\pi} \cdot \|f\|_2^2 - \frac{1}{\pi} \cdot \|f - S_n\|_2^2 \leq \frac{1}{\pi} \cdot \|f\|_2^2 .$$

Passing to the limit for $n \to +\infty$ in the above inequality, we get the inequality of (2).

Finally, from the inequality of point (2), it turns out that the series $\sum_{n=1}^{\infty} (a_n^2 + b_n^2)$ is convergent; so the its general term tends to 0, and this leads to the equality (3). ■

Remarks 4.4.6

(i) The inequality of (1) shows us that the sequence of partial sums for the Fourier series associated with f gives the best approximation in norm of f among the other trigonometric polynomials. We will demonstrate later that this sequence L^2-converges to f.

(ii) The inequality (2) is called the **Bessel inequality**. As we will show below, this will actually become an equality.

(iii) The condition (3) can still be written:

$$\lim_n \int_{[-\pi,\pi]} f(x) \cos nx d\lambda(x) = 0 = \lim_n \int_{[-\pi,\pi]} f(x) \sin nx d\lambda(x).$$

In the literature, this result is often encountered as the Riemann-Lebesgue lemma.

To prove that the Fourier series associated with a function L^2-converges to that function, we need some helpful results.

Lemma 4.4.7 *Let $(d_n)_{n\in\mathbb{N}}$ be the sequence of real numbers defined by*

$$d_n = \int_{-\pi}^{\pi} \cos^{2n} t dt, \text{ for any } n \in \mathbb{N}.$$

For any $n \in \mathbb{N}$, and for every $x, y \in \mathbb{R}$, let $D_n(x, y) = \frac{1}{d_n} \cdot \cos^{2n} \frac{x-y}{2}$.

Then, for every $r \in]0, \pi[$:

(1) $\lim_n \int_{y-r}^{y+r} D_n(x, y)dx = 1$, *uniformly with respect to $y \in [-\pi + r, \pi - r]$.*

(2) $\int_{-\pi}^{\pi} D_n(x, y)dx = 1$, *for any $n \in \mathbb{N}$ and for every $y \in \mathbb{R}$.*

Proof

We remark that, for any $n \geq 1$,

$$d_n = \int_{-\pi}^{\pi} \cos^{2n-1} t(\sin t)'dt = \cos^{2n-1} t \sin t|_{-\pi}^{\pi} +$$

$$+(2n-1) \int_{-\pi}^{\pi} \cos^{2n-2} t \sin^2 t dt = (2n-1)d_{n-1} - (2n-1)d_n.$$

From here results the recurrence relation:

$$d_n = \frac{2n-1}{2n} d_{n-1}, \text{ for any } n \geq 1.$$

In the previous relation, we give values to n, from 1 to a number $m \in \mathbb{N}^*$, and, if we multiply the relations, we found:

$$d_m = \frac{(2m-1)!!}{(2m)!!} d_0 = \frac{(2m-1)!!}{(2m)!!} \cdot 2\pi, \tag{1}$$

where $(2m-1)!! = 1 \cdot 3 \cdot 5 \cdots (2m-1)$ and $(2m)!! = 2 \cdot 4 \cdot 6 \cdots (2m)$.

We remark that, for any $m \in \mathbb{N}^*$,

$$\frac{2m-2}{2m-1} < \frac{2m-1}{2m}. \tag{2}$$

In inequality (2), we give values to m, from 2 to a number $n \geq 2$, and by multiplying the relations, we found

$$\frac{(2n-2)!!}{(2n-1)!!} < 2 \cdot \frac{(2n-1)!!}{(2n)!!}. \tag{3}$$

By amplifying the inequality (3) with $\frac{(2n-1)!!}{(2n)!!}$, it stands out:

$$\frac{1}{2n} < 2 \cdot \left[\frac{(2n-1)!!}{(2n)!!}\right]^2, \tag{4}$$

or, equivalent to

$$\frac{(2n-1)!!}{(2n)!!} > \frac{1}{2\sqrt{n}}, \quad \text{for any } n \geq 2. \tag{5}$$

From (1) and (5), finally, results the inequality:

$$d_n \geq \frac{\pi}{\sqrt{n}}, \quad \text{for any } n \geq 2. \tag{6}$$

(1) For any $n \in \mathbb{N}$, let $I_n = \int_{y-r}^{y+r} D_n(x, y)dx$. We make the change of variable $\frac{x-y}{2} = t$, and we obtain

$$I_n = \frac{2}{d_n} \int_{-\frac{r}{2}}^{\frac{r}{2}} \cos^{2n} t dt = \frac{4}{d_n} \int_0^{\frac{r}{2}} \cos^{2n} t dt = \frac{1}{d_n} \left[d_n - 4 \cdot \int_{\frac{r}{2}}^{\frac{\pi}{2}} \cos^{2n} t dt\right] =$$

$$= 1 - \frac{4}{d_n} \int_{\frac{r}{2}}^{\frac{\pi}{2}} \cos^{2n} t dt.$$

But, from (6), we deduce

$$\left|\frac{4}{d_n} \int_{\frac{r}{2}}^{\frac{\pi}{2}} \cos^{2n} t dt\right| \leq \frac{4}{d_n} \cos^{2n} \frac{r}{2} < \frac{4\sqrt{n}}{\pi} \cos^{2n} \frac{r}{2}.$$

As $\lim_n \sqrt{n} \left(\cos \frac{r}{2}\right)^{2n} = 0$, we obtain $I_n \to 1$ and then

$$\lim_n \int_{y-r}^{y+r} D_n(x, y)dx = 1, \text{ uniformly with respect to } y \in [-\pi + r, \pi - r].$$

(2) Let $n \in \mathbb{N}$ and $y \in \mathbb{R}$; then

$$\int_{-\pi}^{\pi} D_n(x, y)dx = \frac{1}{d_n} \int_{-\pi}^{\pi} \cos^{2n} \frac{x-y}{2} dx = \frac{2}{d_n} \int_{-\frac{\pi}{2}-\frac{y}{2}}^{\frac{\pi}{2}-\frac{y}{2}} \cos^{2n} t dt.$$

The function $f : \mathbb{R} \to \mathbb{R}$, $f(t) = \cos^{2n} t$, for any $t \in \mathbb{R}$, is periodic with the period π. It follows that over any interval of length equal to the period, the integral is the same and therefore

$$\int_{-\pi}^{\pi} D_n(x, y)dx = \frac{2}{d_n} \int_{-\frac{\pi}{2}}^{\frac{\pi}{2}} \cos^{2n} t dt = \frac{1}{d_n} \int_{-\pi}^{\pi} \cos^{2n} t dt = 1.$$

■

Lemma 4.4.8 (L. Fejér) *Let $f : [-\pi, \pi] \to \mathbb{R}$ be a continuous function; then*

$$\int_{-\pi}^{\pi} D_n(x, \cdot) f(x)dx \xrightarrow[{[a,b]}]{\text{u}} f, \text{ for every } [a, b] \subseteq]-\pi, \pi[.$$

Proof

The function f is continuous on $[-\pi, \pi]$; according to Weierstrass theorem, f is bounded. Let $M \geq 1$ such that

$$|f(x)| \leq M, \text{ for every } x \in [-\pi, \pi] \tag{1}$$

The function f, being continuous over the closed and bounded interval $[-\pi, \pi]$, is uniformly continuous (Cantor's theorem). It follows that, for every $\varepsilon \in]0, 1[$, there exists $\delta > 0$ such that

$$|f(x) - f(y)| < \frac{\varepsilon}{4M}, \text{ for every } x, y \in [-\pi, \pi] \text{ with } |x - y| < \delta. \tag{2}$$

Let $[a, b] \subseteq]-\pi, \pi[$ be an arbitrary closed interval; we can choose a positive number δ small enough that $-\pi + \delta \leq a < b \leq \pi - \delta$ ($\delta \leq \min\{\pi + a, \pi - b\}$); then $\delta \in (0, \pi)$, and, applying Lemma 4.4.7 with $r = \delta$, it follows that

$$\lim_n \int_{y-\delta}^{y+\delta} D_n(x, y)dx = 1, \text{ uniformly with respect to } y \in [-\pi + \delta, \pi - \delta].$$

Since $[a, b] \subseteq [-\pi + \delta, \pi - \delta]$, $\lim_n \int_{y-\delta}^{y+\delta} D_n(x, y)dx = 1$, uniformly with respect to $y \in [a, b]$. Therefore there exists $n_\varepsilon \in \mathbb{N}$ such that

$$\left| \int_{y-\delta}^{y+\delta} D_n(x, y)dx - 1 \right| < \frac{\varepsilon}{4M}, \forall n \geq n_\varepsilon, \forall y \in [a, b]. \tag{3}$$

Furthermore, from the previous lemma,

$$\int_{-\pi}^{\pi} D_n(x, y)dx = 1, \text{ for any } n \in \mathbb{N} \text{ and for every } y \in \mathbb{R}. \tag{4}$$

Then, for every $y \in [a, b]$, $[y - \delta, y + \delta] \subseteq [-\pi, \pi]$, and, using the conditions (1)–(4), we obtain

$$\left| \int_{-\pi}^{\pi} D_n(x, y) f(x)dx - f(y) \right| = \left| \int_{-\pi}^{\pi} D_n(x, y) [f(x) - f(y)] dx \right| \leq$$

$$\leq \int_{-\pi}^{\pi} D_n(x,y)|f(x)-f(y)|dx \leq \int_{-\pi}^{y-\delta} D_n(x,y)\left(|f(x)|+|f(y)|\right)dx +$$

$$+\int_{y-\delta}^{y+\delta} D_n(x,y)|f(x)-f(y)|dx + \int_{y+\delta}^{\pi} D_n(x,y)\left(|f(x)|+|f(y)|\right)dx \leq$$

$$\leq 2M\int_{-\pi}^{y-\delta} D_n(x,y)dx + \frac{\varepsilon}{4M}\int_{y-\delta}^{y+\delta} D_n(x,y)dx + 2M\int_{y+\delta}^{\pi} D_n(x,y)dx$$

$$= 2M\left(1-\int_{y-\delta}^{y+\delta} D_n(x,y)dx\right) + \frac{\varepsilon}{4M}\int_{y-\delta}^{y+\delta} D_n(x,y)dx <$$

$$< 2M\cdot\frac{\varepsilon}{4M} + \frac{\varepsilon}{4M}\left(1+\frac{\varepsilon}{4M}\right) < \frac{\varepsilon}{2} + \frac{\varepsilon}{4}(1+\varepsilon) < \frac{\varepsilon}{2}+\frac{\varepsilon}{2} = \varepsilon.$$

Therefore $\int_{-\pi}^{\pi} D_n(x,\cdot)f(x)dx \xrightarrow[{[a,b]}]{u} f$ ■

Theorem 4.4.9
For every $f \in L^2([-\pi,\pi])$, the Fourier series associated with f L^2-converges to f.

Proof
For every $f \in L^2([-\pi,\pi])$ and any $n \geq 1$, let $a_0 = \frac{1}{\pi}\int_{-\pi}^{\pi} f(x)dx$, $a_n = \frac{1}{\pi}\int_{-\pi}^{\pi} f(x)\cos nx dx$, $b_n = \frac{1}{\pi}\int_{-\pi}^{\pi} f(x)\sin nx dx$ (Definition 4.4.2), and let $S_n = \frac{a_0}{2}f_0 + \sum_{k=1}^{n}(a_k f_{2k-1} + b_k f_{2k})$.

Let us show that $S_n \xrightarrow[{[-\pi,\pi]}]{\|\cdot\|_2} f$.

As $\overline{C([-\pi,\pi])}^2 = L^2([-\pi,\pi])$ (see Theorem 4.2.3), for every $\varepsilon > 0$, there exists $g \in C([-\pi,\pi])$ such that $\|f-g\|_2 < \varepsilon$. The function g being continuous on $[-\pi,\pi]$, there exists $M > 0$ such that $|g(x)| \leq M$, for every $x \in [-\pi,\pi]$.

For any $n \in \mathbb{N}$, let $T_n : [-\pi,\pi] \to \mathbb{R}$,

$$T_n(y) = \int_{-\pi}^{\pi} D_n(x,y)g(x)dx = \frac{1}{d_n}\int_{-\pi}^{\pi} \cos^{2n}\frac{x-y}{2}g(x)dx.$$

We notice that, for every $y \in [-\pi,\pi]$ and for any $n \in \mathbb{N}$,

$$T_n(y) = \frac{1}{2^n d_n}\int_{-\pi}^{\pi}(1+\cos x\cos y + \sin x \sin y)^n g(x)dx.$$

We can easily demonstrate by induction that, for any $n \in \mathbb{N}$, and any $k = 0,1,\ldots,n$, there exist two continuous functions $c_k^n, d_k^n : \mathbb{R} \to \mathbb{R}$, such that

$$(1+\cos x\cos y + \sin x\sin y)^n = \sum_{k=0}^{n}(c_k^n(x)\cos ky + d_k^n(x)\sin ky), \forall x, y \in \mathbb{R}.$$

Then, for any $n \in \mathbb{N}$ and for every $y \in \mathbb{R}$,

$$T_n(y) = \frac{1}{2^n d_n}\sum_{k=0}^{n}\left(\int_{-\pi}^{\pi} c_k^n(x)g(x)dx\right)\cos ky+$$

$$+\frac{1}{2^n d_n}\sum_{k=0}^{n}\left(\int_{-\pi}^{\pi} d_k^n(x)g(x)dx\right)\sin ky.$$

For any $n \in \mathbb{N}$, and any $k = 0, 1, \ldots, n$, we denote

$$\alpha_k^n = \frac{1}{2^n d_n}\int_{-\pi}^{\pi} c_k^n(x)g(x)dx \text{ and } \beta_k^n = \frac{1}{2^n d_n}\int_{-\pi}^{\pi} d_k^n(x)g(x)dx.$$

Then

$$T_n = \alpha_0^n f_0 + \sum_{k=1}^{n}(\alpha_k^n f_{2k-1} + \beta_k^n f_{2k}).$$

From (1) of Theorem 4.4.5, for any $n \in \mathbb{N}$,

$$\|f - S_n\|_2 \le \|f - T_n\|_2 \le \|f - g\|_2 + \|g - T_n\|_2. \qquad (*)$$

On the other hand, from Lemma 4.4.8, for every $[a,b] \subseteq (-\pi,\pi)$, $T_n \xrightarrow[{[a,b]}]{\text{u}} g$.

It follows that $|g - T_n|^2 \xrightarrow[{[-\pi,\pi]}]{\cdot} \underline{0}$.
Then, for any $n \in \mathbb{N}$ and for every $y \in \mathbb{R}$,

$$|T_n(y)| \le \int_{-\pi}^{\pi} D_n(x,y)|g(x)|dx \le M\int_{-\pi}^{\pi} D_n(x,y)dx = M \text{ and so,}$$

$$|T_n - g|^2 \le (|T_n| + |g|)^2 \le (M+M)^2 = 4M^2.$$

So we can apply to $\left(|T_n - g|^2\right)_{n\in\mathbb{N}}$ the bounded convergence theorem (Corollary 3.3.10). Then $\lim_n \int_{-\pi}^{\pi} |T_n - g|^2 d\lambda = 0$; in other words $T_n \xrightarrow{\|\cdot\|_2} g$.

Using the relation $(*)$, it follows that $\limsup_n \|f - S_n\|_2 \le \varepsilon$, for every $\varepsilon > 0$, from where $S_n \xrightarrow{\|\cdot\|_2} f$, which concludes the demonstration. ■

Corollary 4.4.10 (Parseval Equality) *For every $f \in L^2([-\pi,\pi])$, let $(a_n)_{n\in\mathbb{N}}$, $(b_n)_{n\in\mathbb{N}}$ be the Fourier coefficients associated with f; then*

$$\frac{a_0^2}{2} + \sum_{n=1}^{\infty}(a_n^2 + b_n^2) = \frac{1}{\pi}\int_{-\pi}^{\pi} f^2(x)dx.$$

Proof

According to the previous theorem, if $f \in L^2([-\pi, \pi])$, then the Fourier series associated with f L^2-converges to f. The result follows by passing to the limit in the relation $(*)$ of the proof of Theorem 4.4.5. ■

Corollary 4.4.11

(1) *Every function $f \in L^2([-\pi, \pi])$, for which all the Fourier coefficients are zero, is zero almost everywhere.*

(2) *If two functions $f, g \in L^2([-\pi, \pi])$ have the same Fourier coefficients, then $f = g$ is almost everywhere.*

Proof

(1) If $a_n = b_n = 0$, for any $n \in \mathbb{N}$, then the Fourier series associated has partial sums S_n nules, for any $n \in \mathbb{N}$.

As $S_n \xrightarrow{\|\cdot\|_2} f$, $\|f\|_2 = \|S_n - f\|_2 \to 0$. Then $\int_A |f|^2 d\lambda = 0$, and so $f = \underline{0}$ a.e. (see 1) of Theorem 3.3.3).

(2) If $f, g \in L^2([-\pi, \pi])$ have the same Fourier coefficients, then $f - g$ has all the Fourier coefficients zero, and so, according to point (1), $f - g = 0$ a.e. ■

Examples 4.4.12

(i) Let $f : [-\pi, \pi] \to \mathbb{R}$, $f(x) = \text{sign}x = \begin{cases} -1, & x < 0, \\ 0, & x = 0, \\ 1, & x > 0. \end{cases}$

It is obvious that $f \in L^2([-\pi, \pi])$. Since f is an odd function, $a_n = 0$, for any $n \in \mathbb{N}$:

$$b_n = \frac{1}{\pi} \cdot \int_{-\pi}^{\pi} f(x) \cdot \sin nx dx = \frac{-2}{n\pi} \left[(-1)^n - 1\right].$$

Parseval equality is

$$\sum_{n=1}^{\infty} b_n^2 = \frac{1}{\pi} \cdot \int_{-\pi}^{\pi} f^2(x) dx = 2,$$

from where

$$\sum_{n=1}^{\infty} \frac{1}{(2n-1)^2} = \frac{\pi^2}{8}.$$

As $\sum_{n=1}^{\infty} \frac{1}{n^2} = \sum_{n=1}^{\infty} \frac{1}{(2n-1)^2} + \frac{1}{4} \cdot \sum_{n=1}^{\infty} \frac{1}{n^2}$, it follows that

$$\sum_{n=1}^{\infty} \frac{1}{n^2} = \frac{\pi^2}{6}.$$

(ii) Let $f : [-\pi, \pi] \to \mathbb{R}$, $f(x) = |x|$. Since f is an even function, for all $n \in \mathbb{N}^*$, $b_n = 0$. $a_0 = \frac{1}{\pi} \cdot \int_{-\pi}^{\pi} |x| dx = \pi$, and, for any $n \geq 1$, $a_n = \frac{1}{\pi} \cdot \int_{-\pi}^{\pi} |x| \cos nx dx = \frac{2}{n^2\pi}[(-1)^n - 1]$.

Parseval equality is

$$\frac{a_0^2}{2} + \sum_{n=1}^{\infty} a_n^2 = \frac{1}{\pi} \cdot \int_{-\pi}^{\pi} x^2 dx = \frac{2\pi^2}{3},$$

from where

$$\sum_{n=1}^{\infty} a_{2n-1}^2 = \frac{\pi^2}{6} \text{ or } \frac{16}{\pi^2} \sum_{n=1}^{\infty} \frac{1}{(2n-1)^4} = \frac{\pi^2}{6}.$$

Considering that $\sum_{n=1}^{\infty} \frac{1}{n^4} = \sum_{n=1}^{\infty} \frac{1}{(2n-1)^4} + \frac{1}{16} \sum_{n=1}^{\infty} \frac{1}{n^4}$, it follows that

$$\sum_{n=1}^{\infty} \frac{1}{n^4} = \frac{\pi^4}{90}.$$

4.5 Abstract Setting

Let $(X, \mathcal{A})$ be a measurable space, and let $\gamma : \mathcal{A} \to \bar{\mathbb{R}}_+$ be a complete and σ-finite measure on the σ-algebra $\mathcal{A}$. As above, we will limit to describe the essential points and to state the main theorems (without demonstration) for the theory of spaces L^p in this abstract framework.

Definition 4.5.1

Let $p \geq 1$; a function $\mathcal{A}$-measurable $f : X \to \mathbb{R}$ is said to be p-integrable if $|f|^p \in \mathfrak{L}_+^1(X)$. Let $\mathfrak{L}^p(X)$ be the set of all p-integrable functions. We notice that, according to Theorem 3.6.7, $\mathfrak{L}^1(X)$ is accurate to the set of all integrable functions on X.

Let $f = \sum_{i=1}^{n} a_i \chi_{A_i}$ be a simple function: $\{a_1, \ldots, a_p\} \subseteq \mathbb{R}$ and $\{A_1, \ldots, A_p\}$ is a $\mathcal{A}$-measurable partition of X. Since $|f|^p = \sum_{i=1}^{n} |a_i|^p \chi_{A_i}$, $f \in \mathfrak{L}^p(X)$ if and only if $a_i = 0$, for any i for which $\gamma(A_i) = +\infty$. Therefore $f \in \mathfrak{L}^p(X)$ if and only if $f \in \mathcal{E}^1(X) \subseteq \mathfrak{L}^1(X)$ ($\mathcal{E}^1(X)$ is the set of all integrable $\mathcal{A}$-simple functions).

Let $\|\cdot\|_p : \mathfrak{L}^p(X) \to \mathbb{R}_+$, defined by $\|f\|_p = \left(\int_X |f|^p d\gamma\right)^{\frac{1}{p}}$.

The inequalities of Hölder and Minkowski (see Theorems 4.1.4 and 4.1.5) can also be demonstrated in this context:

Theorem 4.5.2 (Hölder Inequality)
Let $p, q > 1$ such that $\frac{1}{p} + \frac{1}{q} = 1$. For every $f \in \mathfrak{L}^p(X)$ and every $g \in \mathfrak{L}^q(X)$, $f \cdot g \in \mathfrak{L}^1(X)$ and

$$\|f \cdot g\|_1 = \int_X |f \cdot g| d\gamma \leq \left(\int_X |f|^p d\gamma\right)^{\frac{1}{p}} \cdot \left(\int_X |g|^q d\gamma\right)^{\frac{1}{q}} = \|f\|_p \cdot \|g\|_q.$$

Theorem 4.5.3 (Minkowski Inequality)
For any $p \geq 1$ and for every $f, g \in \mathfrak{L}^p(X)$, $f + g \in \mathfrak{L}^p(X)$ and $\|f + g\|_p =$

$$= \left(\int_X |f + g|^p d\gamma\right)^{\frac{1}{p}} \leq \left(\int_X |f|^p d\gamma\right)^{\frac{1}{p}} + \left(\int_X |g|^p d\gamma\right)^{\frac{1}{p}} = \|f\|_p + \|g\|_p.$$

We can also show that space $(\mathfrak{L}^p(X), \|\cdot\|_p)$ is a complete seminormed space. Theorem 4.1.12 is still valid:

Theorem 4.5.4
If $\gamma(X) < +\infty$ and if $1 \leq p < r$, then $\mathfrak{L}^r(X) \subseteq \mathfrak{L}^p(X)$ and

$$\|f\|_p \leq [\gamma(X)]^{\frac{r-p}{rp}} \cdot \|f\|_r, \text{ for every } f \in \mathfrak{L}^r(X).$$

The relation defined by $f \sim g$ if and only if $f = g$ γ-almost everywhere is an equivalence relation on $\mathfrak{L}^p(X)$; then we denote by $\mathfrak{L}^p(X)|_\sim = L^p(X)$ the quotient set. We can define correctly the mapping: $\|\cdot\|_p : L^p(X) \to \mathbb{R}_+$, $\|[f]\|_p = \|f\|_p$, for every equivalence class $[f] \in L^p(X)$. The space $(L^p(X), \|\cdot\|_p)$ is a Banach space.

The spaces $(L^p(X), \|\cdot\|_p)$, $p \geq 1$ are called classic Banach spaces.

Remark 4.5.5 If $X = \mathbb{N}$ is the set of natural numbers and if γ is the counting measure (see (ii) of 1.4.6), then $\mathfrak{L}^p(\mathbb{N}) = L^p(\mathbb{N}) = \ell^p$ is the space of all real p-summable sequences ($x = (x_n)_{n\in\mathbb{N}}$, $\|x\|_p = \left(\sum_{n=0}^{\infty} |x_n|^p\right)^{\frac{1}{p}} < +\infty$).

We can also find some density results in $\mathfrak{L}^p(X)$.

Theorem 4.5.6

(1) $\mathcal{E}^1(X)$ *is dense in* $(\mathfrak{L}^p(X), \|\cdot\|_p)$.
(2) *If* (X, τ) *is a normal topological space,* $\mathcal{A}$ *is a* σ*-algebra on* X *such that* $\tau \subseteq \mathcal{A}$*, and if* γ *is a complete* σ*-finite regular measure on* $(X, \mathcal{A})$*, then* $C_p(X) = C(X) \cap \mathfrak{L}^p(X)$ *is dense in* $(\mathfrak{L}^p(X), \|\cdot\|_p)$*. Moreover, if* (X, τ) *is compact and if* γ *is a finite measure, then the set of all continuous functions on* X*,* $C(X)$*, is dense in* $(\mathfrak{L}^p(X), \|\cdot\|_p)$.

We can also define $\mathfrak{L}^\infty(X)$ as the set of all measurable functions $f : X \to \mathbb{R}$ for which $\|f\|_\infty = \inf\{\alpha \in \overline{\mathbb{R}}_+ : |f| \leq \alpha\ \gamma - \text{a.e.}\} < +\infty$. The space $(\mathfrak{L}^\infty(X), \|\cdot\|_\infty)$ is a complete seminormed space, and the quotient space $(L^\infty(X), \|\cdot\|_\infty)$ is a Banach space. If $\gamma(X) < +\infty$, then we can prove the Riesz theorem (Theorem 4.3.11).

4.6 Exercises

(1) Let $f_n :]0, 1] \to \mathbb{R}$, $f_n(x) = \dfrac{n}{1 + n\sqrt{x}}$. Show that
  (a) $(f_n) \subseteq \mathfrak{L}^2(]0, 1])$.
  (b) There exists $f :]0, 1] \to \mathbb{R}$ such that $f_n(x) \to f(x)$, for every $x \in]0, 1]$.
  (c) (f_n) don't converge in $(\mathfrak{L}^2(]0, 1]), \|\cdot\|_2)$ to f.

(2) Let f, g be two Riemann integrable functions on $[a, b]$; show that

$$\left(\int_a^b f(x)g(x)dx\right)^2 \leq \int_a^b f^2(x)dx \cdot \int_a^b g^2(x)dx.$$

(3) Let $f \in \mathfrak{L}^1(\mathbb{R})$, $f(x) > 0$, for every $x \in \mathbb{R}$; show that $\dfrac{1}{f} \notin \mathfrak{L}^1(\mathbb{R})$.

Indication: One use Hölder inequality of with the functions $f^{-\frac{1}{2}}, f^{\frac{1}{2}}$ and $p = q = \frac{1}{2}$.

(4) Let $f : [0, 1] \to \mathbb{R}_+$ such that $\sqrt{f} \in \mathfrak{L}^1([0, 1])$; show that

$$\int_{[0,1]} \sqrt{f} d\lambda \leq \sqrt{\int_{[0,1]} f d\lambda}.$$

Compare with Jensen's inequality.

(5) Let $f :]0, +\infty[\to \mathbb{R}$, $f(x) = \dfrac{1}{\sqrt{x} \cdot e^x}$.
Show that $f \in \mathfrak{L}^1(]0, +\infty[) \setminus \mathfrak{L}^2(]0, +\infty[)$.

(6) Let $(f_n) \subseteq \mathfrak{L}^p(A)$ and $f \in \mathfrak{L}^p(A)$ such that $f_n \underset{A}{\rightarrow} f$.
Show that $f_n \xrightarrow[A]{\|\cdot\|_p} f$ if and only if $\|f_n\|_p \to \|f\|_p$.

Indication: For the implication "$\Longleftarrow$" it will first be shown that, for all $a, b \in \mathbb{R}$, and for any $p \geq 1$, $|a+b|^p \leq 2^{p-1}(|a|^p + |b|^p)$; then Fatou's lemma will be applied to the sequence $(g_n)_n$ where $g_n = 2^{p-1}(|f|^p + |f_n|^p) - |f - f_n|^p$.

(7) Let $p > 1, q = \dfrac{p}{p-1}$, $(f_n)_n \subseteq \mathfrak{L}^p(A)$ and $f \in \mathfrak{L}^p(A)$ such that $f_n \xrightarrow[A]{\|\cdot\|_p} f$. Show that, for every $g \in \mathfrak{L}^q(A)$,

$$\int_A (f_n \cdot g) d\lambda \to \int_A (f \cdot g) d\lambda.$$

(8) Prove the following properties in $\mathfrak{L}^2(A)$:
(*a*) $|(f, g)| \leq \|f\|_2 \cdot \|g\|_2$, for every $f, g \in \mathfrak{L}^2(A)$.
(*b*) If $f_n \xrightarrow[A]{\|\cdot\|_2} f$, then $(f_n, g) \to (f, g)$, where $(f_n)_n \subseteq \mathfrak{L}^2(A)$, $f, g \in \mathfrak{L}^2(A)$.
(*c*) $\|f+g\|_2^2 + \|f-g\|_2^2 = 2 \cdot (\|f\|_2^2 + \|g\|_2^2)$, for every $f, g \in \mathfrak{L}^2(A)$.
(*d*) If $f \perp g_n$, for any $n \in \mathbb{N}$ and $g_n \xrightarrow[A]{\|\cdot\|_2} g$, then $f \perp g$.

(9) Write Parseval equality for functions:
(*a*) $f : [-\pi, \pi] \to \mathbb{R}$, $f(x) = x^2$.
(*b*) $f : [-\pi, \pi] \to \mathbb{R}$, $f(x) = \chi_{[0,\alpha]}$, $\alpha \in]0, \pi]$; see the case $\alpha = \frac{\pi}{2}$.

(10) Show that the series $\displaystyle\sum_{n=1}^{\infty} \frac{\sin nx}{\sqrt{n}}$ is pointwise convergent on $[-\pi, \pi]$, but it cannot be the Fourier series for any function $f \in \mathfrak{L}^2([-\pi, \pi])$.

Lebesgue Integral on $\mathbb{R}^2$

Let $\mathbb{R} \times \mathbb{R} = \mathbb{R}^2 = \{(x_1, x_2) : x_1, x_2 \in \mathbb{R}\}$ and $\mathbb{R} \times \mathbb{R} \times \mathbb{R} = \mathbb{R}^3 = \{(x_1, x_2, x_3) : x_1, x_2, x_3 \in \mathbb{R}\}$; by the usual operations of addition and multiplication with real scalars (operations defined on the coordinates), these sets are organized like real vector spaces.

The mappings defined by $(x_1, x_2) \mapsto \|(x_1, x_2)\| = \sqrt{x_1^2 + x_2^2}$ and respectively $(x_1, x_2, x_3) \mapsto \|(x_1, x_2, x_3)\| = \sqrt{x_1^2 + x_2^2 + x_3^2}$, are norms on these spaces, that is to say that the following properties are satisfied:

(1) $\|x\| = 0$ if and only if $x = \underline{0}$ ($\underline{0} = (0, 0)$ or $\underline{0} = (0, 0, 0)$).
(2) $\|c \cdot x\| = |c| \cdot \|x\|$, for any $c \in \mathbb{R}$ and for every $x \in \mathbb{R}^2$ (or $x \in \mathbb{R}^3$).
(3) $\|x + y\| \leq \|x\| + \|y\|$, for every $x, y \in \mathbb{R}^2$ (or $x, y \in \mathbb{R}^3$).

Any norm on a vector space defines a metric on this space; the mapping $(x, y) \mapsto d(x, y) = \|x - y\|$ is a metric on $\mathbb{R}^2$ ($\mathbb{R}^3$); it has the following properties:

(1) $d(x, y) = 0$ if and only if $x = y$.
(2) $d(x, y) = d(y, x)$, for every $x, y \in \mathbb{R}^2$ (or $x, y \in \mathbb{R}^3$).
(3) $d(x, y) \leq d(x, z) + d(y, z)$, for every $x, y, z \in \mathbb{R}^2$ (or $x, y, z \in \mathbb{R}^3$).

Let $x \in \mathbb{R}^2(\mathbb{R}^3)$ and let $r > 0$; we call **open ball** centered at x and of radius r the set $S(x, r) = \{y \in \mathbb{R}^2(\mathbb{R}^3) : d(x, y) < r\}$. The set $T(x, r) = \{y \in \mathbb{R}^2(\mathbb{R}^3) : d(x, y) \leq r\}$ is the **closed ball** centered at x and of radius r.

5.1 Lebesgue Measure on $\mathbb{R}^2$

First, we will define some topological notions in $\mathbb{R}^2$. With slight adaptations, they can be easily rewritten for $\mathbb{R}^3$.

L. C. Florescu, *Lebesgue Integral*, Compact Textbooks in Mathematics,
https://doi.org/10.1007/978-3-030-60163-8_5

Definition 5.1.1

A set $D \subseteq \mathbb{R}^2$ is **open** set if $D = \emptyset$ or, in the case where $D \neq \emptyset$, for every $x \in D$, there exists $r > 0$ such that $S(x, r) \subseteq D$.
A set $F \subseteq \mathbb{R}^2$ is **closed** if this complement, F^c, is open.
It is said that the family of open sets of $\mathbb{R}^2$ is the **usual topology** of $\mathbb{R}^2$; it is denoted by τ_u^2. The family of closed sets is denoted by $\mathcal{F}^2$.

Using the above definition, it is easy to prove the following proposition:

Proposition 5.1.2 *The usual topology has the following properties:*
(1) *For every* $D, G \in \tau_u^2$, $D \cap G \in \tau_u^2$.
(2) *For every* $\{D_i : i \in I\} \subseteq \tau_u^2$, $\bigcup_{i \in I} D_i \in \tau_u^2$.
(3) $\mathbb{R}^2, \emptyset \in \tau_u^2$.

The family of closed sets has the dual properties:
(1′) *For every* $F, H \in \mathcal{F}^2$, $F \cup H \in \mathcal{F}^2$.
(2′) *For every* $\{F_i : i \in I\} \subseteq \mathcal{F}^2$, $\bigcap_{i \in I} F_i \in \mathcal{F}^2$.
(3′) $\mathbb{R}^2, \emptyset \in \mathcal{F}^2$.

Definition 5.1.3

A neighborhood of $x \in \mathbb{R}^2$ is a set $V \subseteq \mathbb{R}^2$ which contains an open ball: $S(x, r) \subseteq V$.
Let $\mathcal{V}(x)$ be the family of all neighborhoods of x.
A point $x \in \mathbb{R}^2$ is said to be **adherent** to $A \subseteq \mathbb{R}^2$ when every neighborhood V of x meets A ($V \cap A \neq \emptyset$), in other words, every open ball $S(x, r)$ meets A. Let $\bar{A}$ be the set of all adherent points of A; set $\bar{A}$ is said to be the **closure** of A.
A point $x \in \mathbb{R}^2$ is called **interior** point of $A \subseteq \mathbb{R}^2$ when A is a neighborhood of x.
The **interior** of A, A°, is the set of all interior points of A.
A set $A \subseteq \mathbb{R}^2$ is **bounded** if there is $r > 0$ such that $A \subseteq S(\underline{0}, r)$.
A sequence $(x_n)_n \subseteq \mathbb{R}^2$ **converges** to $x \in \mathbb{R}^2$ if $\|x_n - x\| \to 0$.
A set $K \subseteq \mathbb{R}^2$ is **compact** if it is bounded and closed.

Remarks 5.1.4

(i) $x \in \bar{A}$ if and only if there exists a sequence $(x_n)_n \subseteq A$, $x_n \to x$.
(ii) F is closed set if and only if $F = \bar{F}$.
(iii) Let $(x_n)_n \subseteq \mathbb{R}^2$, $x_n = (x_1^n, x_2^n)$, for any $n \in \mathbb{N}$ and let $x = (x_1, x_2) \in \mathbb{R}^2$. Then $x_n \to x$ if and only if $x_1^n \to x_1$ and $x_2^n \to x_2$.
(iv) A set $K \subseteq \mathbb{R}^2$ is compact if and only if every sequence in K has a subsequence convergent to a point of K.
(v) A set $K \subseteq \mathbb{R}^2$ is compact if and only if from any open cover of K, we can extract a finite subcover (the proof is similar to that of Lemma 7.1.6).

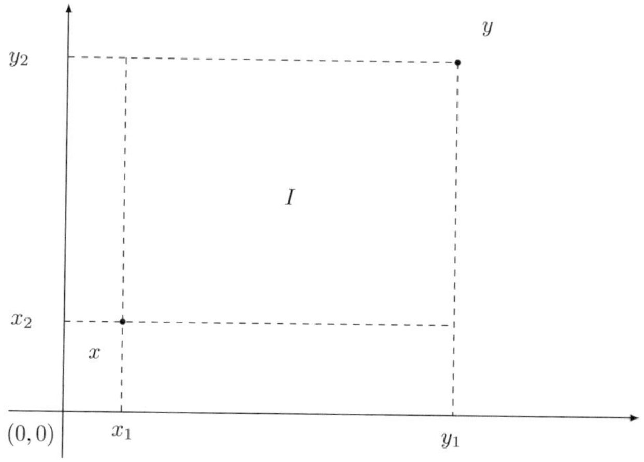

Fig. 5.1 A two-dimensional open interval

Definition 5.1.5

Let $x = (x_1, x_2), y = (y_1, y_2) \in \mathbb{R}^2$; then $x \leq y$ if $x_1 \leq y_1$ and $x_2 \leq y_2$. This is a partial order on $\mathbb{R}^2$ (it is not possible to compare any two elements of $\mathbb{R}^2$).
If $x = (x_1, x_2) \leq y = (y_1, y_2)$, then we can define the **two-dimensional open interval**:

$$I =]x, y[=]x_1, y_1[\times]x_2, y_2[= \{(z_1, z_2) \in \mathbb{R}^2 : x_1 < z_1 < y_1, x_2 < z_2 < y_2\}$$

The above Fig. 5.1 shows us that the interval I is the interior of a rectangle whose sides are parallel to the coordinate axes.
We will denote with $\mathcal{I}(\mathbb{R}^2)$, or with $\mathcal{I}$ if there is no danger of confusion, the family of open intervals; it's obvious that $\mathcal{I} \subseteq \tau_u^2$.
We can also define the **two-dimensional closed intervals**: $[x, y] = [x_1, y_1] \times [x_2, y_2]$; every closed interval is a closed set. We can also define the other types of intervals, generically marked by $J = |x, y| = |x_1, y_1| \times |x_2, y_2|$ (the vertical bar can be a closed bracket or an open bracket). We will denote with $\mathcal{J}(\mathbb{R}^2)$, or with $\mathcal{J}$ if there is no risk of confusion, the family of all the intervals in $\mathbb{R}^2$.

In the plan, we will consider only bounded intervals!

For every interval $J = |x, y| = |x_1, y_1| \times |x_2, y_2| \in \mathcal{J}$, the **measure** of J is $|J| = (y_1 - x_1) \cdot (y_2 - x_2)$; $|J|$ is the area of the rectangle J.
For every $J = |x_1, y_1| \times |x_2, y_2| \in \mathcal{J}$ and every $z = (z_1, z_2) \in \mathbb{R}^2$, $z + J = |x_1 + z_1, y_1 + z_1| \times |x_2 + z_2, y_2 + z_2| \in \mathcal{J}$ is the image of J by the translation with z; it is obvious that $|z + J| = |J|$.

(Continued)

Definition 5.1.5 (continued)

A partial order relation is defined similarly in $\mathbb{R}^3$. You can also define three-dimensional intervals (these will be parallelepipeds with edges parallel to the coordinate axes). The measure of such an interval will be the volume of the parallelepiped. We also observe that the translation of an interval is an interval of the same type and that the measure is invariant to translations.

We will present below the construction of Lebesgue's measure on $\mathbb{R}^2$; the concepts and results can be easily adapted to $\mathbb{R}^3$.

The method of construction of the measure on $\mathbb{R}^2$ follows the same steps as in the case of $\mathbb{R}$: the measure of open sets will be defined, then the outer measure will be constructed in the plane, and the Lebesgue measurable sets will be defined in $\mathbb{R}^2$.

Recall that, in the definition of the open sets measure of $\mathbb{R}$, the theorem of the open sets structure plays a major role: any open set of $\mathbb{R}$ is uniquely written as a countable union of pairwise disjoint open intervals (Theorem 1.1.3).

In $\mathbb{R}^2$, we no longer have a representation of this type; however, we can give a theorem of representation of open sets as a countable union of the two-dimensional nonoverlapping (without common interior points) closed intervals (we have given a similar theorem in the case of $\mathbb{R}$, Theorem 1.1.8). To be able to demonstrate such a result, we need some preparatory lemmas.

Lemma 5.1.6 *Let $J, J_1, J_2, \ldots, J_n \in \mathcal{J}$ such that $J \supseteq \bigcup_{k=1}^{n} J_k$ and, for any $k \neq l$, J_k and J_l do not have common interior points; then $|J| \geq \sum_{k=1}^{n} |J_k|$.*

Proof

Let $J_1, J_2, \ldots, J_n \in \mathcal{J}$ be arbitrary two-dimensional nonoverlapping intervals so that their union is included in the interval J. For any $k \in \{1, \ldots, n\}$, we extend the sides of the rectangle J_k to the border of J. We thus obtain a partition of J into nonoverlapping rectangles: $K_1, \ldots, K_m$ and a partition $N_0, N_1, \ldots, N_n$ of the set $\{1, \ldots, m\}$ such that, for any $k \in \{1, \cdots, n\}$, $J_k = \cup_{i \in N_k} K_i$ and $\cup_{i \in N_0} K_i = J \setminus \cup_{k=1}^{n} J_k$. We suppose that $J = |a, b| \times |c, d|$; the intersection points of the sides extensions of the intervals J_k with the sides of J determine two partitions $a = a_0 < a_1 < \ldots < a_p = b$ and $c = c_0 < c_1 < \ldots < c_q = d$ of intervals $|a, b|$, respectively $|c, d|$ ($m = p \cdot q$). Then

$$|J| = (b-a) \cdot (d-c) = (b-a) \cdot \sum_{j=0}^{q-1} (c_{j+1} - c_j) =$$

$$= \sum_{i=0}^{p-1} \sum_{j=0}^{q-1} (a_{i+1} - a_i) \cdot (c_{j+1} - c_j) = \sum_{l=1}^{m} |K_l| = \sum_{i \in N_0} |K_i| + \sum_{k=1}^{n} \sum_{i \in N_k} |K_i|.$$

Similarly, it is shown that, for any $k \in \{1, \ldots, n\}$, $|J_k| = \sum_{i \in N_k} |K_i|$. It follows that

$$|J| \geq \sum_{k=1}^{n} \sum_{i \in N_k} |K_i| = \sum_{k=1}^{n} |J_k|.$$

■

From the above demonstration, we note that if $N_0 = \emptyset$, then $J = \bigcup_{k=1}^{n} J_k$ and $|J| = \sum_{k=1}^{n} |J_k|$.

In Fig. 5.2, we have imagined a possible situation to illustrate the previous lemma. The intervals delimited by continuous lines are the intervals J_k (the large rectangle is J). The extensions of the sides of the rectangles J_k were marked by dotted lines. Then, $J \supset \cup_{k=1}^{4} J_k$, $N_0 = \{1, 6, 7, 9\}$, $N_1 = \{4\}$, $N_2 = \{2, 5\}$, $N_3 = \{3\}$, $and N_4 = \{8\}$. According to this partition, $J_1 = K_4$, $J_2 = K_2 \cup K_5$, $J_3 = K_3$, $and J_4 = K_8$.

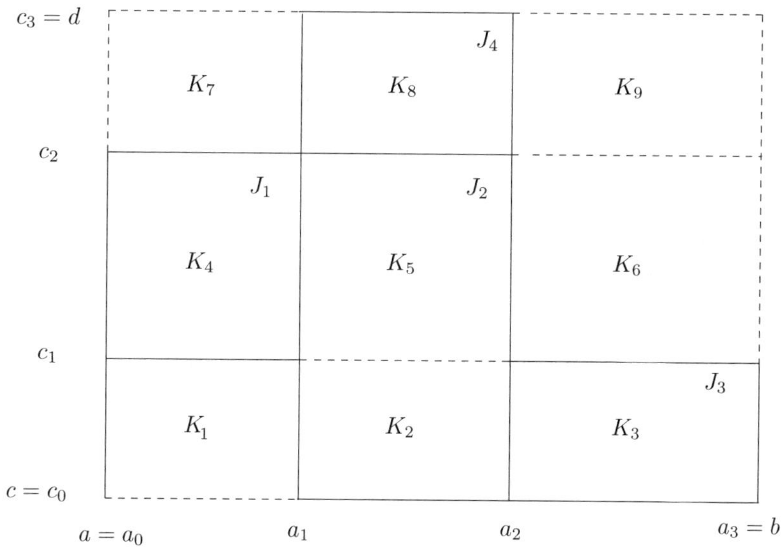

Fig. 5.2 A possible situation for the interval $J = [a, b] \times [c, d]$ and the intervals $J_1, \cdots, J_n$

In the following lemma, we will no longer consider the intervals J_k with disjoint interiors.

Lemma 5.1.7 *Let $J_1, \ldots, J_n \in \mathcal{J}$ such that $J = \bigcup_{k=1}^{n} J_k \in \mathcal{J}$; then*

$$|J| \leq \sum_{k=1}^{n} |J_k|$$

Proof

We proceed as in the previous lemma demonstration by extending the sides of the intervals J_k and obtaining a new partition of J: $J = \bigcup_{l=1}^{m} K_l$ and $\{N_1, \ldots, N_n\}$ a cover (but not a partition!) of $\{1, \ldots, m\}$ such that, for any $k = 1, \cdots, n$, $J_k = \bigcup_{i \in N_k} K_i$ (in this case $N_0 = \emptyset$). Then

$$|J| = \sum_{l=1}^{m} |K_l| \leq \sum_{k=1}^{n} \sum_{i \in N_k} |K_i| = \sum_{k=1}^{n} |J_k|. \qquad \blacksquare$$

Lemma 5.1.8 *Let $J_1, \ldots, J_n, K_1, \ldots, K_p \in \mathcal{J}$ s.t. $\bigcup_{i=1}^{n} J_i \subseteq \bigcup_{j=1}^{p} K_j$. If, for any $i \neq l$, J_i and J_l don't have common interior points, then*

$$\sum_{i=1}^{n} |J_i| \leq \sum_{j=1}^{p} |K_j|.$$

Proof

Let $L_{ij} = J_i \bigcap K_j \in \mathcal{J}$; then, for any $i \neq l$, L_{ij} and L_{lj} do not have common interior points and are included in K_j. According to Lemma 5.1.6, $\sum_{i=1}^{n} |L_{ij}| \leq |K_j|$, for any $j = 1, \ldots, p$, from where

$$\sum_{i=1}^{n} \sum_{j=1}^{p} |L_{ij}| = \sum_{j=1}^{p} \sum_{i=1}^{n} |L_{ij}| \leq \sum_{j=1}^{p} |K_j|.$$

But $J_i = \bigcup_{j=1}^{p} L_{ij}$ and then, according to Lemma 5.1.7, $|J_i| \leq \sum_{j=1}^{p} |L_{ij}|$ from where

$$\sum_{i=1}^{n} |J_i| \leq \sum_{i=1}^{n} \sum_{j=1}^{p} |L_{ij} \leq \sum_{j=1}^{p} |K_j|. \qquad \blacksquare$$

Lemma 5.1.9 *Let $(J_n)_n$ and $(K_m)_m$ be two sequences of **closed** intervals such that $\bigcup_{n=1}^{\infty} J_n \subseteq \bigcup_{m=1}^{\infty} K_m$. If, for any $l \neq n$, J_l and J_n don't have common interior points, then*

$$\sum_{n=1}^{\infty} |J_n| \leq \sum_{m=1}^{\infty} |K_m|.$$

Proof

Let us suppose that $\sum_{n=1}^{\infty} |J_n| > \sum_{m=1}^{\infty} |K_m|$. Then there exist $N \in \mathbb{N}$ and $\varepsilon > 0$ such that $\sum_{n=1}^{N} |J_n| > \sum_{m=1}^{\infty} |K_m| + \varepsilon$. For any $m \in \mathbb{N}^*$, let I_m be an open interval such that $K_m \subseteq I_m$ and $|I_m| < |K_m| + \dfrac{\varepsilon}{2^m}$.

The set $C = \bigcup_{n=1}^{N} J_n$ is compact (bounded and closed) and $C \subseteq \bigcup_{m=1}^{\infty} I_m$. Therefore, $\{I_m : m \in \mathbb{N}^*\}$ is an open cover of the compact set C; from this open cover, we can extract a finite subcover (see (v) of 5.1.4). Let $M \in \mathbb{N}^*$ such that $C \subseteq \bigcup_{m=1}^{M} I_m$.

As $\bigcup_{n=1}^{N} J_n \subseteq \bigcup_{m=1}^{M} I_m$, we can use Lemma 5.1.8:

$$\sum_{n=1}^{N} |J_n| \le \sum_{m=1}^{M} |I_m| \le \sum_{m=1}^{\infty} |I_m| < \sum_{m=1}^{\infty} |K_m| + \varepsilon,$$

which contradicts the choice of ε. ∎

Definition 5.1.10

An interval **dyadic** of $\mathbb{R}^2$ is a closed two-dimensional interval of the form

$$\left[\frac{k}{2^n}, \frac{k+1}{2^n}\right] \times \left[\frac{l}{2^n}, \frac{l+1}{2^n}\right],$$

where k, l are integers and $n \in \mathbb{N}$.

Proposition 5.1.11 *If the dyadic intervals J_1 and J_2 have a common interior point, then $J_1 \subseteq J_2$ or $J_2 \subseteq J_1$.*

Proof

Let $n, m \in \mathbb{N}$, $k_1, l_1, k_2, l_2 \in \mathbb{Z}$ and let

$$J_1 = \left[\frac{k_1}{2^n}, \frac{k_1+1}{2^n}\right] \times \left[\frac{l_1}{2^n}, \frac{l_1+1}{2^n}\right] \text{ and } J_2 = \left[\frac{k_2}{2^m}, \frac{k_2+1}{2^m}\right] \times \left[\frac{l_2}{2^m}, \frac{l_2+1}{2^m}\right].$$

We suppose that $x = (x_1, x_2)$ is a common interior point, that is to say

$$x \in \left(\left]\frac{k_1}{2^n}, \frac{k_1+1}{2^n}\right[\times \left]\frac{l_1}{2^n}, \frac{l_1+1}{2^n}\right[\right) \cap \left(\left]\frac{k_2}{2^m}, \frac{k_2+1}{2^m}\right[\times \left]\frac{l_2}{2^m}, \frac{l_2+1}{2^m}\right[\right).$$

We will show that, if $n \le m$, then $J_2 \subseteq J_1$. Indeed, from the above assumptions, we can deduce the inequalities $\frac{k_2}{2^m} < x_1 < \frac{k_2+1}{2^m}$ and

$$\frac{2^{m-n} \cdot k_1}{2^m} = \frac{k_1}{2^n} < x_1 < \frac{k_1+1}{2^n} = \frac{2^{m-n} \cdot k_1 + 2^{m-n}}{2^m}.$$

It follows that

$$k_2 < 2^{m-n} \cdot k_1 + 2^{m-n} \text{ and } 2^{m-n} \cdot k_1 < k_2 + 1 \text{ or}$$

$$k_2 + 1 \le 2^{m-n} \cdot k_1 + 2^{m-n} \text{ and } 2^{m-n} \cdot k_1 \le k_2.$$

According to the above inequalities, it follows that, for all $y = (y_1, y_2) \in J_2$,

$$\frac{k_1}{2^n} = \frac{2^{m-n} \cdot k_1}{2^m} \leq \frac{k_2}{2^m} \leq y_1 \leq \frac{k_2+1}{2^m} \leq \frac{2^{m-n} \cdot k_1 + 2^{m-n}}{2^m} = \frac{k_1+1}{2^n}.$$

With a similar calculation, $\frac{l_1}{2^n} \leq y_2 \leq \frac{l_1+1}{2^n}$, from where $y \in J_1$ and so $J_2 \subseteq J_1$.

If $m \leq n$, then we demonstrate in a similar way that $J_1 \subseteq J_2$. ■

Theorem 5.1.12

Any non-empty open set $D \in \tau_u^2$ is a countable union of two-dimensional nonoverlapping closed intervals.

If $D = \bigcup_{n=1}^{\infty} J_n = \bigcup_{m=1}^{\infty} K_m$, which are two countable unions of two-dimensional nonoverlapping closed intervals, then

$$\sum_{n=1}^{\infty} |J_n| = \sum_{m=1}^{\infty} |K_m|.$$

Proof

Let $D \in \tau_u^2$, $D \neq \emptyset$.

For $n = 0$, we consider the network of lines parallel to the coordinate axes passing through integer points: $\ldots, -3, -2, -1, 0, 1, 2, 3, \ldots$; let J_0 be the union of the closed squares of the 1 sides thus created which are included in D. Therefore, $J_0 \subseteq D$ and $D \setminus J_0$ contains no square of this form.

For $n = 1$, we consider the network of lines parallel to the coordinate axes passing through points: $\frac{k}{2}, k \in \mathbb{Z}$; let J_1 the union of the closed squares of the $\frac{1}{2}$ sides which are included in D and whose interiors are disjoint from J_0 and so on.

In the step n, we consider the network of lines parallel to the coordinate axes passing through the points of the form $\dfrac{k}{2^n}, k \in \mathbb{Z}$; let J_n be the union of the closed squares of the $\frac{1}{2^n}$ sides which are included in D and whose interiors are disjoint from $\bigcup_{i=0}^{n-1} J_i$.

We will continue this construction in the same way.

Then $\bigcup_{n=1}^{\infty} J_n$ is a countable union of closed squares whose interiors are disjoint and $\bigcup_{n=1}^{\infty} J_n \subseteq D$. Let us show that the previous inclusion is in fact an equality.

For every $x = (x_1, x_2) \in D$, there exists $\varepsilon > 0$ such that the open square $P =]x_1 - \varepsilon, x_1 + \varepsilon[\times]x_2 - \varepsilon, x_2 + \varepsilon[\subseteq D$. Let $n \in \mathbb{N}$ such that $\frac{1}{2^n} < \varepsilon$ and let $k = [2^n x_1], l = [2^n x_2] \in \mathbb{Z}$, where $[a]$ is the least integer greater than or equal to a. Then, the closed square $K = \left[\dfrac{k}{2^n}, \dfrac{k+1}{2^n}\right] \times \left[\dfrac{l}{2^n}, \dfrac{l+1}{2^n}\right]$ (obtained in the network $\frac{k}{2^n}$) contains x and is contained in P. K is either contained in one of the sets $J_m, m < n$, or is one of the squares constituting the set J_n. It follows that $x \in \bigcup_{n=1}^{\infty} J_n$ and then $D = \bigcup_{n=1}^{\infty} J_n$.

If D admits two of these representations $D = \bigcup_{n=1}^{\infty} J_n = \bigcup_{m=1}^{\infty} K_m$ of which, for each, the interiors are disjoint, then according to Lemma 5.1.9, $\sum_{n=1}^{\infty} |J_n| \le \sum_{m=1}^{\infty} |K_m|$ and $\sum_{m=1}^{\infty} |K_m| \le \sum_{n=1}^{\infty} |J_n|$, from where the equality requested. ■

From the proof of Theorem 5.1.12, it follows that any non-empty open set can be represented as a countable union of nonoverlapping dyadic intervals.

We can now define the measure of open sets in $\mathbb{R}^2$.

Definition 5.1.13

Let $D \in \tau_u^2$; if $D = \emptyset$, then $\mu(D) = 0$. If $D \neq \emptyset$, then, according to the previous theorem, $D = \bigcup_{n=1}^{\infty} J_n$, where J_n are two-dimensional nonoverlapping closed intervals; we say that this writing of D is a **representation** of the open set D. We will then define the **measure** of D by $\mu(D) = \sum_{n=1}^{\infty} |J_n|$.

The definition is consistent because, using Theorem 5.1.12 again, the above sum does not depend on the representation of the set D.

So we defined an application $\mu : \tau_u^2 \to \bar{\mathbb{R}}_+$; in the following theorem, we will give some of the properties of open sets measure.

Theorem 5.1.14

(1) $\mu(\emptyset) = 0$, $\mu(\mathbb{R}^2) = +\infty$.
(2) $\mu(x + D) = \mu(D)$, *for every* $x \in \mathbb{R}^2$ *and every* $D \subset \tau_u^2$.
(3) $\mu(D) \le \mu(G)$, *for every* $D, G \in \tau_u^2$ *with* $D \subseteq G$.
(4) $\mu(\bigcup_{n=1}^{\infty} D_n) = \sum_{n=1}^{\infty} \mu(D_n)$, *for every* $(D_n)_n \subseteq \tau_u^2$ *pairwise disjoint.*
(5) $\mu(\bigcup_{n=1}^{\infty} D_n) \le \sum_{n=1}^{\infty} \mu(D_n)$, *for every* $(D_n)_n \subseteq \tau_u^2$.
(6) $\mu(I) = |I|$, *for every* $I \in \mathcal{I}$

Proof

(1) $\mathbb{R}^2$ admits the following representation as a counting union of nonoverlapping closed squares:

$$\mathbb{R}^2 = \bigcup_{k,l \in \mathbb{Z}} ([k, k+1] \times [l, l+1]) .$$

It follows that the measure of $\mathbb{R}^2$ is an infinite sum of 1 (the area of these squares), and then it is $+\infty$.

(2) If $D = \bigcup_{n=1}^{\infty} J_n$ is a representation of D, then $x+D = \bigcup_{n=1}^{\infty}(x+J_n)$ is a representation of $x + D$ (since J_n are closed intervals without common interior points, so are the intervals $x + J_n$. It is obvious that $\mu(x + D) = \sum_{n=1}^{\infty} |x + J_n| = \sum_{n=1}^{\infty} |J_n| = \mu(D)$.

(3) Let $D = \bigcup_{n=1}^{\infty} J_n$ and let $G = \bigcup_{m=1}^{\infty} K_m$ be the representations of the open sets D and G. Since $D \subseteq G$, we can use Lemma 5.1.9; then $\mu(D) = \sum_{n=1}^{\infty} |J_n| \le \sum_{m=1}^{\infty} |K_m| = \mu(G)$.

(4) For any $n \in \mathbb{N}^*$, let $D_n = \bigcup_{k=1}^{\infty} J_k^n$ be a representation of open set D_n. The set $D = \bigcup_{n=1}^{\infty} D_n$ is open and $D = \bigcup_{n=1}^{\infty} \bigcup_{k=1}^{\infty} J_k^n$ is a representation of it. Indeed, J_k^n are closed intervals. Two distinct intervals J_k^n and J_l^m have no common interior points because, if $m \ne n$, then $J_k^n \cap J_l^m = \emptyset$ (are included in the disjointed sets D_n and D_m), and, if $m = n$, then $k \ne l$ and so J_k^n and J_l^n have no common interior points. It follows that

$$\mu(D) = \sum_{n=1}^{\infty} \sum_{k=1}^{\infty} |J_k^n| = \sum_{n=1}^{\infty} \mu(D_n).$$

(5) Let $D = \bigcup_{k=1}^{\infty} J_k$ be a representation of open set $D = \bigcup_{n=1}^{\infty} D_n$ and, for any $n \in \mathbb{N}^*$, let $D_n = \bigcup_{k=1}^{\infty} J_k^n$ a representation of D_n. Since $\bigcup_{k=1}^{\infty} J_j \subseteq \bigcup_{n=1}^{\infty} \bigcup_{k=1}^{\infty} J_k^n$, we can use Lemma 5.1.9 and therefore

$$\mu(D) = \sum_{k=1}^{\infty} |J_k| \le \sum_{n=1}^{\infty} \sum_{k=1}^{\infty} |J_k^n| = \sum_{n=1}^{\infty} \mu(D_n).$$

(6) Let $I =]x, y[\in \mathcal{I} \subseteq \tau_u^2$, where $x = (x_1, x_2)$ and $y = (y_1, y_2)$; then $I =]x_1, y_1[\times]x_2, y_2[$. Consider the sequences: $x_1^n \downarrow x_1$, $y_1^m \uparrow y_1$ with $x_1^0 < y_1^0$ and the sequences $x_2^p \downarrow x_2$, $y_2^q \uparrow y_2$ with $x_2^0 < y_2^0$. Then, the open intervals $]x_1, y_1[$ and $]x_2, y_2[$ can be written as countable unions of closed intervals:

$$]x_1, y_1[= [x_1^0, y_1^0] \cup \bigcup_{n=0}^{\infty} [x_1^{n+1}, x_1^n] \cup \bigcup_{m=0}^{\infty} [y_1^m, y_1^{m+1}] \text{ and}$$

$$]x_2, y_2[= [x_2^0, y_2^0] \cup \bigcup_{p=0}^{\infty} [x_2^{p+1}, x_2^p] \cup \bigcup_{q=0}^{\infty} [y_2^q, y_2^{q+1}],$$

so that their Cartesian product will be a countable union of two-dimensional nonoverlapping closed intervals:

$$I = \left([x_1^0, y_1^0] \times [x_2^0, y_2^0]\right) \cup$$

$$\cup \bigcup_{p=0}^{\infty} \left([x_1^0, y_1^0] \times [x_2^{p+1}, x_2^p]\right) \cup \bigcup_{q=0}^{\infty} \left([x_1^0, y_1^0] \times [y_2^q, y_2^{q+1}]\right) \cup$$

$$\cup \bigcup_{n=0}^{\infty} \left([x_1^{n+1}, x_1^n] \times [x_2^0, y_2^0]\right) \cup \bigcup_{n,p=0}^{\infty} \left([x_1^{n+1}, x_1^n] \times [x_2^{p+1}, x_2^p]\right) \cup$$

$$\cup \bigcup_{n,q=0}^{\infty} \left([x_1^{n+1}, x_1^n] \times [y_2^q, y_2^{q+1}] \right) \cup \bigcup_{m=0}^{\infty} \left([y_1^m, y_1^{m+1}] \times [x_2^0, y_2^0] \right) \cup$$

$$\cup \bigcup_{m,p=0}^{\infty} \left([y_1^m, y_1^{m+1}] \times [x_2^{p+1}, x_2^p] \right) \cup \bigcup_{m,q=0}^{\infty} \left([y_1^m, y_1^{m+1}] \times [y_2^q, y_2^{q+1}] \right).$$

This being a representation of the open set I, $\mu(I)$ will be the sum of the surfaces of the closed rectangles on the components. If we consider that

$$\sum_{n=0}^{\infty}(x_1^n - x_1^{n+1}) = x_1^0 - x_1, \sum_{m=0}^{\infty}(y_1^{m+1} - y_1^m) = y_1 - y_1^0,$$

$$\sum_{p=0}^{\infty}(x_2^p - x_2^{p+1}) = x_2^0 - x_2 \text{ and } \sum_{q=0}^{\infty}(y_2^{q+1} - y_2^q) = y_2 - y_2^0,$$

a simple calculation show that

$$\mu(I) = (y_1 - x_1) \cdot (y_2 - x_2) = |I|. \qquad \blacksquare$$

Once the measure is defined for the open sets, we will proceed as in the case of $\mathbb{R}$: we will define the Lebesgue outer measure in the plan, and then we will define the measurable sets and the Lebesgue measure in the plan. The demonstrations being identical, we will limit ourselves to presenting the main definitions and results concerning the Lebesgue measure on $\mathbb{R}^2$.

Definition 5.1.15

The mapping $\mu^* : \mathcal{P}(\mathbb{R}^2) \to \bar{\mathbb{R}}_+$, defined by

$$\mu^*(E) = \inf\{\mu(D) : D \in \tau_u^2, E \subseteq D\}, \text{ for every } E \subseteq \mathbb{R}^2,$$

is the **Lebesgue outer measure** on $\mathbb{R}^2$.

The outer measure has the following properties:

Theorem 5.1.16

(1) $\mu^*(\emptyset) = 0$,
(2) $\mu^*(E) \leq \mu^*(F)$ *and* $E \subseteq F$,
(3) $\mu^*(\bigcup_{n=1}^{\infty} E_n) \leq \sum_{n=1}^{\infty} \mu^*(E_n)$, *for every* $(E_n)_n \subseteq \mathcal{P}(\mathbb{R}^2)$.

Remarks 5.1.17

(i) $\mu^*(D) = \mu(D)$, for every $D \in \tau_u^2$.
(ii) $\mu^*(\{x\}) = 0$, for every $x \in \mathbb{R}^2$.
(iii) $\mu^*(J) = |J|$, for every $J \in \mathcal{J}$.
(iv) $\mu^*(\bigcup_{k=1}^n E_k) \leq \sum_{k=1}^n \mu^*(E_k)$, for all $n \in \mathbb{N}^*$ and every $E_1, \cdots, E_n \in \mathcal{P}(\mathbb{R}^2)$.
(v) $\mu^*(x + E) = \mu^*(E)$, for every $x \in \mathbb{R}^2$ and every $E \subseteq \mathbb{R}^2$.

Definition 5.1.18

A set $E \subseteq \mathbb{R}^2$ is a **null set** or **Lebesgue-negligible** if $\mu^*(E) = 0$.
According to the definition, E is a null set if and only if, for all $\varepsilon > 0$, there exists a sequence of nonoverlapping closed intervals $(J_n)_n$ such that $E \subseteq \bigcup_{n=0}^\infty J_n$ and $\sum_{n=0}^\infty |J_n| < \varepsilon$.
A property P is satisfied **almost everywhere** on the set $E \subseteq \mathbb{R}^2$ (P is μ-**almost everywhere**, or **a.e.** accomplished) if the set $E_P = \{x \in E : x \text{ does not have the property } P\}$ is a null set, which means that $\mu^*(A_P) = 0$.

Definition 5.1.19

A set $E \subseteq \mathbb{R}^2$ is **measurable** (in the sense of Lebesgue) if, for every $\varepsilon > 0$, there exists $D \in \tau_u^2$ such that $E \subseteq D$ and $\mu^*(D \setminus E) < \varepsilon$.
Let $\mathscr{L}(\mathbb{R}^2)$ be the family of all measurable sets of $\mathbb{R}^2$; the mapping $\mu = \mu^*|_{\mathcal{L}(\mathbb{R}^2)}$ is σ-additive. μ is the **Lebesgue measure** on $\mathbb{R}^2$.
If $E \in \mathscr{L}(\mathbb{R}^2)$, then $\mathscr{L}(E) = \{F \subseteq E : F \in \mathscr{L}(\mathbb{R}^2)\}$ is the family of all measurable subsets of E.
We can easily prove that $E \in \mathscr{L}(\mathbb{R}^2)$ if and only if, for every $\varepsilon > 0$, there exists a closed set F and an open one D such that $F \subseteq E \subseteq D$ and $\mu(D \setminus F) < \varepsilon$.

Remarks 5.1.20

(i) $\tau_u^2 \subseteq \mathscr{L}(\mathbb{R}^2)$. It follows that μ is the extension of the measure of open sets, and then the notation performed does not give rise to confusion.
(ii) $E \in \mathscr{L}(\mathbb{R}^2)$, for every $E \subseteq \mathbb{R}^2$ with $\mu^*(E) = 0$.
(iii) $\bigcup_{n=1}^\infty E_n \in \mathscr{L}(\mathbb{R}^2)$, for every $(E_n)_n \subseteq \mathscr{L}(\mathbb{R}^2)$.
(iv) For every $E \in \mathscr{L}(\mathbb{R}^2)$ with $\mu(E) = 0$ and every $F \subseteq E$, $F \in \mathscr{L}(\mathbb{R}^2)$.
(v) Every interval $J \in \mathcal{J}$ is measurable and $\mu(J) = |J|$.

It follows that $\mathscr{L}(\mathbb{R}^2)$ is a σ-algebra on $\mathbb{R}^2$ and that μ is a complete measure on $\mathscr{L}(\mathbb{R}^2)$. Then μ will have all the properties of Theorem 1.3.11.

At the end of this paragraph, we will give two results showing how the measure of a set of $\mathscr{L}(\mathbb{R}^2)$ can be calculated using the Lebesgue measure on $\mathbb{R}$.

Theorem 5.1.21
Let $A, B \in \mathscr{L}(\mathbb{R})$; then $A \times B \in \mathscr{L}(\mathbb{R}^2)$ and $\mu(A \times B) = \lambda(A) \cdot \lambda(B)$.

Proof

(1) At the beginning, we assume that $\lambda(A) < +\infty$ and $\lambda(B) < +\infty$.

(a) If A and B are intervals, then, according to (v) of 5.1.20,

$$A \times B \in \mathcal{J}(\mathbb{R}^2) \text{ and } \mu(A \times B) = |A \times B| = |A| \cdot |B| = \lambda(A) \cdot \lambda(B).$$

(b) Let $A, B \in \tau_u$; then, according to the theorem of structure of open sets of $\mathbb{R}$ (Theorem 1.1.3), $A = \bigcup_{n=1}^{\infty} I_n$ and $B = \bigcup_{m=1}^{\infty} J_m$, where $(I_n)_n$, $(J_m)_m$ are sequences of open interval, pairwise disjoint.

$A \times B = \cup_n \cup_m (I_n \times J_m)$ and, since $I_n \times J_m$ are open two-dimensional intervals disjoint,

$$\mu(A \times B) = \sum_{n,m} |I_n \times J_m| = \sum_{n,m} |I_n| \cdot |J_m| = \lambda(A) \cdot \lambda(B).$$

(c) Let now the case where $A, B \in \mathscr{L}(\mathbb{R})$. Using the characterization given in Exercise 10) of 1.5, for every $\varepsilon > 0$, there exist the closed sets F_1, F_2 and the open sets D_1, D_2 of $\mathbb{R}$ such that $F_1 \subseteq A \subseteq D_1$, $F_2 \subseteq B \subseteq D_2$, and $\lambda(D_1 \setminus F_1) < \varepsilon_1$, $\lambda(D_2 \setminus F_2) < \varepsilon_2$, where $\varepsilon_1 = \min\left\{1, \frac{\varepsilon}{2(\lambda(B)+1)}\right\}$ and $\varepsilon_2 = \min\left\{1, \frac{\varepsilon}{2(\lambda(A)+1)}\right\}$. Then $F_1 \times F_2$ is closed, $D_1 \times D_2$ is open in $\mathbb{R}^2$, and $F_1 \times F_2 \subseteq A \times B \subseteq D_1 \times D_2$. Since $(D_1 \times D_2) \setminus (F_1 \times F_2) \subseteq [(D_1 \setminus F_1) \times D_2] \bigcup [D_1 \times (D_2 \setminus F_2)]$, we obtain que

$$\mu[(D_1 \times D_2) \setminus (F_1 \times F_2)] \leq \mu[(D_1 \setminus F_1) \times D_2] + \mu[D_1 \times (D_2 \setminus F_2)] =$$

$$= \lambda(D_1 \setminus F_1) \cdot \lambda(D_2) + \lambda(D_1) \cdot \lambda(D_2 \setminus F_2) < \varepsilon_1 \cdot \lambda(D_2) + \lambda(D_1) \cdot \varepsilon_2 <$$

$$< \varepsilon_1 \cdot (\lambda(B) + \varepsilon_2) + \varepsilon_2 \cdot (\lambda(A) + \varepsilon_1) \leq \frac{\varepsilon}{2} + \frac{\varepsilon}{2} = \varepsilon.$$

It follows that $A \times B \in \mathscr{L}(\mathbb{R}^2)$.

On the one hand, $\mu(A \times B) \leq \mu(D_1 \times D_2) = \lambda(D_1) \cdot \lambda(D_2) < (\lambda(A) + \varepsilon_1) \cdot (\lambda(B) + \varepsilon_2) = \lambda(A) \cdot \lambda(B) + \varepsilon_1 \cdot (\lambda(B) + 1) + \varepsilon_2 \cdot (\lambda(A) + 1) < \lambda(A) \cdot \lambda(B) + \varepsilon$.

On the other hand, $\mu(A \times B) \geq \mu(F_1 \times F_2) = \mu(D_1 \times D_2) - \mu((D_1 \times D_2) \setminus (F_1 \times F_2)) > \lambda(D_1) \cdot \lambda(D_2) - \varepsilon \geq \lambda(A) \cdot \lambda(B) - \varepsilon$.

Since ε is arbitrarily positive, it follows that $\mu(A \times B) = \lambda(A) \cdot \lambda(B)$.

(2) Now we consider the case where A or B can have the measure $+\infty$. For any $n \in \mathbb{N}$, let $A_n = A \cap [-n, n]$ and $B_n = B \cap [-n, n]$. Then $A = \bigcup_{n=0}^{\infty} A_n$, $B = \bigcup_{n=0}^{\infty} B_n$, and,

since $(A_n)_n$ and $(B_n)_n$ are increasing sequences, $A \times B = \bigcup_{n=0}^{\infty}(A_n \times B_n)$. Since A_n and B_n have a finite measure, it appears from (1) that $A_n \times B_n \in \mathscr{L}(\mathbb{R}^2)$, for any $n \in \mathbb{N}$; therefore $A \times B \in \mathscr{L}(\mathbb{R}^2)$.

The sequences $(A_n)_n$ and $(B_n)_n$ are increasing and then the sequence $(A_n \times B_n)_n$ is also increasing; using the property of continuity form below of the measure (property 6) of Theorem 1.4.7) and point (1) of this proof, it follows that

$$\mu(A \times B) = \lim_n \mu(A_n \times B_n) = \lim_n \lambda(A_n) \cdot \lambda(B_n) = \lambda(A) \cdot \lambda(B).$$ ■

For every $E \subseteq \mathbb{R}^2$ and every $x, y \in \mathbb{R}$, let

$$E_x = \{y \in \mathbb{R} : (x, y) \in E\}$$

be the **section of** E **at** x and

$$E^y = \{x \in \mathbb{R} : (x, y) \in E\}$$

be the section of E at the second variable y.

Theorem 5.1.22

Let $E \in \mathscr{L}(\mathbb{R}^2)$. Then, for almost every $x \in \mathbb{R}$, $E_x \in \mathscr{L}(\mathbb{R})$. Let us define $f : \mathbb{R} \to [0, +\infty]$ letting $f(x) = \lambda(E_x)$ ($f(x) = 0$ at the points where $E_x \notin \mathscr{L}(\mathbb{R})$), and let $Z = f^{-1}(+\infty) = \{x \in \mathbb{R} : \lambda(E_x) = +\infty\}$; then $Z \in \mathscr{L}(\mathbb{R})$, f is measurable and positive on $\mathbb{R}$, and

$$\mu(E) = \int_{\mathbb{R}} \lambda(E_x) d\lambda(x) = \begin{cases} \int_{\mathbb{R}\setminus Z} \lambda(E_x) d\lambda(x), & \lambda(Z) = 0, \\ +\infty, & \lambda(Z) > 0. \end{cases}$$

Proof

(1) We will first deal with the case where E is bounded; therefore E_x is bounded, for every $x \in \mathbb{R}$ and so $Z = \emptyset$.

(a) Let $E = [a, b] \times [c, d]$ be a closed interval of $\mathbb{R}^2$. For every $x \in \mathbb{R}$, $E_x = \begin{cases} \emptyset, & x \notin [a, b] \\ [c, d], & x \in [a, b] \end{cases} \in \mathscr{L}(\mathbb{R})$ and $\lambda(E_x) = (d - c) \cdot \chi_{[a, b]}(x)$. It follows that $x \mapsto \lambda(E_x)$ is a measurable function and

$$\int_{\mathbb{R}} \lambda(E_x) d\lambda(x) = \int_{[a,b]} (d - c) d\lambda = (b - a) \cdot (d - c) = \mu(E).$$

(b) Let $E \in \tau_u^2$; according to the theorem of structure of open sets of $\mathbb{R}^2$ (Theorem 5.1.12), $E = \bigcup_{n=1}^{\infty} J^n$, where J^n are two-dimensional nonoverlapping closed intervals. For every $x \in \mathbb{R}$, $E_x = \bigcup_{n=1}^{\infty} J_x^n$. Since J_x^n are closed intervals in $\mathbb{R}$, it follows that E_x is a Borel set of $\mathbb{R}$ and therefore $E_x \in \mathscr{L}(\mathbb{R})$.

The intervals J_x^n have no common interior points; then $\lambda(E_x) = \sum_{n=1}^{\infty} \lambda(J_x^n)$ (see Theorem 1.1.8). Then the function $x \mapsto \lambda(E_x)$ is the pointwise limit of the sequence of functions $x \mapsto \sum_{k=1}^{n} \lambda(J_x^k)$; according to point a), this sequence is a sequence of measurable functions, and so its pointwise limit is measurable also (see 5) of Theorem 2.1.18). Now we use the theorem of Beppo Levi (Corollary 3.1.10) and point a) to obtain

$$\int_{\mathbb{R}} \lambda(E_x) d\lambda(x) = \sum_{n=1}^{\infty} \int_{\mathbb{R}} \lambda(J_x^n) d\lambda(x) = \sum_{n=1}^{\infty} \mu(J_n) = \mu(E).$$

(c) Now let E be a closed set of $\mathbb{R}^2$. Since E is bounded, there exists an open interval $I =]a, b[\times]c, d[\subseteq \mathbb{R}^2$ such that $E \subseteq I$. The set $D = I \setminus E$ is open and $\mu(E) = \mu(I) - \mu(D)$. From b), $\mu(D) = \int_{\mathbb{R}} \lambda(D_x) d\lambda(x) = \int_{\mathbb{R}} [\lambda(I_x) - \lambda(E_x)] d\lambda(x) = (b - a) \cdot (d - c) - \int_{\mathbb{R}} \lambda(E_x) d\lambda(x)$. It follows that $\mu(E) = \mu(I) - \mu(D) = \int_{\mathbb{R}} \lambda(E_x) d\lambda(x)$.

(d) Let $E \in \mathscr{L}(\mathbb{R}^2)$ be any measurable bounded set and let $0 < \delta < 1$ be an arbitrary number.

- From the characterization mentioned by Definition 5.1.19, there exists an open set D^1 and a closed one F^1 such that $F^1 \subseteq E \subseteq D^1$ and $\mu(D^1 \setminus F^1) < \frac{1}{4} \cdot \delta^2$. The set $U^1 = D^1 \setminus F^1$ is open and then, using point b), we obtain

$$\mu(U^1) = \int_{\mathbb{R}} \lambda(U_x^1) d\lambda(x) < \frac{1}{4} \cdot \delta^2.$$

We will prove, by reducing to the absurd, that there is a set $E_1^\delta \in \mathscr{L}(\mathbb{R})$ such that $\{x \in \mathbb{R} : \lambda(U_x^1) \geq \frac{1}{2} \cdot \delta\} \subseteq E_1^\delta$ and and $\lambda(E_1^\delta) < \frac{1}{2} \cdot \delta$. Indeed, if it was not the case, then the set $E_1 = \{x \in \mathbb{R} : \lambda(U_x^1) \geq \frac{1}{2} \cdot \delta\}$ would have the measure $\lambda(E_1) \geq \frac{1}{2} \cdot \delta$ and then

$$\mu(U^1) \geq \int_{E_1} \lambda(U_x^1) d\lambda(x) \geq \frac{1}{2} \cdot \delta \cdot \lambda(E_1) \geq \frac{1}{4} \cdot \delta^2,$$

which is a contradiction.

- Let D^2 be an open set and let F^2 be a closed one such that $F^2 \subseteq E \subseteq D^2$ and $\mu(D^2 \setminus F^2) < \frac{1}{4^2} \cdot \delta^2$. We denote by $U^2 = D^2 \setminus F^2 \in \tau_u^2$; as in the previous paragraph, we prove that there are $E_2^\delta \in \mathscr{L}(\mathbb{R})$ such that $\{x \in \mathbb{R} : \lambda(U_x^2) \geq \frac{1}{2^2} \cdot \delta\} \subseteq E_2^\delta$ and $\lambda(E_2^\delta) < \frac{1}{2^2} \cdot \delta$.

  We continue inductively this reasoning.
- Let D^i be an open set and let F^i be a closed one such that $F^i \subseteq E \subseteq D^i$ and $\mu(D^i \setminus F^i) < \frac{1}{4^i} \cdot \delta^2$; the set $U^i = D^i \setminus F^i$ is open, and, as above, there is a set $E_i^\delta \in \mathscr{L}(\mathbb{R})$ such that $\{x \in \mathbb{R} : \lambda(U_x^i) \geq \frac{1}{2^i} \cdot \delta\} \subseteq E_i^\delta$ si $\lambda(E_i^\delta) < \frac{1}{2^i} \cdot \delta$.
- ....

Let $N_\delta = \bigcup_{i=1}^{\infty} E_i^\delta$; then $\lambda(N_\delta) < \delta$.

Since δ is arbitrarily positive, we make it take the values $\{\frac{1}{n} : n \in \mathbb{N}^*\}$ and we denote by $N = \bigcap_{n=1}^{\infty} N_{\frac{1}{n}}$; then $\lambda(N) \le \lambda(N_{\frac{1}{n}}) \le \frac{1}{n}$, for any $n \in \mathbb{N}^*$. Therefore, $\lambda(N) = 0$.

Let now $x \in \mathbb{R} \setminus N$; then there is $n \in \mathbb{N}^*$ such that $x \in \mathbb{R} \setminus N_{\frac{1}{n}} = \bigcap_{i=1}^{\infty} (\mathbb{R} \setminus E_i^{\frac{1}{n}})$.

For every $\varepsilon > 0$, there is i such that $\frac{1}{2^i} \cdot \frac{1}{n} < \varepsilon$; since $x \notin E_i^{\frac{1}{n}}$, $\lambda(U_x^i) < \frac{1}{2^i} \cdot \frac{1}{n} < \varepsilon$.

We have shown that, for every $x \in \mathbb{R} \setminus N$ and every $\varepsilon > 0$, there exist a closed set F_x^i and an open one D_x^i in $\mathbb{R}$ such that $F_x^i \subseteq E_x \subseteq D_x^i$ and $\lambda(D_x^i \setminus F_x^i) < \varepsilon$. It means that $E_x \in \mathscr{L}(\mathbb{R})$, for every $x \in \mathbb{R} \setminus N$. Therefore, $E_x \in \mathscr{L}(\mathbb{R})$ for almost every $x \in \mathbb{R}$.

The function $x \mapsto \lambda(E_x)$ is therefore well defined on $\mathbb{R}$ (in the points of N, we agree to give of this function the value 0). Moreover, $\lambda(E_x) = \lim_{i\to\infty} \lambda(D_x^i)$; then the above function is the sum of a series of measurable functions (see point b)), and so it is measurable and positive.

Since $\lambda(F_x^i) \le \lambda(E_x) \le \lambda(D_x^i)$, for every $x \in \mathbb{R}$ and any i, we can use points b) and c) to obtain

$$\mu(F^i) = \int_{\mathbb{R}} \lambda(F_x^i) d\lambda(x) \le \int_{\mathbb{R}} \lambda(E_x) d\lambda(x) \le \int_{\mathbb{R}} \lambda(D_x^i) d\lambda(x) = \mu(D^i) <$$

$$< \mu(F^i) + \frac{1}{4^i} \cdot \delta^2, \text{ for any } i \in \mathbb{N}^* \text{ and}$$

$$\mu(F^i) \le \mu(E) \le \mu(D^i) < \mu(F^i) + \frac{1}{4^i} \cdot \delta^2, \text{ for any } i \in \mathbb{N}^*.$$

It follows that

$$\left| \mu(E) - \int_{\mathbb{R}} \lambda(E_x) d\lambda(x) \right| < \frac{1}{4^i} \cdot \delta^2, \text{ for any } i \in \mathbb{N}^*;$$

therefore, $\mu(E) = \int_{\mathbb{R}} \lambda(E_x) d\lambda(x)$.

(2) If E is not bounded, then, for any $n \in \mathbb{N}$, let the bounded measurable sets be $E^n = E \cap ([-n, n] \times [-n, n])$.

According to point (1),

$$\mu(E^n) = \int_{\mathbb{R}} \lambda(E_x^n) d\lambda(x), \text{ for any } n \in \mathbb{N}. \tag{1}$$

The sequence $(E^n)_n$ is increasing and then

$$\mu(E) = \mu\left(\bigcup_{n=1}^{\infty} E_x^n\right) = \lim_n \mu(E^n). \tag{2}$$

On the other hand, the sequence $(f_n)_n$, where, for any $n \in \mathbb{N}$, $f_n(x) = \lambda(E_x^n)$, is a sequence of measurable and positive functions on $\mathbb{R}$. We note that

$$f_n(x) = \lambda(E_x^n) = \begin{cases} \lambda(E_x \cap [-n, n]), & |x| \le n, \\ 0, & |x| > n. \end{cases}$$

Therefore, $(f_n)_n$ is increasing and $f_n \uparrow f$, where $f : \mathbb{R} \to [0, +\infty]$, $f(x) = \lambda(E_x)$. We remark that the function f can also take the value $+\infty$ and then let $Z = f^{-1}(+\infty)$. According to Corollary 3.1.8, $Z \in \mathscr{L}(\mathbb{R})$, f is measurable on $\mathbb{R}$, and

$$\lim_n \int_{\mathbb{R}} f_n d\lambda = \int_{\mathbb{R}} f d\lambda = \begin{cases} \int_{\mathbb{R}\setminus Z} \lambda(E_x) d\lambda(x), & \lambda(Z) = 0, \\ +\infty, & \lambda(Z) > 0. \end{cases} \tag{3}$$

By (1), (2), and (3), we obtain the conclusion of the theorem. ■

In Fig. 5.3, we have represented a set $E \subseteq \mathbb{R}^2$ and two of it sections, one at x and one at y; the integrals on $\mathbb{R}$ of their measures give the measure of the set (the "area" of E).

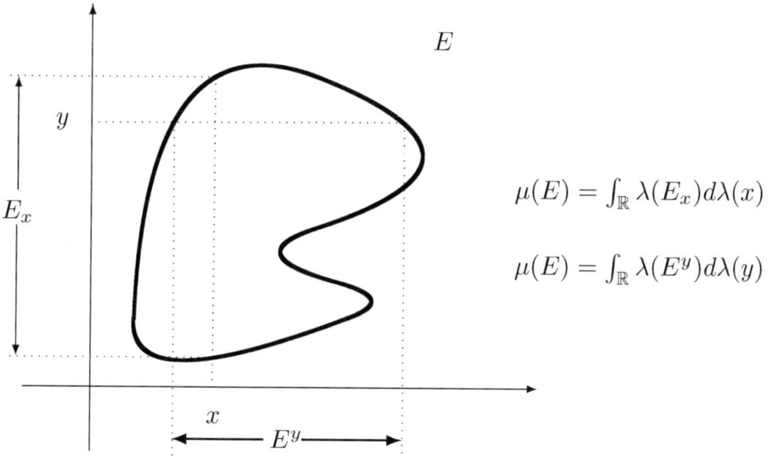

Fig. 5.3 The sections E_x and E^y of a set $E \in \mathfrak{L}(\mathbb{R}^2)$

Remarks 5.1.23

(i) In general, the sections of a measurable set in $\mathbb{R}^2$ are not all measurable in $\mathbb{R}$. For example, we consider a set $N \in \mathcal{P}(\mathbb{R}) \setminus \mathscr{L}(\mathbb{R})$ (in (ii) of Remark 1.3.17, we have justified the existence of such sets), and let $E = \{0\} \times N \subseteq \mathbb{R}^2$; $E \subseteq \{0\} \times \mathbb{R}$ and $\mu(\{0\} \times \mathbb{R}) = 0$ (see Exercise 1) of 5.5). Since the measure μ is complete, $E \in \mathscr{L}(\mathbb{R}^2)$. In the section of this set at 0, $E_0 = N$ and so it is not measurable in $\mathbb{R}$.

(ii) Similarly, we can show that

$$\mu(E) = \int_{\mathbb{R}} \lambda(E^y)d\lambda(y) = \begin{cases} \int_{\mathbb{R}\setminus Z} \lambda(E^y)d\lambda(y), & \lambda(Z) = 0, \\ +\infty, & \lambda(Z) > 0. \end{cases}, \text{ where}$$

$Z = \{y \in \mathbb{R} : \lambda(E^y) = +\infty\}$.

With this extended form of the integral in mind, we will write

$$\mu(E) = \int_{\mathbb{R}} \lambda(E_x)d\lambda(x) = \int_{\mathbb{R}} \lambda(E^y)d\lambda(y).$$

(iii) If $A, B \in \mathscr{L}(\mathbb{R})$ and if $E \in \mathscr{L}(A \times B)$, then

$$\mu(E) = \int_{A} \lambda(E_x)d\lambda(x) = \int_{B} \lambda(E^y)d\lambda(y).$$

Indeed, in this case, for every $x \in \mathbb{R} \setminus A$ and every $y \in \mathbb{R} \setminus B$, $E_x = E^y = \emptyset$.

(iv) The formula given in the preceding theorem also allows a quick demonstration of the **principle of Cavalieri in plane**:

If two plane figures are bounded by two parallel lines and if each other line parallel to these two lines cuts the two figures into segments of the same length, then the two figures have the same area.

Indeed, let $E, F \subseteq \mathbb{R}^2$ be two measurable sets so that any straight line parallel to Oy intersects them after two linear sets of the same length, so $\lambda(E_x) = \lambda(F_x)$, for all $x \in \mathbb{R}$ ("length" of a set of $\mathbb{R}$ is here the Lebesgue measure of this set). The previous theorem assures us that $\mu(E) = \mu(F)$, and, if the "surface" of a set of planes is its Lebesgue measure, we obtain the conclusion of the principle.

(v) The process of building the Lebesgue measure on $\mathbb{R}^2$ can be easily adapted for $\mathbb{R}^3$ (and more generally for $\mathbb{R}^n, n \geq 1$). Thus, the σ-algebra of Lebesgue measurable subsets $\mathscr{L}(\mathbb{R}^3)$ can be constructed in $\mathbb{R}^3$, and then a complete measure, θ, can be constructed on this σ-algebra; this measure has all the properties of the Lebesgue measure on $\mathbb{R}^2$. For every $M \in \mathscr{L}(\mathbb{R}^3)$ and for every $x \in \mathbb{R}$, we can define the section of M at x:

$$M_x = \{(y, z) \in \mathbb{R}^2 : (x, y, z) \in M\} \in \mathscr{L}(\mathbb{R}^2).$$

We can show that, λ-almost for every $x \in \mathbb{R}$, $M_x \in \mathscr{L}(\mathbb{R}^2)$. The function $x \mapsto \mu(M_x)$ is measurable and positive and

$$\theta(M) = \int_{\mathbb{R}} \mu(M_x)d\lambda(x).$$

(vi) The formula given in the previous point provides a quick demonstration of the **principle of Cavalieri in space**:

If two solids are bounded by two parallel planes and if all the intersections of these solids with a plane parallel to the first two have the same area, then the solids have the same volume.

Indeed, let $M, N \subseteq \mathbb{R}^3$ be measurable, so that any plane parallel to yOz intersects them after sets with the same area, so $\mu(M_x) = \mu(N_x)$, for all $x \in \mathbb{R}$ ("area" of a set of $\mathbb{R}^2$ is here Lebesgue's measure of this set). According to point (v), $\theta(M) = \theta(N)$, and, as "volume" of a set of $\mathbb{R}^3$ is the Lebesgue measure of this set, we obtain the conclusion of the principle.

The following theorem is an immediate consequence of Theorem 5.1.22.

Theorem 5.1.24
Let $A \in \mathscr{L}(\mathbb{R})$ and let $f \in \mathfrak{L}^1(A)$. Then the function $t \mapsto \lambda(|f| > t)$ is integrable on $[0, +\infty[$ and

$$\int_A |f| d\lambda = \int_{[0,+\infty[} \lambda(|f| > t) d\lambda(t).$$

Proof
Let us denote $E = \{(x, t) \in A \times [0, +\infty[: 0 \le t < |f(x)|\}$. Let's show that $E \in \mathscr{L}(A \times [0, +\infty[)$.

If f is a positive simple function, then $f = \sum_{k=1}^{n} a_k \chi_{A_k}$, where $\{a_1, \cdots, a_n\} \subseteq \mathbb{R}_+$ and $\{A_1, \cdots, A_n\}$ is a $\mathscr{L}$-partition of A. It is immediately noticed that $E = \bigcup_{k=1}^{n} (A_k \times [0, a_k[)$ and, according to Theorem 5.1.21, $E \in \mathscr{L}(A \times [0, +\infty[)$.

Let now $f \in \mathfrak{L}^1(A)$; according to Theorem 2.3.3, there exists a strictly increasing sequence of positive simple functions $(f_n)_n \subseteq \mathcal{E}_+(A)$ such that $f_n \uparrow |f|$ (why ?). Based on the theorem proven above, $E_n = \{(x, t) \in A \times [0, +\infty[: 0 \le t < f_n(x)\} \in \mathscr{L}(A \times [0, +\infty[)$. Then $E = \liminf_n E_n = \bigcup_{n=1}^{\infty} \bigcap_{k=n}^{\infty} E_k \in \mathscr{L}(A \times [0, +\infty[)$.

Therefore, according to Theorem 5.1.22 and to (iii) of Remark 5.1.23, we obtain

$$\mu(E) = \int_A \lambda(E_x) d\lambda(x) = \int_{[0,+\infty[} \lambda(E^t) d\lambda(t).$$

The demonstration ends if we notice that $E_x = [0, |f(x)|[$ and that $E^t = (|f| > t)$. ∎

5.2 Lebesgue Multiple Integrals

In the previous paragraph, we built a complete measure $\mu : \mathscr{L}(\mathbb{R}^2) \to \bar{\mathbb{R}}_+$.

Definition 5.2.1

Let $E \in \mathscr{L}(\mathbb{R}^2)$ and let $f : E \to \mathbb{R}$; the function of two variables f is said to be μ-**measurable** on E, or simply **measurable**, if, for every $a \in \mathbb{R}$,

$$f^{-1}(]-\infty, a[) = \{(x, y) \in E : f(x, y) < a\} \in \mathscr{L}(\mathbb{R}^2)$$

(Continued)

Definition 5.2.1 (continued)

We denote with $\mathfrak{L}(E)$ the set of all measurable functions on E and with $\mathfrak{L}_+(E)$ the subset of measurable and positive functions.

As in the one-dimensional case, we can show that $\mathfrak{L}(E)$ is, concerning the usual operations of addition and multiplication with scalars, a real vector space, containing all the real continuous functions on E, $C(E)$.

Definition 5.2.2

A function $f : E \to \mathbb{R}$ is said to be **simple** on $E \in \mathscr{L}(\mathbb{R}^2)$ if it takes a finite number of values on measurable subsets of E, so it is of form $f = \sum_{k=1}^{p} a_k \chi_{E_k}$, where $\{E_1, \cdots, E_p\}$ is a measurable partition of E and $\chi_{E_k}(x, y) = \begin{cases} 1, & (x, y) \in E_k, \\ 0, & (x, y) \in E \setminus E_k. \end{cases}$

Let $\mathcal{E}(E)$ be the set of all simple functions on E; we will notice that $\mathcal{E}(E) \subseteq \mathfrak{L}(E)$.

We can define the μ-almost everywhere (a.e.) limit of a sequence of measurable functions, and we can prove that it is a measurable function. In addition, the composition of a continuous function with a measurable function is measurable.

As in the case of $\mathbb{R}$, we can define the almost uniform convergence and the convergence in measure; the relationships between them and the convergence a.e. are the same as in the one-dimensional case.

As in the one-dimensional case, the Riesz theorem (any convergent in measure sequence admits an almost uniform convergent subsequence) and the theorem of Egorov (convergence a.e. on sets of finite measures leads to almost uniform convergence) are true.

We can also prove a theorem of simple functions approximation of measurable functions:

Theorem 5.2.3

Let $E \in \mathscr{L}(\mathbb{R}^2)$ and $f : E \to \mathbb{R}$.

(1) *$f \in \mathfrak{L}_+(E)$ if and only if there is a sequence $(f_n)_n \subseteq \mathcal{E}_+(E)$ such that $f_n \uparrow f$.*

(2) *$f \in \mathfrak{L}(E)$ if and only if there is a sequence $(f_n)_n \subseteq \mathcal{E}(E)$ pointwise convergent to f on E.*

(3) *If the function $f \in \mathfrak{L}(E)$ is bounded, then there exists a sequence $(f_n)_n \subseteq \mathcal{E}(E)$ uniform convergent on E to f.*

The integral of the functions of two variables is introduced by following the same steps as in the case of the functions of a variable. Let $E \in \mathscr{L}(\mathbb{R}^2)$ and $f : E \to \mathbb{R}$.

(1) If $f = \sum_{k=1}^{p} a_k \chi_{E_k} \in \mathcal{E}_+(E)$, then

$$\iint_E f d\mu = \iint_E f(x, y) d\mu(x, y) = \sum_{k=1}^{p} a_k \mu(E_k).$$

(2) If $f \in \mathfrak{L}_+(E)$, then

$$\iint_E f d\mu = \sup\left\{\iint_E \varphi d\mu : \varphi \in \mathcal{E}_+(E), \varphi \leq f\right\}.$$

(3) If $f \in \mathfrak{L}(E)$ and if at least one of the integrals of its positive part $f^+ = \sup\{f, \underline{0}\}$ or of its negative part $f^- = \sup\{-f, \underline{0}\}$ is finite, then

$$\iint_E f d\mu = \iint_E f^+ d\mu - \iint_E f^- d\mu \in [-\infty. + \infty].$$

A function $f \in \mathfrak{L}(E)$ is **μ-integrable** if its integral is finite (the integrals of f^+ and of f^- are finite); the **integral** of f is

$$\iint_E f d\mu = \iint_E f^+ d\mu - \iint_E f^- d\mu \in \mathbb{R}.$$

Let $\mathfrak{L}^1(E)$ be the set of all integrable functions on E and let $\mathfrak{L}^1_+(E)$ be the subset of positive and integrable functions. $\mathfrak{L}^1(E)$ is a vector subspace of $\mathfrak{L}(E)$ and the integral is a linear operator on this subspace.

As in the one-dimensional case, we can show that $f \in \mathfrak{L}^1(E)$ if and only if $|f| \in \mathfrak{L}^1_+(E)$ and that

$$\left|\iint_E f d\mu\right| \leq \iint_E |f| d\mu.$$

We can also prove here the monotone convergence theorem:

$$\iint_E f_n d\mu \uparrow \iint_E f d\mu, \text{ if } (f_n)_n \subseteq \mathfrak{L}_+(E) \text{ and } f_n \uparrow f,$$

and the Fatou's lemma: if $(f_n)_n \subseteq \mathfrak{L}_+(E)$ and if $\liminf_n f_n : E \to \mathbb{R}$, then

$$\iint_E \liminf_n f_n d\mu \leq \liminf_n \iint_E f_n d\mu.$$

We can also formulate the Beppo Levi theorem.

The integral is σ-additive with respect to the domain of integration:

Let $(E_n)_n \subseteq \mathscr{L}(\mathbb{R}^2)$ with $E_n \cap E_m = \emptyset$, for $n \neq m$; if $f \in \mathfrak{L}^1(\cup_{n=1}^{\infty} E_n)$, then $f \in \mathfrak{L}^1(E_n)$, for any $n \in \mathbb{N}$ and

$$\iint_{\cup_{n=1}^{\infty} E_n} f d\mu = \sum_{n=1}^{\infty} \iint_{E_n} f d\mu.$$

The mapping $\|\cdot\|_1 : \mathfrak{L}^1(E) \to \mathbb{R}_+$, defined by $\|f\|_1 = \iint_E |f| d\mu$, for every $f \in \mathfrak{L}^1(E)$, is a seminorm and the space $(\mathfrak{L}^1(E), \|\cdot\|_1)$ is complete.

As in the one-dimensional case, the quotient space $L^1(E) = \mathfrak{L}^1(E)|_{\dot{=}}$ is a Banach space with respect to the norm $\|\cdot\|_1$, defined by $\|[f]\|_1 = \|f\|_1$ (here $\dot{=}$ is the equality μ-a.e.).

We can formulate and prove the dominated convergence theorem.

Theorem 5.2.4

Let $E \in \mathscr{L}(\mathbb{R}^2)$, $(f_n)_n \subseteq \mathfrak{L}(E)$ and $g \in \mathfrak{L}^1(E)$ such that

(1) $f_n \xrightarrow[E]{a.e.} f$ *and*

(2) $|f_n| \leq g$, *a.e. on E, for any $n \in \mathbb{N}$.*

Then $(f_n)_n \subseteq \mathfrak{L}^1(E)$, $f \in \mathfrak{L}^1(E)$ and

$$\iint_E f_n d\mu \to \iint_E f d\mu.$$

We will mention a variable change formula similar to that of the functions of one variable (see Theorem 3.5.6).

Theorem 5.2.5

Let $g = (g_1, g_2) : \mathbb{R}^2 \to \mathbb{R}^2$ be an injective function of class C^1 (there exist all partial derivatives of g_1 and g_2 and they are continuous on $\mathbb{R}^2$); then, for every $E \in \mathscr{L}(\mathbb{R}^2)$, $g(E) \in \mathscr{L}(\mathbb{R}^2)$ and

$$\mu(g(E)) = \iint_E \left|det J_g(x, y)\right| d\mu(x, y).$$

For every Borel function $f : \mathbb{R}^2 \to \mathbb{R}$ and for every set $E \in \mathscr{L}(\mathbb{R}^2)$, $f \in \mathfrak{L}^1(g(E))$ if and only if $(f \circ g) \cdot \left|det J_g\right| \in \mathfrak{L}^1(E)$ and then

$$\int_{g(E)} f(x, y) d\mu(x, y) = \int_E (f(g(x, y)) \cdot \left|det J_g(x, y)\right| d\mu(x, y).$$

We have denoted with $\det J_g$ the determinant of the Jacobian matrix of g: $J_g = \begin{pmatrix} \frac{\partial g_1}{\partial x} & \frac{\partial g_1}{\partial y} \\ \frac{\partial g_2}{\partial x} & \frac{\partial g_2}{\partial y} \end{pmatrix}$. The proof in the one-dimensional case was essentially based on the monotonicity of g. In the two-dimensional case, in which it is unable to use the monotonicity, the demonstration turns out to be much more difficult; we will indicate, for those interested, the Theorem 3.7.1 of the monograph [1].

Finally, the spaces L^p, $1 \leq p \leq +\infty$ can be defined in the same way in $\mathbb{R}^2$, and properties similar to those of the one-dimensional case can be demonstrated.

Remark 5.2.6 All this construction of the integral can be easily adapted for $\mathbb{R}^3$. Using the integral of $\mathbb{R}^2$, we can still give a formula to calculate the measure of the sets of $\mathbb{R}^3$ (further, from the formula presented in point (v) de 5.1.23).

For every $x, y \in \mathbb{R}$, the section of $M \in \mathscr{L}(\mathbb{R}^3)$ at (x, y) is

$$M_{(x,y)} = \{z \in \mathbb{R} : (x, y, z) \in M\}.$$

Then $M_{(x,y)} \in \mathscr{L}(\mathbb{R})$, for μ-almost every $(x, y) \in \mathbb{R}^2$. The function $(x, y) \mapsto \lambda(M_{(x,y)})$ is measurable and positive and

$$\theta(M) = \iint_{\mathbb{R}^2} \lambda(M_{(x,y)}) d\mu(x, y).$$

5.3 Fubini's Theorem

In this section, we will examine conditions under which the double integral can be reduced to two iterated integrals.

Definition 5.3.1

Let $A, B \in \mathscr{L}(\mathbb{R})$ and let $f \in \mathfrak{L}(A \times B)$. The following integrals, if they exist, will be called **iterated integrals** of the function f:

$$\int_A \left(\int_B f(x, y) d\lambda(y) \right) d\lambda(x), \int_B \left(\int_A f(x, y) d\lambda(x) \right) d\lambda(y).$$

In Theorem 5.1.22, we have shown that, for all $E \in \mathscr{L}(\mathbb{R}^2)$, $E_x \in \mathscr{L}(\mathbb{R})$, for almost every $x \in \mathbb{R}$, the function $x \mapsto \lambda(E_x)$ is measurable and positive on $\mathbb{R}$ and $\mu(E) = \int_{\mathbb{R}} \lambda(E_x) d\lambda(x)$. By replacing E with χ_E and noting that, for all $x, y \in \mathbb{R}$, $\chi_{E_x}(y) = \chi_E(x, y)$, we can rewrite the formula of measure calculation of E:

$$\iint_{\mathbb{R}^2} \chi_E d\mu = \int_{\mathbb{R}} \left(\int_{\mathbb{R}} \chi_E(x, y) d\lambda(y) \right) d\lambda(x).$$

If $A, B \in \mathscr{L}(\mathbb{R})$ and $E \subseteq A \times B$, then the above formula becomes

$$\iint_{A\times B} \chi_E d\mu = \int_A \left(\int_B \chi_E(x, y) d\lambda(y) \right) d\lambda(x).$$

We will show that the above relation works for any measurable and positive application.

Theorem 5.3.2 (Tonelli)
If $f \in \mathfrak{L}_+(A \times B)$, then, λ-almost for every $x \in A$, the function $v_x = f(x, \cdot) : B \to \mathbb{R}_+$ is measurable and positive on B; the function $u : A \to [0, +\infty]$, defined by

$$u(x) = \begin{cases} \int_B v_x d\lambda = \int_B f(x, y) d\lambda(y), \ v_x \in \mathfrak{L}^1(B) \\ +\infty, \ v_x \notin \mathfrak{L}^1(B) \end{cases},$$

is measurable and positive on A; and

$$\iint_{A\times B} f d\mu = \int_A u d\lambda \in \bar{\mathbb{R}}_+ \text{ or}$$

$$\iint_{A\times B} f(x, y) d\mu(x, y) = \int_A \left(\int_B f(x, y) d\lambda(y) \right) d\lambda(x) \in \bar{\mathbb{R}}_+.$$

Proof

(1) We suppose first that $f = \chi_E \in \mathfrak{L}_+(A \times B)$. Then $E \in \mathscr{L}(A \times B)$, $v_x = \chi_{E_x}$, $u(x) = \lambda(E_x)$, almost for every $x \in \mathbb{R}$, and so the proof is a consequence of (iii) of Remark 5.1.23.

(2) Let $f = \sum_{i=1}^p a_i \chi_{E_i} \in \mathcal{E}_+(A \times B) \subseteq \mathfrak{L}_+(A \times B)$.
$v_x = \sum_{i=1}^p a_i \chi_{E_{ix}}$ is then measurable for almost every $x \in A$, and $u(x) = \sum_{i=1}^p a_i \lambda(E_{ix})$ is also measurable and

$$\iint_{A\times B} f d\mu(E) = \sum_{i=1}^p a_i \mu(E_i) = \sum_{i=1}^p a_i \int_A \lambda(E_{ix}) d\lambda(x) =$$

$$= \int_A \left(\sum_{i=1}^p a_i \lambda(E_{ix}) \right) d\lambda(x) = \int_A \left[\int_B \left(\sum_{i=1}^p a_i \chi_{E_{ix}}(y) \right) d\lambda(y) \right] d\lambda(x) =$$

$$= \int_A \left(\int_B f(x, y) d\lambda(y) \right) d\lambda(x).$$

(3) Let now $f \in \mathfrak{L}_+(A \times B)$; then there exists an increasing sequence of positive simple functions, $(f_n)_n \subseteq \mathcal{E}_+(A \times B)$, which converges to f (see 5.2.3). From (2), for any $n \in \mathbb{N}$, almost for every $x \in A$, $v_x^n = f_n(x, \cdot)$ is measurable. It follows that, almost for every $x \in A$, the sequence $(v_x^n)_n = (f_n(x, \cdot))_n$ is a sequence of measurable functions (a countable union of null sets is a null set), and therefore its limit, $v_x = f(x, \cdot)$, is measurable on B. Since $v_x^n = f_n(x, \cdot) \uparrow f(x, \cdot) = v_x$, we can apply the monotone convergence theorem (Theorem 3.1.7), and so we obtain

$$u_n(x) = \int_B v_x^n d\lambda = \int_B f_n(x, \cdot) d\lambda \uparrow \int_B f(x.\cdot) d\lambda = \int_B v_x d\lambda = u(x).$$

Because $(u_n)_n$ is a sequence of measurable functions on A, it follows that u is measurable and positive on A but which can take the value $+\infty$ (see (iv) of Remark 2.1.3). Applying again the monotone convergence theorem, the result of point (2), and Corollary 3.1.8, we obtain

$$\iint_{A\times B} f d\mu = \lim_n \iint_{A\times B} f_n d\mu = \lim_n \int_A \left(\int_B f_n(x, y) d\lambda(y) \right) d\lambda(x) =$$

$$= \lim_n \int_A u_n d\lambda = \int_A u d\lambda = \int_A \left(\int_B f(x, y) d\lambda(y) \right) d\lambda(x).$$ ■

Remark 5.3.3 With a similar proof, we obtain, in the hypotheses of Tonelli's theorem,

$$\iint_{A\times B} f(x, y) d\mu(x, y) = \int_B \left(\int_A f(x, y) d\lambda(x) \right) d\lambda(y).$$

The following theorem shows that, if the function f is integrable on $A \times B$, the iterated integrals exist and are equal to the integral of the function.

Theorem 5.3.4 (Fubini)
Let $A, B \in \mathscr{L}(\mathbb{R})$ and $f \in \mathfrak{L}^1(A \times B)$; then, λ-almost for every $x \in A$, the function $v_x = f(x, \cdot) : B \to \mathbb{R}$ is integrable on B; the function $u : A \to \mathbb{R}$, defined by $u(x) = \int_B f(x, y) d\lambda(y)$, is integrable on A (at the points $x \in A$ where v_x is not integrable, we consider $u(x) = 0$); and

$$\iint_{A\times B} f d\mu = \int_A u d\lambda \text{ or}$$

$$\iint_{A\times B} f(x, y) d\mu(x, y) = \int_A \left(\int_B f(x, y) d\lambda(y) \right) d\lambda(x).$$

Proof

The positive part of f is $f^+ = \sup\{f, 0\}$ and the negative part $f^- = \sup\{-f, 0\}$. Since $f \in \mathfrak{L}^1(A \times B)$, $f^+, f^- \in \mathfrak{L}^1_+(A \times B) \subseteq \mathfrak{L}_+(A \times B)$.

From Tonelli's theorem, it follows that

$$\iint_{A\times B} f^+(x, y)d\mu(x, y) = \int_A \left(\int_B f^+(x, y)d\lambda(y)\right) d\lambda(x) < +\infty \text{ and}$$

$$\iint_{A\times B} f^-(x, y)d\mu(x, y) = \int_A \left(\int_B f^-(x, y)d\lambda(y)\right) d\lambda(x) < +\infty.$$

It follows that, almost for every $x \in A$, $v_x^+ = f^+(x, \cdot)$, $v_x^- = f^-(x, \cdot) \in \mathfrak{L}^1_+(B)$; therefore $v_x \in \mathfrak{L}^1(B)$. and also $u^+, u^- \in \mathfrak{L}^1_+(A)$ and then $u \in \mathfrak{L}^1(A)$ and

$$\iint_{A\times B} f d\mu = \iint_{A\times B} f^+ d\mu - \iint_{A\times B} f^- d\mu =$$

$$= \int_A \left(\int_B (f^+(x, y) - f^-(x, y))d\lambda(y)\right) d\lambda(x) =$$

$$= \int_A \left(\int_B f(x, y)d\lambda(y)\right) d\lambda(x).$$

■

Remark 5.3.5 With a similar demonstration, we obtain in the hypotheses of Fubini's theorem

$$\iint_{A\times B} f(x, y)d\mu(x, y) = \int_B \left(\int_A f(x, y)d\lambda(x)\right) d\lambda(y).$$

We conclude this section with another iteration result, an immediate consequence of the theorems of Tonelli and Fubini.

Theorem 5.3.6

Let $f \in \mathfrak{L}(A \times B)$; if one of the integrals

$$\iint_{A\times B} |f| d\mu, \int_A \left(\int_B |f(x, y)| d\lambda(y)\right) d\lambda(x), \int_B \left(\int_A |f(x, y)| d\lambda(x)\right) d\lambda(y)$$

is finite, then

$$\iint_{A\times B} f d\mu = \int_A \left(\int_B f(x, y)d\lambda(y)\right) d\lambda(x) = \int_B \left(\int_A f(x, y)d\lambda(x)\right) d\lambda(y).$$

Proof

Tonelli's theorem applied to the function $|f|$ gives $\iint_{A\times B} |f| d\mu =$

$$= \int_A \left(\int_B |f(x, y)| d\lambda(y) \right) d\lambda(x) = \int_B \left(\int_A |f(x, y)| d\lambda(x) \right) d\lambda(y).$$

From the hypothesis, one of the above integrals is finite, so all of them are finite. It follows that f is integrable on $A \times B$; using Fubini's theorem, we get the desired result. ■

5.4 Abstract Setting

Let $(X, \mathcal{A})$, $(Y, \mathcal{B})$ be two measurable spaces; the set of measurable rectangles

$$\mathcal{R} = \{A \times B : A \in \mathcal{A}, B \in \mathcal{B}\}$$

is not a σ-algebra on the Cartesian product $X \times Y$. Let then $\mathcal{C} = \mathcal{A} \otimes \mathcal{B}$ be the σ-algebra generated by $\mathcal{R}$; we recall that $\mathcal{C}$ is the smalest σ-algebra on $X \times Y$ which contains $\mathcal{R}$. $\mathcal{C}$ is the **product** of σ-algebras $\mathcal{A}$ and $\mathcal{B}$.

For every function $f : X \times Y \to \mathbb{R}$, the sections by the two variables $x \in X$ and $y \in Y$ are $f_x : Y \to \mathbb{R}$, $f_x(v) = f(x, v)$ and $f^y : X \to \mathbb{R}$, $f^y(u) = f(u, y)$. We can show that, for any function $\mathcal{C}$-measurable, $f : X \times Y \to \mathbb{R}$, every $x \in X$ and every $y \in Y$, f_x is $\mathcal{B}$-measurable, and f^y is $\mathcal{A}$-measurable.

If $C \subseteq X \times Y$, $x \in X$, and $y \in Y$, the sections of C are $C_x = \{v \in Y : (x, v) \in C\}$, respectively $C^y = \{u \in X : (u, y) \in C\}$; it is obvious that $(\chi_C)_x = \chi_{C_x}$ and $(\chi_C)^y = \chi_{C^y}$. If $C \in \mathcal{C}$, then $C_x \in \mathcal{B}$ and $C^y \in \mathcal{A}$.

We suppose that $(X, \mathcal{A})$ and $(Y, \mathcal{B})$ are provided with two measures σ-finite, γ and respectively ν.

Theorem 5.4.1

For every $C \in \mathcal{C}$ let $f_C : X \to \mathbb{R}_+$, $f_C(x) = \nu(C_x)$ and $g_C : Y \to \mathbb{R}_+$, $g_C(y) = \gamma(C^y)$. Then, f_C is $\mathcal{A}$-measurable, g_C is $\mathcal{B}$-measurable, and $\int_X f_C d\gamma = \int_Y g_C d\nu$. The set function $\eta : \mathcal{C} \to \mathbb{R}_+$, defined by

$$\eta(C) = \int_X f_C d\gamma = \int_Y g_C d\nu, \text{ for every } C \in \mathcal{C},$$

is a measure σ-finite on $(X \times Y, \mathcal{C})$.
For every $A \times B \in \mathcal{R} \subseteq \mathcal{C}$,

$$\eta(A \times B) = \gamma(A) \cdot \nu(B). \qquad (*)$$

η is the only measure on the product space $(X \times Y, \mathcal{C})$ verifying $(*)$.

Definition 5.4.2

η is called the **product** of measures γ and ν, and we denote $\eta = \gamma \otimes \nu$.

Remarks 5.4.3 Let λ be the Lebesgue measure on $(\mathbb{R}, \mathscr{L})$ and let μ be the Lebesgue measure on $(\mathbb{R}^2, \mathscr{L}(\mathbb{R}^2))$.

(i) We have shown in Theorem 5.1.21 that, for every $A, B \in \mathscr{L}$, $A \times B \in \mathscr{L}(\mathbb{R}^2)$ and that $\mu(A \times B) = \lambda(A) \cdot \lambda(B)$. It follows that $\mathscr{L} \otimes \mathscr{L} \subseteq \mathscr{L}(\mathbb{R}^2)$ and that $\mu|_{\mathscr{L}\otimes\mathscr{L}} = \lambda \otimes \lambda$.

(ii) In general, the product measure is not complete even if the two factors are complete measures. Indeed, let $V \subseteq \mathbb{R}$ be Vitali's set (1.2.7 and 1.3.17). $(\lambda \otimes \lambda)(\mathbb{R} \times \{0\}) = 0$, $V \times \{0\} \subseteq \mathbb{R} \times \{0\}$ but $C = V \times \{0\} \notin \mathscr{L} \otimes \mathscr{L}$ $(C^0 = V \notin \mathscr{L})$. Since μ is complete on $\mathscr{L}(\mathbb{R}^2)$, and $\mu(\mathbb{R} \times \{0\}) = \mu(\cup_{n=1}^{\infty}([-n, n] \times \{0\})) = \lim_n \mu([-n, n] \times \{0\}) = \lim_n(2n \cdot 0) = 0$, $V \times \{0\} \in \mathscr{L}(\mathbb{R}^2)$. We deduce from here that $\mathscr{L} \otimes \mathscr{L} \subsetneq \mathscr{L}(\mathbb{R}^2)$. In fact the Lebesgue measure on $\mathbb{R}^2$, μ, is the completion of $\lambda \otimes \lambda$ (see Exercise 18) of 1.5).

Let $\mathfrak{L}^1(X \times Y)$ be the space of all integrable functions on $(X \times Y, \mathcal{A} \otimes \mathcal{B}, \gamma \otimes \nu)$ (see ▶ Sect. 3.6). If $f \in \mathfrak{L}^1(X \times Y)$, then the integral of f is

$$\int_{X\times Y} f(x, y) d\eta(x, y) = \int_{X\times Y} f(x, y) d(\gamma \otimes \nu)(x, y).$$

We find in this abstract framework the theorems of Tonelli and Fubini.

Theorem 5.4.4 (Tonelli)

If $f : X \times Y \to \mathbb{R}_+$ is C-measurable and positive, then the functions $x \mapsto \int_Y f_x d\nu$ and $y \mapsto \int_Y f^y d\gamma$ are measurable and

$$\int_{X\times Y} f d\eta = \int_X \left(\int_Y f_x d\nu \right) d\gamma = \int_Y \left(\int_X f^y d\gamma \right) d\nu \in \bar{\mathbb{R}}_+.$$

Theorem 5.4.5 (Fubini)

Let $f \in \mathfrak{L}^1(X \times Y)$; then, γ-almost for every $x \in X$, $f_x \in \mathfrak{L}^1(Y)$ and ν-almost for every $y \in Y$, $f^y \in \mathfrak{L}^1(X)$.

Let the functions $u : X \to \mathbb{R}$, $u(x) = \int_Y f_x d\nu$, and $v : Y \to \mathbb{R}$, $v(y) = \int_X f^y d\gamma$. Then $u \in \mathfrak{L}^1(X)$, $v \in \mathfrak{L}^1(Y)$, and

$$\int_{X\times Y} f d\nu = \int_X u d\gamma = \int_Y v d\nu, \text{ or}$$

$$\int_{X\times Y} f d\eta = \int_X \left(\int_Y f(x, y) d\nu(y) \right) d\gamma(x) = \int_Y \left(\int_X f(x.y) d\gamma(x) \right) d\nu(y).$$

5.5 Exercises

(1) Let $A, B \subseteq \mathbb{R}$; if $\lambda^*(A) = 0$, then $\mu^*(A \times B) = 0$ and so $A \times B \in \mathscr{L}(\mathbb{R}^2)$.

Indication: First of all it will be considered the case where $B = [a, b]$ is a closed bounded interval; it will then be taken into account that $\mathbb{R}$ can be represented as a countable union of such intervals.

(2) Show that any straight line in the plane is a null set in $\mathbb{R}^2$.

Indication: A straight line parallel to the axis Oy or to the Ox has the following form: $\{a\} \times \mathbb{R}$ or respectively $\mathbb{R} \times \{b\}$, and the result follows from the previous point.

Let now be a line of equation $y = ax + b$; we will assume that $a > 0$. We will first show that the set $E = \{(x, ax + b) : x \geq 0\}$ is a null set. For every $\varepsilon > 0$, let $\varepsilon_1 = \sqrt{\dfrac{6\varepsilon}{a\pi^2}}$ and let the sequence $(x_n^\varepsilon)_n$ where $x_0^\varepsilon = 0$ and $x_n^\varepsilon = \varepsilon_1 \cdot \left(1 + \frac{1}{2} + \cdots + \frac{1}{n}\right)$, for any $n \in \mathbb{N}^*$. We are going to build closed intervals $J_n = [x_n^\varepsilon, x_{n+1}^\varepsilon] \times [ax_n^\varepsilon + b, ax_{n+1}^\varepsilon + b]$, for any $n \in \mathbb{N}$. We show that $E \subseteq \bigcup_{n=0}^\infty J_n$ and that $\sum_{n=0}^\infty |J_n| = \varepsilon$.

(3) Let $E = [0, 1] \times \{0\}$, $F = \{0\} \times [0, 1] \subseteq \mathbb{R}^2$; show that $\mu(E) = \mu(F) = 0$ and $\mu(E + F) = 1$ $(E + F = \{x + y : x \in E, y \in F\})$.

(4) Using the formula given in Theorem 5.1.22, calculate $\mu(E)$, where $E = \{(x, y) \in \mathbb{R}^2 : x^2 + y^2 = r^2\}$.

(5) For every $E \in \mathscr{L}(\mathbb{R}^2)$, $x = (x_1, x_2) \in \mathbb{R}^2$ and every $a \in \mathbb{R}$ let $x + E = \{x + y = (x_1+y_1, x_2+y_2) : y = (y_1, y_1) \in E\}$ and $a \cdot E = \{ay = (ay_1, ay_2) : y = (y_1, y_2) \in E\}$. Show that $x + E, a \cdot E \in \mathscr{L}(\mathbb{R}^2)$ and that $\mu(x + E) = \mu(E), \mu(a \cdot E) = a^2 \cdot \mu(E)$.

Indication: It will be used the first part of Theorem 5.2.5 for the following variable changes $g, h : \mathbb{R}^2 \to \mathbb{R}^2$, $g(y) = x + y$, $h(y) = a \cdot y$. A direct demonstration can also be given: the proof of the sets measurability is similar to that of Theorem 1.3.15 and for the measure calculation is used Theorem 5.1.22.

(6) Calculate $\theta(M)$, where $M = \{(x, y, z) \in \mathbb{R}^3 : x^2 + y^2 + z^2 \leq r^2\}$.

(7) Show that if $A, B \in \mathscr{L}(\mathbb{R})$, $f \in \mathfrak{L}^1(A)$, and $g \in \mathfrak{L}^1(B)$, then the function $h : A \times B \to \mathbb{R}$, $h(x, y) = f(x) \cdot g(y)$, for every $(x, y) \in A \times B$, is integrable on $A \times B$ and

$$\iint_{A\times B} f(x) \cdot g(y) d\mu(x, y) = \left(\int_A f(x) d\lambda(x)\right) \cdot \left(\int_B g(y) d\lambda(y)\right).$$

(8) Let $f : [-1, 1] \times [-1, 1] \to \mathbb{R}$ be defined by

$$f(x, y) = \begin{cases} \frac{xy}{(x^2+y^2)^2}, & (x, y) \neq (0, 0) \\ 0, & (x, y) = (0, 0) \end{cases}.$$

Show that $f \notin \mathfrak{L}^1([-1, 1] \times [-1, 1])$ but

$$\int_{[-1,1]} \left(\int_{[-1,1]} f(x, y) d\lambda(y)\right) d\lambda(x) = \int_{[-1,1]} \left(\int_{[-1,1]} f(x, y) d\lambda(x)\right) d\lambda(y).$$

Signed Measures

In Theorem 3.3.1, we have seen that the integral of a function can be considered as a set function σ-additive with respect to the set on which we integrate. This justifies the interest in σ-additives set functions with real values. We will discuss in this chapter some interesting properties of these set functions. We will show that such a function is a difference of two positive measures, and we will present the conditions under which such set functions can be represented as integrals of certain measurable functions. The last paragraph of the chapter presents some connections between the integral and the derivative.

6.1 Decomposition Theorems

In this chapter, we will use the concepts and results of paragraphs 1.4, 2.4, and 3.6 generically titled "Abstract Setting."

Let $(X, \mathcal{A})$ be a measurable space where X is an abstract set and $\mathcal{A}$ is a σ-algebra on X. For every $B \in \mathcal{A}$, we denote by $\mathcal{A}(B)$ the family of sets $C \subseteq B, C \in \mathcal{A}$. $\mathcal{A}(B)$ is a σ-algebra on B. Let $\gamma : \mathcal{A} \to [0, +\infty] = \bar{\mathbb{R}}_+$ be a complete, σ-finite positive measure on X. Let us denote by $\mathcal{M}(X)$, the vector space of all real measurable functions on X; by $\mathcal{M}_+(X)$, the subset of measurable and positive functions; and by $\mathcal{E}_+(X)$, the subset of positive simple functions.

In paragraph 3.6, we defined the integral of every $f \in \mathcal{M}_+(X)$ by

$$\int_X f d\gamma = \sup \left\{ \int_X \varphi d\gamma : \varphi \in \mathcal{E}_+(X), \varphi \leq f \right\} \in [0, +\infty].$$

$\mathfrak{L}^1_+(X)$ denotes the set of measurable and positive functions which have a finite integral.

Finally, for all $f \in \mathcal{M}(X)$ for which $f^- = \sup\{-f, \underline{0}\} \in \mathfrak{L}^1_+(X)$, let $\nu_f : \mathcal{A} \to]-\infty, +\infty]$ defined by

$$\nu_f(B) = \int_B f d\gamma = \int_B f^+ d\gamma - \int_B f^- d\gamma, \text{ for every } B \in \mathcal{A}.$$

L. C. Florescu, *Lebesgue Integral*, Compact Textbooks in Mathematics,
https://doi.org/10.1007/978-3-030-60163-8_6

If $f^+ \in \mathfrak{L}^1_+(X)$, then ν_f takes the values in $\mathbb{R}$; in this case the function f is integrable on X. Let $\mathfrak{L}^1(X)$ be the vector space of integrable functions on X. According to (4) of Theorem 3.6.14, the integral is σ-additive with respect to the set on which we integrate. It follows that ν_f is a σ-additive set function. This example justifies the following extension of the concept of measure.

Definition 6.1.1

Let $\nu : \mathcal{A} \to [-\infty, +\infty]$; ν is called a **signed measure** if it satisfies:

(1) $\nu(\emptyset) = 0$,

(2) $\nu\left(\bigcup_{n=1}^{\infty} B_n\right) = \sum_{n=1}^{\infty} \nu(B_n)$, for every sequence of pairwise disjoint sets $(B_n)_n \subseteq \mathcal{A}$.

(3) $\nu(\mathcal{A}) \subseteq]-\infty, +\infty]$ or $\nu(\mathcal{A}) \subseteq [-\infty, +\infty[$ (ν assumes at most one of the values $-\infty, +\infty$).

In the following, we will consider that the signed measures take values in $]-\infty, +\infty]$. The results of this chapter also apply to measures taking values in $[-\infty, +\infty[$.

A measure ν taking values in $\mathbb{R} =]-\infty, +\infty[$ is said to be a **finite** signed measure. ν is said to be σ**-finite** if there is a countable partition of X, $\{X_n : n \in \mathbb{N}\} \subseteq \mathcal{A}$, such that $\nu(X_n) \in \mathbb{R}$, for every $n \in \mathbb{N}$.

As we noted above, if γ is a positive, complete, and σ-finite measure on X, for every $f \in \mathcal{M}(X)$ with $f^- \in \mathfrak{L}^1_+(X)$, $\nu_f : \mathcal{A} \to]-\infty, +\infty]$, defined by $\nu_f(B) = \int_B f d\gamma$, is a signed measure; we say that ν_f is the **measure generated by** f; ν_f is also noted as $f \cdot \gamma$. If $f \in \mathfrak{L}^1(X)$, then ν_f is a finite signed measure.

A signed measure retains certain properties of a positive measure (see Theorem 1.4.7).

Proposition 6.1.2 *Let $\nu : \mathcal{A} \to]-\infty, +\infty]$ be a signed measure; then:*

(1) $\nu(B \cup C) = \nu(B) + \nu(C)$, *for every $B, C \in \mathcal{A}$ with $B \cap C = \emptyset$ (ν is finitely additive).*

(2) $\nu(B \setminus C) = \nu(B) - \nu(C)$, $\forall B, C \in \mathcal{A}$ *with $C \subseteq B$ and $\nu(C) < +\infty$.*

(3) $\nu(\bigcup_{n=1}^{\infty} B_n) = \lim_n \nu(B_n)$, *for every increasing sequence $(B_n)_n \subseteq \mathcal{A}$ (ν is continuous from below).*

(4) $\nu(\bigcap_{n=1}^{\infty} B_n) = \lim_n \nu(B_n)$, *for every decreasing sequence $(B_n)_n \subseteq \mathcal{A}$, with $\nu(B_1) < +\infty$ (ν is continuous from above).*

Proof

(1) Let $B_1 = B$, $B_2 = C$, and $B_n = \emptyset$, for any $n \geq 3$; then $\nu(B \cup C) = \nu(\bigcup_{n=1}^{\infty} B_n) = \sum_{n=1}^{\infty} \nu(B_n) = \nu(B) + \nu(C) + \nu(\emptyset) + \cdots + \nu(\emptyset) + \cdots = \nu(B) + \nu(C)$.

(2) $B = (B \setminus C) \cup C$; using the finite additivity of ν, it follows that $\nu(B) = \nu(B \setminus C) + \nu(C)$ and, since $\nu(C) < +\infty$, $\nu(B \setminus C) = \nu(B) - \nu(C)$.

(3) Let $B = \bigcup_{n=1}^{\infty} B_n$. If there is $n_0 \in \mathbb{N}$ such that $\nu(B_{n_0}) = +\infty$, then, for any $n > n_0$, $\nu(B_n) = \nu(B_n \setminus B_{n_0}) + \nu(B_{n_0}) = +\infty$ and so $\nu(B) = \nu(B \setminus B_{n_0}) + \nu(B_{n_0}) = +\infty = \lim_n \nu(B_n)$.

Let us suppose that $\nu(B_n) < +\infty$, for any $n \in \mathbb{N}^*$. Then, since $B = B_1 \cup (B_2 \setminus B_1) \cup \cdots \cup (B_n \setminus B_{n-1}) \cup \cdots$, it follows that $\nu(B) = \nu(B_1) + \nu(B_2 \setminus B_1) + \cdots + \nu(B_n \setminus B_{n-1}) + \cdots = \nu(B_1) + \sum_{n=2}^{\infty}[\nu(B_n) - \nu(B_{n-1})] =$
$= \lim_n [\nu(B_1) + \sum_{k=2}^{n}(\nu(B_k) - \nu(B_{k-1}))] = \lim_n \nu(B_n)$.

(4) Let $B = \bigcap_{n=1}^{\infty} B_n$; since $\nu(B_1) = \nu(B_1 \setminus B_n) + \nu(B_n)$, for any $n \in \mathbb{N}^*$ and $\nu(B_1) = \nu(B_1 \setminus B) + \nu(B)$, it follows that $\nu(B) < +\infty$ and $\nu(B_n) < +\infty$, for any $n \in \mathbb{N}^*$.

The sequence $(B_1 \setminus B_n)_n$ is increasing and $\bigcup_{n=1}^{\infty}(B_1 \setminus B_n) = B_1 \setminus B$. Using properties (2) and (3), $\nu(B_1) - \nu(B) = \nu(B_1 \setminus B) = \lim_n \nu(B_1 \setminus B_n) = \nu(B_1) - \lim_n \nu(B_n)$, from where it follows the property of continuity from above.

■

Remark 6.1.3 Signed measures do not verify all the properties of the positive measures of Theorem 1.4.7. In general, a signed measure is not monotonic. Indeed, let $f : [0, 2] \to \mathbb{R}$ be the function defined by $f(x) = \begin{cases} 1 \ , x \in [0, 1] \\ -1 \ , x \in]1, 2] \end{cases}$; $f \in \mathfrak{L}^1([0, 2])$. The signed measure generated by f is defined by $\nu_f(B) = \lambda(B \cap [0, 1]) - \lambda(B \cap]1, 2])$, for every $B \in \mathscr{L}([0, 2])$ (see Definition 6.1.1). The set $[0, 1] \in \mathscr{L}([0, 2])$, $\nu_f([0, 1]) = 1$, and $\nu_f([0, 2]) = 0$.

The property of finite subadditivity (and therefore that of σ-subadditivity) is not verified either by the signed measures. For the above function f, we consider the measurable sets $B =]\frac{2}{3}, 2]$ and $C = [0, \frac{3}{2}[$; then $\nu_f(B \cup C) = \nu_f([0, 2]) = 0 > -\frac{1}{6} = \nu_f(B) + \nu_f(C)$.

Definition 6.1.4

Let $\nu : \mathcal{A} \to (-\infty, +\infty]$ be a signed measure. A set $B \in \mathcal{A}$ is said to be **ν-positive** (**ν-negative**) if, for every $C \in \mathcal{A}(B)$ ($C \in \mathcal{A}$ and $C \subseteq B$), $\nu(C) \geq 0$ (respectively $\nu(C) \leq 0$).

If B is ν-positive (ν-negative), then $\nu(B) \geq 0$ ($\nu(B) \leq 0$). The converse is not satisfied; indeed, in the example given in the above remark, $\nu_f([0, 2]) = 0$, but the set $[0, 2]$ is neither ν_f-positive ($\nu_f(]1, 2]) = -1 < 0$) nor ν_f -negative ($\nu_f([0, 1]) = 1 > 0$).

The set $B \in \mathcal{A}$ is said to be **ν-null** if it is both ν-positive and ν-negative. B is ν-null if and only if $\nu(C) = 0$, for every $C \in \mathcal{A}(B)$.

The empty set $\emptyset$ is ν-positive and ν-negative, hence ν-null, for every signed measure ν. If ν is a positive measure on $\mathcal{A}$, then every set of $\mathcal{A}$ is ν-positive; since the positive measures are monotonic, a set $B \in \mathcal{A}$ is ν-null if and only if $\nu(B) = 0$.

Example 6.1.5

Let $f \in \mathcal{M}(X)$ with $f^- \in \mathfrak{L}^1_+(X)$ and let ν_f be the signed measure generated by f. The set $X^+ = (f \geq 0)$ is ν_f-positive, $X^- = (f \leq 0)$ is ν_f-negative, and $(f = 0)$ is ν_f-null. Let us remember that $(f \geq 0) = \{x \in X : f(x) \geq 0\}$; $(f \leq 0)$ and $(f = 0)$ are similarly defined.

Moreover, X^+ is a maximal set with the property mentioned in the sense that for any other ν_f-positive set $B \in \mathcal{A}$, $\nu_f(B \setminus X^+) = 0$. Indeed, because $B \setminus X^+ \in \mathcal{A}$, $B \setminus X^+ \subseteq B$, $\nu_f(B \setminus X^+) \geq 0$; if we suppose that $\nu_f(B \setminus X^+) > 0$, then

$$0 < \nu_f(B \setminus X^+) = \int_{B \setminus X^+} f d\gamma = \int_{B \setminus X^+} (-f^-) d\gamma \leq 0,$$

which is absurd. We can even notice that $B \setminus X^+$ is ν_f-null set.

A similar property of maximality can be formulated for the set X^-: for every ν_f-negative set $B \in \mathcal{A}$, $\nu_f(B \setminus X^-) = 0$.

Proposition 6.1.6 *Let $\nu : \mathcal{A} \to (-\infty, +\infty]$ be a signed measure and let $E \in \mathcal{A}$ with $\nu(E) < 0$. There exists a ν-negative set $F \in \mathcal{A}$, $F \subseteq E$ such that $\nu(F) < 0$.*

Proof

If E is ν-negative, then the set $F = E$ verifies the conclusion of the proposition.

Let us suppose that E is not ν-negative; then E has strictly positive measure subsets. Let n_1 be the smallest natural number for which there exists $E_1 \in \mathcal{A}(E)$ with $\nu(E_1) \geq \frac{1}{n_1}$. If $F = E \setminus E_1$ is ν-negative, then, since $\nu(E) = \nu(E_1) + \nu(F) < 0$, it follows that $\nu(F) < 0$, and therefore F verifies the conclusion.

If $E \setminus E_1$ is not ν-negative, then it has strictly positive measure subsets. Let n_2 be the smallest natural number for which there exists $E_2 \in \mathcal{A}(E \setminus E_1)$ with $\nu(E_2) \geq \frac{1}{n_2}$ and so on.

If $F_k = E \setminus (E_1 \cup \cdots \cup E_{k-1})$ is ν-negative, then $F = F_k$ verifies the conclusion.

If F_k is not ν-negative, then it has strictly positive measure subsets, and we consider n_k the smallest natural number for which there exists $E_k \in \mathcal{A}(F_k)$ with $\nu(E_k) \geq \frac{1}{n_k}$ and so on.

If the above construction does not produce a solution F of the problem after a finite number of steps, then we consider $F = \bigcap_{n=1}^{\infty} F_k = E \setminus \left(\bigcup_{n=1}^{\infty} E_k \right)$. Since $0 > \nu(E) > -\infty$, $\sum_{n=1}^{\infty} \nu(E_k) \geq 0$, and $\nu(E) = \nu(F) + \sum_{n=1}^{\infty} \nu(E_k)$, it follows that $\nu(F) \in]-\infty, 0[$. Then $\sum_{k=1}^{\infty} \frac{1}{n_k} \leq \sum_{k=1}^{\infty} \nu(E_k) < +\infty$, from where $n_k \to +\infty$.

Let us show that F verifies the conclusion of proposition.

If we suppose that F is not ν-negative, then there exists $G \in \mathcal{A}(F)$ such that $\nu(G) > 0$. Let $N \in \mathbb{N}$ such that $\nu(G) > \frac{1}{N}$ and $k \in \mathbb{N}$ with $N < n_k$. The fact that $G \subseteq F_k = E \setminus (E_1 \cup \cdots \cup E_{k-1})$ and $\nu(G) > \frac{1}{N}$ contradicts the property that n_k is the smallest natural number for which there exists $E_k \subseteq F_k$ with $\nu(E_k) \geq \frac{1}{n_k}$.

Therefore, F is ν-negative and, since $\nu(F) < 0$, F verifies the conclusion of proposition.

■

The previous proposition proves its usefulness in the demonstration of the Hahn decomposition theorem.

Theorem 6.1.7 (Hahn Decomposition Theorem)
For every signed measure $\nu : \mathcal{A} \to (-\infty, +\infty]$, there exists two sets $E, F \in \mathcal{A}$, E ν-negative, F ν-positive such that $X = E \cup F$ and $E \cap F = \emptyset$.
What would be another pair of sets E', F' with the same properties, the set $E \Delta E' = F \Delta F'$ is ν-null.

Proof
Let $a = \inf\{\nu(B) : B \in \mathcal{A}, B\ \nu\text{-negative}\}$ and let $(B_n)_{n\geq 1} \subseteq \mathcal{A}$ be a sequence of ν-negative sets such that $\nu(B_n) \downarrow a$. Let us denote $E = \bigcup_{n=1}^{\infty} B_n$, and let us consider $(C_n)_{n\geq 1}$ the disjoint sequence associated with $(B_n)_{n\geq 1}$: $C_1 = B_1, C_n = B_n \setminus (B_1 \cup \cdots \cup B_{n-1})$, for any $n \geq 2$. The sequence $(C_n)_{n\geq 1}$ is composed of pairwise disjoint ν-negative sets and $E = \bigcup_{n=1}^{\infty} C_n$.

For every $C \in \mathcal{A}(E)$, $\nu(C) = \nu(C \cap \bigcup_{n=1}^{\infty} C_n) = \sum_{n=1}^{\infty} \nu(C \cap C_n) \leq 0$. Therefore, E is ν-negative. Then, for any $n \in \mathbb{N}^*$, $\nu(E \setminus B_n) \leq 0$ and so $\nu(E) = \nu(B_n) + \nu(E \setminus B_n) \leq \nu(B_n)$. Passing to the limit for $n \to +\infty$, we obtain $\nu(E) \leq \lim_n \nu(B_n) = a$. According to the definition of a, it follows that $\nu(E) = a$.

Let $F = A \setminus E$; if we suppose that F is not ν-positive, then there exists $B \in \mathcal{A}(F)$ such that $\nu(B) < 0$. Now we use Proposition 6.1.6; there exists $C \in \mathcal{A}(B)$ ν-negative such that $\nu(C) < 0$. Then, $E \cup C$ is ν-negative and $\nu(E \cup C) = \nu(E) + \nu(C) = a + \nu(C) < a$, which contradicts the definition of a.

Therefore, F is ν-positive and then the pair of sets E, F verifies the conclusion of the theorem.

Let now E', F' another pair of sets, E' ν-negative, F' ν-positive, $X = E' \cup F'$, and $E' \cap F' = \emptyset$.

It is clear that $E \setminus E' = F^c \setminus (F')^c = F^c \cap F' = (F') \setminus F$ ($F^c = X \setminus F$ is the complement of F). For every $B \in \mathcal{A}$, $B \subseteq E \setminus E' = (F') \setminus F$, $\nu(B) \leq 0$ (E is ν-negative) and $\nu(B) \geq 0$ (F' is ν-positive). It follows that $\nu(B) = 0$. Similarly, $F \setminus F' = (E') \setminus E$ and, for every $B \in \mathcal{A}$, $B \subseteq F \setminus F' = (E') \setminus E$, $\nu(B) = 0$. Hence, $E \Delta E' = F \Delta F'$ is a ν-null set. ■

Definition 6.1.8

Let $\nu : \mathcal{A} \to]-\infty, +\infty]$ be a signed measure; a pair of sets $(E, F) \in \mathcal{A} \times \mathcal{A}$ is said to be a **Hahn decomposition** of X with respect to ν if E is ν-negative, F is ν-positive, $X = E \cup F$, and $E \cap F = \emptyset$.
The previous theorem showed us that the set X admits a Hahn decomposition with respect to any signed measure ν; two decompositions coincide up to a ν-null set.
Let $f \in \mathcal{M}(X)$ with $f^- \in \mathfrak{L}^1_+(X)$; so as we observed it in Example 6.1.5, a Hahn decomposition of X with respect to the generated measure ν_f is (X^-, X^+), where $X^- = (f < \underline{0}) = \{x \in X : f(x) < 0\}$ and $X^+ = (f \geq \underline{0}) = \{x \in X : f(x) \geq 0\}$.

Definition 6.1.9

Two signed measures $\nu_1, \nu_2 : \mathcal{A} \to]-\infty, +\infty]$ on a measurable space (X, $\mathcal{A}$) are said to be mutually **singular** (or simple **singular**) if there exists a set $E \in \mathcal{A}$, such that E is ν_1-null and $E^c = X \setminus E$ is ν_2-null. We denote this by $\nu_1 \perp \nu_2$. If ν_1 and ν_2 are two positive measures, then $\nu_1 \perp \nu_2$ if and only if there exists $E \in \mathcal{A}$ such that $\nu_1(E) = 0 = \nu_2(X \setminus E)$.

Let γ be a positive, complete, and σ-finite measure on X, and let $f, g \in \mathcal{M}_+(X)$ such that $f = \underline{0}$ γ-a.e. on $E \in \mathcal{A}$ and $g = \underline{0}$ γ-a.e. on $X \setminus E$; then $\nu_f \perp \nu_g$, where ν_f and ν_g are the signed measures generated by f, respectively by g.

The following theorem allows us to decompose any real measure into a difference of positive measures.

Theorem 6.1.10 (Jordan Decomposition Theorem)

For every signed measure $\nu : \mathcal{A} \to]-\infty, +\infty]$, there exist two positive measures, $\nu^+, \nu^- : \mathcal{A} \to \bar{\mathbb{R}}_+$, such that $\nu = \nu^+ - \nu^-$, $\nu^+ \perp \nu^-$, and ν^- is finite.

This decomposition of ν as the difference of two positive orthogonal measures is unique.

Proof

Let (E, F) be a Hahn decomposition of the set X with respect to ν; therefore E is ν-negative, F is ν-positive, $X = E \cup F$, and $E \cap F = \emptyset$. Let $\nu^+, \nu^- : \mathcal{A} \to \bar{\mathbb{R}}_+$ be defined by $\nu^+(B) = \nu(F \cap B)$, $\nu^-(B) = -\nu(E \cap B)$, for every $B \in \mathcal{A}$. It is easy to see that ν^+, ν^- are two positive measures on X, that ν^- is finite, and that, for every $B \in \mathcal{A}$, $\nu^+(B) - \nu^-(B) = \nu(F \cap B) + \nu(E \cap B) = \nu((E \cup F) \cap B) = \nu(B)$.

Moreover, $\nu^+(E) = \nu(F \cap E) = 0 = -\nu(E \cap F) = \nu^-(F) = \nu^-(X \setminus E)$; hence $\nu^+ \perp \nu^-$.

Let now $\nu = \nu_1 - \nu_2$ be another decomposition of ν as difference of positive singular measures. Since $\nu_1 \perp \nu_2$, there exists $E_1 \in \mathcal{A}$ such that $\nu_1(E_1) = \nu_2(X \setminus E_1) = 0$. Then $(E_1, X \setminus E_1)$ is another Hahn decomposition of X with respect to ν. Indeed, for every $B \in \mathcal{A}(E_1)$, $\nu(B) = \nu_1(B) - \nu_2(B) = 0 - \nu_2(B) \leq 0$ (ν_1 is a positive measure; hence, it is monotonic: $0 \leq \nu_1(B) \leq \nu_1(E_1) = 0$). Then, E_1 is ν-negative. Similarly, it is shown that $X \setminus E_1$ is ν-positive. According to Theorem 6.1.7, $E \Delta E_1 = F \Delta F_1$ is a ν-null set (we denoted $F_1 = X \setminus E_1$). Then, for every $B \in \mathcal{A}$, $\nu((F \setminus F_1) \cap B) = 0 = \nu((F_1 \setminus F) \cap B)$ from where,

$$\nu^+(B) = \nu(F \cap B) = \nu((F \setminus F_1) \cap B) + \nu((F \cap F_1) \cap B) =$$

$$= \nu((F_1 \cap F) \cap B) = \nu((F_1 \cap F) \cap B) + \nu((F_1 \setminus F) \cap B) =$$

$$= \nu(F_1 \cap B) = \nu_1(F_1 \cap B) - \nu_2(F_1 \cap B) = \nu_1(F_1 \cap B) = \nu_1(B).$$

It is also shown that $\nu^-(B) = \nu_2(B)$. ■

Definition 6.1.11

The decomposition $\nu = \nu^+ - \nu^-$ with ν^+ and ν^- positive, singular measures and ν^- finite is said to be the **Jordan decomposition** of the measure ν; the mapping $|\nu| : \mathcal{A} \to \bar{\mathbb{R}}_+$, defined by $|\nu| = \nu^+ + \nu^-$, is a positive measure, and it is called the **total variation** of ν.

Remark 6.1.12 Note that the measure ν is finite if and only if ν^+ is finite or, equivalently, if and only if $|\nu|$ is finite.

Let $f \in \mathcal{M}(X)$ with $f^- \in \mathfrak{L}^1_+(X)$; a Hahn decomposition of X with respect to the measure generated by f, ν_f, is (X^-, X^+), where $X^- = (f < 0) = \{x \in X : f(x) < 0\}$ and $X^+ = (f \geq 0) = \{x \in X : f(x) \geq 0\}$ (see Definition 6.1.8). Note that $x \in X^+$ if and only if $f(x) = \sup\{f(x), 0\} = f^+(x)$; hence $X^+ = \{x \in X : f(x) = f^+(x)\}$. Then, for every $B \in \mathcal{A}$, $\nu_f^+(B) = \nu_f(X^+ \cap B) = \int_{X^+ \cap B} f d\gamma = \int_B f^+ d\gamma$. Similarly, $\nu_f^-(B) = \int_B f^- d\gamma$. It follows that ν_f^+ is the measure generated by f^+ and ν_f^- is the measure generated by f^-. The total variation of ν_f is $\nu_f^+ + \nu_f^-$—the positive measure generated by $f^+ + f^- = |f|$.

The explanation of the name of total variation can be found in the following theorem.

Theorem 6.1.13

Let $\nu : \mathcal{A} \to]-\infty, +\infty]$ be a signed measure; the total variation of ν, $|\nu|$, has the following properties:

(1) $|\nu|(B) = \sup\left\{\sum_{k=1}^n |\nu(B_k)| : n \in \mathbb{N}^*, \{B_1, \ldots, B_n\} \in \Pi_B\right\}$, *where Π_B is the family of all finite partitions of B with sets of $\mathcal{A}$.*

(2) $\sup\{|\nu(C)| : C \in \mathcal{A}(B)\} \leq |\nu|(B) \leq 2 \cdot \sup\{|\nu(C)| : C \in \mathcal{A}(B)\}$, *for every $B \in \mathcal{A}$.*

(3) *$|\nu|$ is the smallest of the positive measures σ for which $|\nu(B)| \leq \sigma(B)$, for every $B \in \mathcal{A}$.*

Proof

(1) Let $\{B_1, \ldots, B_n\} \in \Pi_B$ be arbitrary; then

$$|\nu|(B) = \sum_{k=1}^n |\nu|(B_k) = \sum_{k=1}^n [\nu^+(B_k) + \nu^-(B_k)] \geq \quad (*)$$

$$\geq \sum_{k=1}^n |\nu^+(B_k) - \nu^-(B_k)| = \sum_{k=1}^n |\nu(B_k)|.$$

On the other hand, if (E, F) is a Hahn decomposition of X with respect to ν, then $\{E \cap B, F \cap B\} \in \Pi_B$ and

$$|\nu(E \cap B)| + |\nu(F \cap B)| = -\nu(E \cap B) + \nu(F \cap B) = \nu^-(B) + \nu^+(B) = |\nu|(B).$$

It follows that

$$|\nu|(B) \le \sup \left\{ \sum_{k=1}^{n} |\nu(B_k)| : n \in \mathbb{N}^*, \{B_1, \dots, B_n\} \in \Pi_B \right\},$$

which, with the inequality $(*)$, demonstrates the equality of point (1).

(2) For every $B \in \mathcal{A}$ and every $C \in \mathcal{A}(B)$, $\{C, B \setminus C\} \in \Pi_B$ and then

$$|\nu|(B) \ge |\nu(C)| + |\nu(B \setminus C)| \ge |\nu(C)|.$$

Therefore, $\sup\{|\nu(C)| : C \in \mathcal{A}(B)\} \le |\nu|(B)$.

If (E, F) is a Hahn decomposition of X, then

$$|\nu|(B) = \nu^+(B) + \nu^-(B) = \nu(B \cap F) - \nu(B \cap E) =$$

$$= |\nu(B \cap E)| + |\nu(B \cap F)| \le 2 \cdot \sup\{|\nu(C)| : C \in \mathcal{A}(B)\}.$$

(3) Using the first inequality from the previous point, it turns out that $|\nu|$ is a positive measure such that $|\nu(B)| \le |\nu|(B)$, for every $B \in \mathcal{A}$.

Let $\sigma : \mathcal{A} \to \bar{\mathbb{R}}_+$ be an arbitrary positive measure with the property $|\nu(B)| \le \sigma(B)$, for every $B \in \mathcal{A}$, and let (E, F) be a Hahn decomposition of X with respect to ν. Then, for every $B \in \mathcal{A}$,

$$|\nu|(B) = \nu(B \cap F) - \nu(B \cap E) = |\nu(B \cap E)| + |\nu(B \cap F)| \le$$

$$\le \sigma(B \cap E) + \sigma(B \cap F) = \sigma(B).$$

Therefore, $|\nu| \le \sigma$.

■

Remark 6.1.14 Let $\nu : \mathcal{A} \to]-\infty, +\infty]$ be a signed measure and let $|\nu|$ be the total variation of ν. Then, a set $B \in \mathcal{A}$ is ν-null if and only if $|\nu|(B) = 0$. Indeed, according to Definition 6.1.4, B is ν-null if and only if $\nu(C) = 0$, for every $C \in \mathcal{A}(B)$, which is equivalent to $\sup\{|\nu(C)| : C \in \mathcal{A}(B)\} = 0$. Point (2) of the previous theorem shows us that it comes down to $|\nu|(B) = 0$.

6.2 Radon-Nikodym Theorem

Let $\mathcal{A}$ be a σ-algebra on X and let $\gamma : \mathcal{A} \to \bar{\mathbb{R}}_+$ be a positive, σ-finite measure on X. For every $f \in \mathcal{M}(X)$ with $f^- \in \mathfrak{L}^1_+(X)$, let $\nu_f : \mathcal{A} \to]-\infty, +\infty]$ be the measure generated by f; we defined $\nu_f(B) = \int_B f d\gamma$ and we observed that $|\nu_f|(B) = \int_B |f| d\gamma$ (see Definition 6.1.11). Point (2) of Theorem 3.6.8 assures us that if $\gamma(B) = 0$, then $|\nu_f|(B) = 0$; according to Remark 6.1.14, B is ν_f-null set. We will show in this paragraph that this property characterizes the measures generated by the measurable functions: any signed measure whose total variation is zero on the γ-null sets is generated by a measurable function.

Definition 6.2.1

A signed measure $\nu : \mathcal{A} \to]-\infty, +\infty]$ is said to be **absolutely continuous** with respect to a positive measure γ if, for every $B \in \mathcal{A}$ with $\gamma(B) = 0$, $\nu(B) = 0$. We denote this by $\nu \ll \gamma$.

Remark 6.2.2 $\nu \ll \gamma$ if and only if $|\nu|(B) = 0$, for every $B \in \mathcal{A}$ with $\gamma(B) = 0$. The sufficiency of condition results from point (2) of Theorem 6.1.13. To show that this condition is necessary, we observe that because γ is a positive measure and $\gamma(B) = 0$, then B is ν-null; Remark 6.1.14 assures us that $|\nu|(B) = 0$.

Proposition 6.2.3 *Let ν be a finite signed measure; $\nu \ll \gamma$ if and only if, for every $\varepsilon > 0$, there exists $\delta > 0$ such that, for every $B \in \mathcal{A}$ with $\gamma(B) < \delta$, $|\nu|(B) < \varepsilon$.*

Proof

The sufficiency: let us suppose that, for every $\varepsilon > 0$, there exists $\delta > 0$ such that, for every $C \in \mathcal{A}$ with $\gamma(C) < \delta$, $|\nu|(C) < \varepsilon$ and let $B \in \mathcal{A}$ with $\gamma(B) = 0$. Then, $\gamma(B) < \delta$ and so $|\nu|(B) < \varepsilon$. Since ε is arbitrary, it follows that $|\nu|(B) = 0$ and therefore $\nu(B) = 0$.

The necessity: We reason by reduction to the absurd; we suppose that $\nu \ll \gamma$ and that there is $\varepsilon_0 > 0$ so that, for any $k \in \mathbb{N}$, there exists $B_k \in \mathcal{A}$ with $\gamma(B_k) < \frac{1}{2^k} = \delta$ and $|\nu|(B_k) \geq \varepsilon_0$. Let $B = \limsup_n B_n = \bigcap_{n=1}^{\infty} \bigcup_{k=n}^{\infty} B_k \in \mathcal{A}$. For any $n \in \mathbb{N}$,

$$\gamma(B) \leq \gamma\left(\bigcup_{k=n}^{\infty} B_k\right) \leq \sum_{k=n}^{\infty} \gamma(B_k) < \sum_{k=n}^{\infty} \frac{1}{2^k} = \frac{1}{2^{n-1}}.$$

It follows that $\gamma(B) = 0$ and, according to the hypothesis, $|\nu|(B) = 0$.

On the other hand, ν being finite, $|\nu|$ is finite and then $|\nu|(\bigcup_{n=1}^{\infty} B_n) < +\infty$; from (9) of Theorem 1.4.7,

$$|\nu|(B) = |\nu|(\limsup_n B_n) \geq \limsup_n |\nu|(B_n) \geq \varepsilon_0,$$

which is a contradiction. ■

Remarks 6.2.4

(i) After the previous proposition, we can say that the reason why ν is absolutely continuous with respect to γ is actually (as the name suggests) a property of continuity; we can write $\nu \ll \gamma$ if and only if $\lim_{\gamma(B)\to 0} \nu(B) = 0$.

(ii) According to the definition, $\nu \ll \gamma$ if and only if $|\nu| \ll \gamma$ which is to say that
$$\begin{cases} \nu^+ \ll \gamma, \\ \nu^- \ll \gamma. \end{cases}$$

(iii) According to the definition and to Theorem 3.6.14, we can remark that if ν_f is the measure generated by the measurable function f, then $\nu_f \ll \gamma$. We will show that the reverse is satisfied.

Theorem 6.2.5 (Radon-Nikodym)
Let $\nu : \mathcal{A} \to \mathbb{R}$ be a finite signed measure absolutely continuous with respect to a σ-finite positive measure γ. Then there exists a function $f \in \mathfrak{L}^1(X, \mathcal{A}, \gamma)$ such that $\nu = \nu_f = f \cdot \gamma$ or

$$\nu(B) = \int_B f d\gamma, \text{ for every } B \in \mathcal{A}.$$

The function f is unique up to a γ-null set.

Proof

Let us first deal with the existence of f. We will demonstrate it in several stages.

(1) First, assume that ν is a finite and positive measure and that $\gamma(X) < +\infty$. Let

$$\mathcal{F} = \left\{ g \in \mathfrak{L}^1_+(X) : \int_B g d\gamma \le \nu(B), \text{ for every } B \in \mathcal{A} \right\}.$$

Since $\underline{0} \in \mathcal{F}$, $\mathcal{F} \neq \emptyset$. Let $L = \sup\{\int_X g d\gamma \,:\, g \in \mathcal{F}\} \le \nu(X) < +\infty$ and let $(g_n)_n \subseteq \mathcal{F}$ such that $\int_X g_n d\gamma \uparrow L$.

We can assume that the sequence $(g_n)_{n\ge 1}$ is increasing. Indeed, we can build a sequence $(h_n)_{n\ge 1}$, defined by $h_n = \max\{g_1, \ldots, g_n\}$, for any $n \in \mathbb{N}^*$. Then $(h_n)_{n\ge 1}$ is obviously increasing. Let us show by induction that $(h_n)_{n\ge 1} \subseteq \mathcal{F}$. $h_1 = g_1 \in \mathcal{F}$. We suppose that $h_1, \ldots, h_n \in \mathcal{F}$; $h_{n+1} = \max\{g_1, \ldots, g_n, g_{n+1}\} = \max\{h_n, g_{n+1}\}$. Let $C = (h_n \ge g_{n+1}) \in \mathcal{A}$; then, since $h_n, g_{n+1} \in \mathcal{F}$, for every $B \in \mathcal{A}$,

$$\int_B h_{n+1} d\gamma = \int_{B\cap C} h_n d\gamma + \int_{B\setminus C} g_{n+1} d\gamma \le \nu(B \cap C) + \nu(B \setminus C) = \nu(B).$$

It follows that $h_{n+1} \in \mathcal{F}$; therefore $(h_n)_{n\ge 1} \subseteq \mathcal{F}$. Since $g_n \le h_n$, for any $n \in \mathbb{N}^*$, $\int_X g_n d\gamma \le \int_X h_n d\gamma$, for any $n \in \mathbb{N}^*$; so $\int_X h_n d\gamma \uparrow L$. Hence, the sequence $(h_n)_n$

can replace the sequence $(g_n)_n$. We can therefore assume from the start that $(g_n)_n$ is increasing.

Let $A = \{x \in X : \lim_n g_n(x) = +\infty\} = \{x \in X : \sup_n g_n(x) = +\infty\}$; we remark that $A = \bigcap_{k=1}^{\infty} \bigcup_{n=1}^{\infty} (g_n > k) \in \mathcal{A}$. If we suppose that $\gamma(A) = a > 0$, then, for any $k \in \mathbb{N}^*$, there is $n_k \in \mathbb{N}^*$ such that $\gamma(g_{n_k} > k) \geq a$ and then

$$k \cdot a \leq k \cdot \gamma((g_{n_k} > k)) \leq \int_{(g_{n_k} > k)} g_{n_k} d\gamma \leq \int_X g_{n_k} d\gamma \leq L, \text{ for any } k \in \mathbb{N}^*,$$

which is absurd. Therefore, $\gamma(A) = 0$ and, since $\nu \ll \gamma$, $\nu(A) = 0$.

Let then $f : X \to \mathbb{R}_+$, defined by $f(x) = \begin{cases} \lim_n g_n(x) \ , x \in X \setminus A, \\ 0 \ , x \in A. \end{cases}$

The function f is measurable (see Exercise (4) of 2.5) and positive; on the set $X \setminus A$, $g_n \uparrow f$. Using the theorem of monotone convergence on $X \setminus A$, we obtain $\int_{X \setminus A} g_n d\gamma \uparrow \int_{X \setminus A} f d\gamma \leq L$. On the other hand, according to Theorem 3.6.8, $\int_A f d\gamma = 0 = \int_A g_n d\gamma$. It follows that $f \in \mathfrak{L}_+^1(X)$ and $\int_X g_n d\gamma \uparrow \int_X f d\gamma$. Therefore, $\int_X f d\gamma = L$.

Using again the theorem of monotone convergence on $B \setminus A$, we obtain

$$\int_B f d\gamma = \int_{B \setminus A} f d\gamma = \lim_n \int_{B \setminus A} g_n d\gamma \leq \nu(B \setminus A) = \nu(B), \text{ for every } B \in \mathcal{A}.$$

Therefore, $f \in \mathcal{F}$.

We prove that f is the desired function.

Let us suppose that there exists $B_0 \in \mathcal{A}$ such that $\int_{B_0} f d\gamma < \nu(B_0)$. Note that $\gamma(B_0) > 0$ (if we suppose that $\gamma(B_0) = 0$, then $\nu(B_0) = 0$ and the above strict inequality could not happen). Then, there is $\varepsilon > 0$ such that

$$\eta(B_0) = \nu(B_0) - \int_{B_0} f d\gamma - \varepsilon \cdot \gamma(B_0) > 0.$$

The mapping $\eta : \mathcal{A} \to \mathbb{R}$, defined by

$$\eta(B) = \nu(B) - \int_B f d\gamma - \varepsilon \cdot \gamma(B), \text{ for every } B \in \mathcal{A},$$

is a signed measure and $\eta(B_0) > 0$. The Hahn decomposition theorem assures us that there exist two sets $E, F \in \mathcal{A}$, such that $E \cap F = \emptyset$ and $X = E \cup F$ and E is η-negative and F is η-positive.

Since $0 < \eta(B_0) = \eta(B_0 \cap E) + \eta(B_0 \cap F)$, it follows that $\eta(B_0 \cap F) > 0$ and then $\eta(F) > 0$.

For every $B \in \mathcal{A}$, $\eta(B \cap F) \geq 0$, hence

$$\int_{B \cap F} f d\gamma + \varepsilon \cdot \gamma(B \cap F) \leq \nu(B \cap F).$$

Then

$$\int_B (f+\varepsilon\cdot\chi_F)d\gamma = \int_B f d\gamma + \varepsilon\cdot\gamma(B\cap F) =$$

$$= \int_{B\setminus F} f d\gamma + \int_{B\cap F} f d\gamma + \varepsilon\cdot\gamma(B\cap F) \leq \nu(B\setminus F)+\nu(B\cap F) = \nu(B).$$

It follows that $f+\varepsilon\cdot\chi_F \in \mathcal{F}$ and then

$$L \geq \int_X (f+\varepsilon\cdot\chi_F)d\gamma = \int_X f d\gamma + \varepsilon\cdot\gamma(F) = L+\varepsilon\cdot\gamma(F).$$

From the above relation, $\gamma(F) = 0$ and, since $\nu \ll \gamma$, $\nu(F) = 0$. It follows that $\eta(F) = 0$, which contradicts $\eta(F) > 0$.

The contradiction we have reached shows that the hypothesis that there is $B_0 \in \mathcal{A}$ such that $\int_{B_0} f d\gamma < \nu(B_0)$ is false.

Therefore, for every $B \in \mathcal{A}$, $\nu(B) = \int_B f d\gamma$ which shows that the function f verifies the conclusion of the theorem: $\nu = \nu_f$.

(2) Now consider the case where ν is a positive finite measure and γ is a positive σ-finite measure. Let $(X_n)_{n\geq 1} \subseteq \mathcal{A}$ be an increasing sequence of sets such that $\bigcup_{n=1}^{\infty} X_n = X$ and $\gamma(X_n) < +\infty$, for any $n \geq 1$. For any $n \geq 1$, we define $\gamma_n, \nu_n : \mathcal{A} \to \mathbb{R}_+$ by $\gamma_n(B) = \gamma(B\cap X_n)$ and $\nu_n(B) = \nu(B\cap X_n)$, for every $B \in \mathcal{A}$. Since $\nu_n \ll \gamma_n$, for any $n \in \mathbb{N}^*$, we can apply the previous point of the demonstration to find the positive and γ_n-integrable functions f_n on X such that

$$\nu_n(B) = \int_B f_n d\gamma_n, \text{ for any } n \in \mathbb{N}^* \text{ and for every } B \in \mathcal{A}.$$

We remark that $\gamma_n(B) = \gamma(B)$, $\nu_n(B) = \nu(B)$, for every $B \in \mathcal{A}$, $B \subseteq X_n$ and $\nu_n(B) = \gamma_n(B) = 0$, for every $B \subseteq X\setminus X_n$; it follows that

$$\int_B (f_{n+1}-f_n)d\gamma = \int_B f_{n+1} d\gamma_{n+1} - \int_B f_n d\gamma_n =$$

$$= \nu_{n+1}(B) - \nu_n(B) = \nu(B) - \nu(B) = 0,$$

for every $B \in \mathcal{A}$, $B \subseteq X_n \subseteq X_{n+1}$. Using point (c) of Theorem 3.6.9, $f_n = f_{n+1}$ γ-a.e. on the set X_n. Every γ_n-integrable function is γ-integrable (see Exercise (21) of 3.7). It follows that $(f_n)_{n\geq 1} \subseteq \mathfrak{L}^1_+(X)$.

We then define consistently $f : X \to \mathbb{R}$ by $f(x) = f_n(x)$, if $x \in X_n$. For any $a \in \mathbb{R}$, $f^{-1}(]-\infty, a[) = \bigcup_{n\geq 1} f_n^{-1}(]-\infty, a[) \in \mathcal{A}$. It follows that the function thus defined is measurable and positive. Using the property of continuity from below of measures on increasing sequences (see (6) of Theorem 1.4.7), we obtain

$$\int_B f d\gamma = \lim_n \int_{B\cap X_n} f d\gamma = \lim_n \int_{B\cap X_n} f_n d\gamma_n = \lim_n \nu_n(B \cap X_n) =$$

$$= \lim_n \nu(B \cap X_n) = \nu(B),$$

for every $B \in \mathcal{A}$. Particularly, $\int_X f d\gamma = \nu(X) < +\infty$ and so $f \in \mathfrak{L}^1_+(X)$.

(3) Let ν be a finite signed measure absolutely continuous with respect to a positive and σ-finite measure, γ. Let $\nu = \nu^+ - \nu^-$ be the Jordan decomposition of ν. According to (ii) of Remark 6.2.4, it follows that $\nu^+ \ll \gamma$ and $\nu^- \ll \gamma$. We can now apply the previous case of the demonstration: there exist $g, h \in \mathfrak{L}^1_+(X)$ such that

$$\nu^+(B) = \int_B g d\gamma \text{ and } \nu^-(B) = \int_B h d\gamma, \text{ for every } B \in \mathcal{A}.$$

The function $f = g - h$ is integrable with respect to γ and verifies the conclusion of the theorem.

Let us now deal with the uniqueness. Let $f_1, f_2 \in \mathfrak{L}^1(X, \mathcal{A}, \gamma)$ be two functions such that, for every $B \in \mathcal{A}$,

$$\nu(B) = \int_B f_1 d\gamma = \int_B f_2 d\gamma.$$

Then, $\int_B (f_1 - f_2) d\gamma = 0$, for every $B \in \mathcal{A}$. Theorem 3.6.9 assures us that $f_1 = f_2$, γ-almost everywhere.

■

Remarks 6.2.6

(i) The theorem can be easily extended to the case where ν is a σ-finite signed measure absolutely continuous with respect to the positive σ-finite measure γ; in this case, however, the function f will have the integral, without necessarily being γ-integrable.

(ii) If we take into account point (iii) of Remark 6.2.4 and the Radon-Nikodym theorem, we can say that if ν is a finite signed measure, then $\nu \ll \gamma$ if and only if there exists $f \in \mathfrak{L}^1(X, \mathcal{A}, \gamma)$ tel que $\nu = f \cdot \gamma$.

Definition 6.2.7

The function $f \in \mathfrak{L}^1(X)$ whose existence has been proven in the Radon-Nikodym theorem is called the **Radon-Nikodym derivative** of ν with respect to γ, and it is denoted by $f = \dfrac{d\nu}{d\gamma}$.

A demonstration similar to that of the Radon-Nikodym theorem can be made for the following Lebesgue decomposition theorem.

Theorem 6.2.8 (Lebesgue Decomposition Theorem)
For every finite signed measure, $\nu : \mathcal{A} \to \mathbb{R}$ *and every* σ*-finite positive measure* γ *there exist two signed measures* $\eta, \rho : \mathcal{A} \to \mathbb{R}$ *such that* $\eta \ll \gamma$, $\rho \perp \gamma$, *and* $\nu = \eta + \rho$.

We conclude this paragraph with a result which expresses the integral with respect to $\nu \ll \gamma$ using the integral with respect to γ.

Theorem 6.2.9
Let $\nu : \mathcal{A} \to \mathbb{R}_+$ *be a finite positive measure, absolutely continuous with respect to a positive* σ*-finite measure* γ, *and let* $f = \dfrac{d\nu}{d\gamma} \in \mathfrak{L}^1_+(X, \mathcal{A}, \gamma)$ *be its Radon-Nikodym derivative. Then,*

$$\int_X g d\nu = \int_X g \cdot f d\gamma = \int_X g \cdot \left(\frac{d\nu}{d\gamma}\right) d\gamma, \text{ for every function } g \in \mathfrak{L}^1(X, \mathcal{A}, \nu).$$

Proof
First consider $g = \chi_B$, where $B \in \mathcal{A}$; then

$$\int_X g d\nu = \nu(B) = \int_B f d\gamma = \int_X \chi_B \cdot f d\gamma = \int_X g \cdot f d\gamma.$$

We can immediately extend this result to any positive simple function $g \in \mathcal{E}_+(X)$.

Let now $g \in \mathcal{M}_+(X)$; then there exists a sequence $(g_n)_n \subseteq \mathcal{E}_+(X)$ such that $g_n \uparrow g$. The monotone convergence theorem allows us to write

$$\int_X g d\nu = \lim_n \int_X g_n d\nu = \lim_n \int_X g_n \cdot f d\gamma = \int_X g \cdot f d\gamma.$$

Let g be a ν-integrable function; then $\int_X g^+ d\nu < +\infty$, $\int_X g^- d\nu < +\infty$, $\int_X g^+ d\nu = \int_X g^+ \cdot f d\gamma$, and $\int_X g^- d\nu = \int_X g^- \cdot f d\gamma$. By subtracting the last two relationships, we get the desired result. ■

6.3 The Integral and the Derivative

In this paragraph, we return to integration on the set of real numbers. The Riemann integral gives a partial answer to the following two questions:

(1) Is the integral differentiable with respect to its upper limit?

If the function $f : [a, b] \to \mathbb{R}$ is continuous on $[a, b]$, then

$$\left(\int_a^x f(t)dt\right)' = f(x), \text{ for every } x \in [a, b].$$

But what if f is only integrable?

We will show (see Theorem 6.3.16) that if the function f is Lebesgue integrable and if we replace the Riemann integral by the Lebesgue integral, then the above formula is satisfied for almost all points in the interval $[a, b]$.

(2) If $f : [a, b] \to \mathbb{R}$ is differentiable at every point of $[a, b]$, then is its derivative f' integrable? If we assume that the derivative f' is integrable, then f can be found by the formula: $f(x) = \int_a^x f'(x)dx + k$?

We recall that if the derivative of f, f', is Riemann integrable on $[a, b]$, then the Leibniz-Newton formula works:

$$f(x) = \int_a^x f'(x)dx + f(a), \text{ for every } x \in [a, b]. \qquad (L - N)$$

The formula remains true and in the more general hypothesis that f' is Lebesgue integrable on $[a, b]$ (see Corollary 6.3.23).

We mention that there are differentiable functions whose derivative, although bounded, is not Riemann integrable but is Lebesgue integrable (see the example of Pompeiu in ▶ Sect. 7.2).

But if the function f is differentiable almost everywhere on $[a, b]$, then the formula $(L - N)$ is no longer true (see Cantor staircase 6.3.17).

We will see (Theorem 6.3.22) that the Leibniz-Newton formula $(L-N)$ remains true in the case where f is absolutely continuous and the Riemann integral will be replaced by the Lebesgue integral.

We denoted by $\mathscr{L}$ the σ-algebra of Lebesgue measurable subsets of $\mathbb{R}$ and by $\lambda : \mathscr{L} \to \bar{\mathbb{R}}_+$ the Lebesgue measure on $\mathbb{R}$. A set $A \subseteq \mathbb{R}$ is a null set if, for every $\varepsilon > 0$, there exists a sequence of open intervals $(I_n)_{n\in\mathbb{N}} \subseteq \mathcal{I}$ such that $A \subseteq \bigcup_{n=0}^{\infty} I_n$ and $\sum_{n=0}^{\infty} |I_n| < \varepsilon$. We recall that any null set is Lebesgue measurable. A property $P(x)$ is satisfied almost for all $x \in A \subseteq \mathbb{R}$ if the set of points $x \in A$ for which $P(x)$ is not satisfied is a null set.

We will present below two important notions: that of function with bounded variation and that of absolutely continuous function.

Definition 6.3.1

(1) Let $f : [a, b] \to \mathbb{R}$ and let $\Delta = \{x_0, x_1, \cdots, x_n\}$ be a partition of the interval $[a, b]$ $(a = x_0 < x_1 < \cdots < x_n = b)$. The variation of f with respect to Δ is

(Continued)

Definition 6.3.1 (continued)

$$V_\Delta(f) = \sum_{k=1}^{n} |f(x_k) - f(x_{k-1})|$$

and the **total variation** of f on $[a, b]$ is

$$V_a^b(f) = \sup\{V_\Delta(f) : \Delta \in \mathcal{D}([a, b])\} \in [0, +\infty],$$

where $\mathcal{D}([a, b])$ is the set of all partitions of $[a, b]$.
We say that f has **bounded variation** on $[a, b]$ if $V_a^b(f) < +\infty$. Let $BV_{[a,b]}$ be the set of all bounded variation functions on $[a, b]$.

(2) The function f is **absolutely continuous** on $[a, b]$ if, for every $\varepsilon > 0$, there exists $\delta > 0$ such that, for every finite family of nonoverlapping subintervals of $[a, b]$, $\{[x_k, y_k] : k = 1, \cdots, n\}$, with $\sum_{k=1}^n (y_k - x_k) < \delta$ we have $\sum_{k=1}^n |f(y_k) - f(x_k)| < \varepsilon$. Let $AC_{[a,b]}$ be the set of all absolutely continuous functions on $[a, b]$. It is obvious that any absolutely continuous function is uniformly continuous; the reverse is not true (see Remark 6.3.21).

Proposition 6.3.2

(1) Every absolutely continuous function on $[a, b]$ has bounded variation on $[a, b]$.

(2) Every monotonic function $f : [a, b] \to \mathbb{R}$ has bounded variation on $[a, b]$ and $V_a^b(f) = |f(b) - f(a)|$.

(3) Every Lipschitz function on $[a, b]$ is absolutely continuous on $[a, b]$.

Proof

(1) The demonstration is the same as that of Proposition 5.3.4 of [1]. Let $f : [a, b] \to \mathbb{R}$ be an absolutely continuous function. We take $\delta > 0$ corresponding to $\varepsilon = 1$ in the definition of absolutely continuous functions. Let $\Delta_0 = \{a_0, a_1, \cdots, a_p\} \in \mathcal{D}([a, b])$ be a fixed partition of mesh $\|\Delta_0\| < \delta$.

For every partition $\Delta = \{x_0, x_1, \cdots, x_n\} \in \mathcal{D}([a, b])$, let $\Delta_1 = \Delta_0 \cup \Delta = \{y_0, y_1, \cdots, y_m\} \in \mathcal{D}([a, b])$; $\Delta \subseteq \Delta_1$ and then $V_\Delta(f) \leq V_{\Delta_1}(f)$. But

$$V_{\Delta_1}(f) = \sum_{j=1}^{p} \sum_{y_{k+1} \in]a_{j-1}, a_j]} |f(y_{k+1}) - f(y_k)|.$$

Since, for any $j = 1, \cdots, p$, $\sum_{y_{k+1} \in]a_{j-1}, a_j]} (y_{k+1} - y_k) \leq a_j - a_{j-1} \leq \|\Delta_0\| < \delta$, $\sum_{y_{k+1} \in]a_{j-1}, a_j]} |f(y_{k+1} - f(y_k)| < 1$ and then $V_\Delta(f) < p$, for every partition $\Delta \in \mathcal{D}([a, b])$. Therefore, $V_a^b(f) \leq p$.

(2) Let $f : [a, b] \to \mathbb{R}$ be an increasing function. For every partition $\Delta = \{x_0, x_1, \cdots, x_n\} \in \mathcal{D}([a, b])$, $V_\Delta(f) = \sum_{k=1}^n |f(x_k) - f(x_{k-1})| = \sum_{k=1}^n (f(x_k) - f(x_{k-1})) = f(b) - f(a) = |f(b) - f(a)|$.

(3) If $f : [a, b] \to \mathbb{R}$ is a L-Lipschitz function, then $|f(x) - f(y)| \leq L \cdot |x - y|$, for every $x, y \in [a, b]$. For every $\varepsilon > 0$, let $\delta = \frac{\varepsilon}{L} > 0$ and let $\{[x_k, y_k] : k = 1, \cdots, n\}$ be a finite family of nonoverlapping subintervals of $[a, b]$ with $\sum_{k=1}^{n}(y_k - x_k) < \delta$. Then, $\sum_{k=1}^{n} |f(y_k) - f(x_k)| \leq L \cdot \sum_{k=1}^{n}(y_k - x_k) < \varepsilon$.

■

An important class of absolutely continuous functions on the interval $[a, b]$ is the set of differentiable functions at every point of the interval and whose derivative is Lebesgue integrable on $[a, b]$. Firstly, we need to prove two lemmas.

Lemma 6.3.3 *Let $A \subseteq [a, b]$ be an arbitrary set, let $f : [a, b] \to \mathbb{R}$ be a function differentiable at every point of A, and let $M = \sup_{x \in A} |f'(x)|$. Then,*

$$\lambda^*(f(A)) \leq M \cdot \lambda^*(A).$$

Proof

Recall that λ^* is the Lebesgue outer measure. If $M = +\infty$, then the inequality is obvious.

Suppose now that $M < +\infty$ and let an arbitrary $\varepsilon > 0$. For every $n \in \mathbb{N}^*$, we denote

$$A_n = \left\{ x \in A : |f(x) - f(y)| \leq (M + \varepsilon) \cdot |x - y|, \forall y \in [a, b], |x - y| < \frac{1}{n} \right\}.$$

Obviously, $A_n \subseteq A_{n+1} \subseteq A$ and then $f(A_n) \subseteq f(A_{n+1}) \subseteq f(A)$, for any $n \in \mathbb{N}^*$. On the other hand, for every $x \in A$, there exists $\lim_{y \to x} \frac{f(y) - f(x)}{y - x} = f'(x)$; hence there is $n \in \mathbb{N}^*$ such that, for every $y \in [a, b]$ with $|y - x| < \frac{1}{n}$, $\left| \frac{f(y) - f(x)}{y - x} - f'(x) \right| < \varepsilon$. From here it follows that

$$\left| \frac{f(y) - f(x)}{y - x} \right| < \varepsilon + |f'(x)| \leq \varepsilon + M \text{ or } |f(y) - f(x)| \leq (M + \varepsilon) \cdot |y - x|$$

and therefore $x \in A_n$. Then, $A = \bigcup_{n=1}^{\infty} A_n$ and then $f(A) = \bigcup_{n=1}^{\infty} f(A_n)$.

Since the outer measure is continuous from below (see Proposition 1.3.13):

$$\begin{cases} \lambda^*(A) = \lim_n \lambda^*(A_n) \text{ and} \\ \lambda^*(f(A)) = \lim_n \lambda^*(f(A_n)). \end{cases} \tag{6.1}$$

According to Proposition 1.2.8, for any $n \in \mathbb{N}^*$, there exist the closed intervals $(I_k^n)_k$ such that

$$A_n \subseteq \bigcup_{k=1}^{\infty} I_k^n \text{ and } \sum_{k=1}^{\infty} |I_k^n| < \lambda^*(A_n) + \varepsilon. \tag{6.2}$$

Eventually replacing I_k^n with $I_k^n \cap [a, b]$, we can assume that $I_k^n \subseteq [a, b]$, for all k and n. Possibly dividing the intervals I_k^n, we can yet assume that $|I_k^n| < \frac{1}{n}$, for all $k, n \in \mathbb{N}^*$.

For all $k, n \in \mathbb{N}^*$ and for every $x, y \in A_n \cap I_k^n$, $|x - y| < \frac{1}{n}$ and then

$$|f(x) - f(y)| \le (M + \varepsilon) \cdot |x - y| \le (M + \varepsilon) \cdot |I_k^n|.$$

From the last inequalities, it follows that

$$\operatorname{diam}(f(A_n \cap I_k^n)) \equiv \sup\{|f(x) - f(y)| : x, y \in A_n \cap I_k^n\} \le (M + \varepsilon) \cdot |I_k^n|,$$

and then

$$\lambda^*(f(A_n \cap I_k^n)) \le (M + \varepsilon) \cdot |I_k^n|, \text{ for all } k, n \in \mathbb{N}^*. \tag{6.3}$$

Using relations (6.3) and (6.2), we obtain

$$\lambda^*(f(A_n)) = \lambda^* \left(\bigcup_{k=1}^{\infty} f(A_n \cap I_k^n) \right) \le \sum_{k=1}^{\infty} \lambda^*(f(A_n \cap I_k^n)) \le$$

$$\le (M + \varepsilon) \cdot \sum_{k=1}^{\infty} |I_k^n| \le (M + \varepsilon) \cdot (\lambda^*(A_n) + \varepsilon).$$

From the last inequality and from (6.1), we obtain

$$\lambda^*(f(A)) \le (M + \varepsilon) \cdot (\lim_n \lambda^*(A_n) + \varepsilon) = (M + \varepsilon) \cdot (\lambda^*(A) + \varepsilon).$$

Because ε is arbitrarily positive, $\lambda^*(f(A)) \le M \cdot \lambda^*(A)$. ■

Remarks 6.3.4

(i) If A is a null set, then $f(A)$ is also a null set. Indeed, for every $n \in N$, let $A_n = \{x \in A : |f'(x)| \le n\}$. Then $\lambda^*(f(A_n)) \le n \cdot \lambda^*(A_n) = 0$, for any $n \in \mathbb{N}$, and so $\lambda^*(f(A)) = \lambda^*(\bigcup_{n=0}^{\infty} f(A_n)) \le \sum_{n=0}^{\infty} \lambda^*(f(A_n)) = 0$.

A function which maps null sets in null sets has the Lusin property. Therefore, any differentiable function has the Lusin property.

(ii) If we replace the differentiability with the differentiability almost everywhere, the Lusin property is no longer preserved. We will present below Cantor's function (see Cantor Staircase 6.3.17). This function has the derivative 0 almost everywhere on [0, 1] but does not have the Lusin property.

Lemma 6.3.5 *Let $A \in \mathscr{L}([a, b])$ be a measurable set, and let $f \in \mathfrak{L}([a, b])$ be a function differentiable at every point of A. Then, $f(A) \in \mathscr{L}$ and*

$$\lambda(f(A)) \le \int_A |f'| d\lambda.$$

Proof

According to Exercise (6) of 2.5, $f' \in \mathfrak{L}([a, b])$ and so there exists the integral $\int_A |f'| d\lambda$.

From the proof of Theorem 1.3.20, it follows that any measurable set A is of the form $A = B \cup N = (\bigcup_{n=1}^{\infty} F_n) \cup N$, where $F_n \subseteq A$ are closed sets and N is a null set. In our case, A is bounded; therefore, F_n are compact sets. f being continuous on A (it is differentiable), $f(F_n)$ is compact, so that $f(F_n) \in \mathscr{L}$, for any $n \in \mathbb{N}$. Moreover, from the (i) of the previous remark, $f(N)$ is negligible, so that it is measurable. So $f(A) = \bigcup_{n=1}^{\infty} f(F_n) \cup f(N) \in \mathscr{L}$.

For every $\varepsilon > 0$ and any $n \in \mathbb{N}^*$, let $A_n = \{x \in A : (n-1)\varepsilon \leq |f'(x)| < n\varepsilon\}$. Then, $A_n \in \mathscr{L}$, $A_n \cap A_k = \emptyset$, for all $n, k \in \mathbb{N}^*$, $n \neq k$, and $A = \bigcup_{n=1}^{\infty} A_n$. It follows that

$$\lambda(A) = \sum_{n=1}^{\infty} \lambda(A_n).$$

Taking into account the previous lemma,

$$\lambda^*(f(A)) = \lambda^*\left(\bigcup_{n=1}^{\infty} f(A_n)\right) \leq \sum_{n=1}^{\infty} \lambda^*(f(A_n)) \leq \sum_{n=1}^{\infty} n\varepsilon \cdot \lambda(A_n) =$$

$$= \sum_{n=1}^{\infty} (n-1)\varepsilon \cdot \lambda(A_n) + \varepsilon \cdot \sum_{n=1}^{\infty} \lambda(A_n) \leq \sum_{n=1}^{\infty} \int_{A_n} |f'| d\lambda + \varepsilon \cdot \lambda(A) =$$

$$= \int_A |f'| d\lambda + \varepsilon \cdot \lambda(A).$$

Because $\lambda(A) < +\infty$ and ε is arbitrarily positive, we obtain the announced result. ■

Theorem 6.3.6

Let $f : [a, b] \to \mathbb{R}$ be a differentiable function at every point of $[a, b]$. If $f' \in \mathfrak{L}^1([a, b])$, then $f \in AC_{[a,b]}$.

Proof

According to the property of absolute continuity of integral (see (2) of Theorem 3.3.1), for every $\varepsilon > 0$, there exists $\delta > 0$ such that

$$\int_B |f'| d\lambda < \varepsilon, \text{ for every } B \in \mathscr{L}([a, b]) \text{ with } \lambda(B) < \delta. \qquad (*)$$

Let $\{[x_k, y_k] : k = 1, \cdots, n\}$ be a family of nonoverlapping subintervals of $[a, b]$ with $\sum_{k=1}^{n} (y_k - x_k) < \delta$, and let $B = \bigcup_{k=1}^{n} [x_k, y_k]$. Then, $\lambda(B) < \delta$ and therefore from $(*)$,

$$\sum_{k=1}^{n} \int_{[x_k, y_k]} |f'| d\lambda = \int_B |f'| d\lambda < \varepsilon. \qquad (**)$$

Since f is continuous on $[a, b]$, for any k for which $f(x_k) \leq f(y_k)$, we have $[f(x_k), f(y_k)] \subseteq f([x_k, y_k])$; if $f(x_k) \geq f(y_k)$, then $[f(y_k), f(x_k)] \subseteq f([x_k, y_k])$. Therefore, from the previous lemma and from $(**)$,

$$\sum_{k=1}^{n} |f(y_k) - f(x_k)| \leq \sum_{k=1}^{n} \lambda^*(f([x_k, y_k])) \leq \sum_{k=1}^{n} \int_{[x_k, y_k]} |f'| d\lambda < \varepsilon.$$

■

Remark 6.3.7 Any differentiable function with bounded derivative on an interval $[a, b]$ is a Lipschitz function; according to (3) of Proposition 6.3.2, it will be absolutely continuous. We note that the result is also a consequence of the previous theorem, because, in this case, the derivative is Lebesgue integrable on $[a, b]$.

The following theorem shows that any functions with bounded variation (and therefore also any absolutely continuous functions) is a difference of two monotonic functions.

Theorem 6.3.8 (Jordan)
Every bounded variation function on $[a, b]$ *is the difference of two increasing functions on* $[a, b]$.

Proof
We suppose that $f : [a, b] \to \mathbb{R}$ has bounded variation and let $g : [a, b] \to \mathbb{R}$ defined by $g(x) = V_a^x(f)$, for every $x \in [a, b]$ (if $f \in BV_{[a,b]}$, then, for any $x \in [a, b]$, $f \in BV_{[a,x]}$). The function g is monotonically increasing on $[a, b]$. Indeed, for every $x, y \in [a, b]$ with $x < y$ and for every partition $\Delta = \{a = y_0, y_1, \cdots, y_n = x\} \in \mathcal{D}([a, x])$, $\Delta_1 = \{a = y_0, y_1, \cdots, y_n = x, y_{n+1} = y\}$ is a partition of $[a, y]$. Then

$$V_\Delta(f) \leq \sum_{k=1}^{n} |f(y_k) - f(y_{k-1})| + |f(y) - f(x)| = V_{\Delta_1}(f) \leq V_a^y(f).$$

Therefore $V_a^x(f) = \sup_{\Delta \in \mathcal{D}([a,x])} V_\Delta(f) \leq V_a^y(f)$ which means that g is monotonically increasing.

Let $h : [a, b] \to \mathbb{R}$, defined by $h(x) = f(x) + V_a^x(f)$; then h is increasing also. Indeed, for every $x, y \in [a, b]$ with $x < y$ we have

$$h(x) - h(y) = f(x) - f(y) + V_a^x(f) - V_a^y(f) = f(x) - f(y) - V_x^y(f) \leq 0.$$

In the above relations we took into account the additivity property of the total variation: $V_a^y(f) = V_a^x(f) + V_x^y(f)$, and the fact that $V_x^y(f) \geq |f(y) - f(x)|$. It is obvious that $f = h - g$. ■

Then, we give some consequences of the Lebesgue derivation theorem (see Theorem 7.3.5), of Proposition 6.3.2 and of the Jordan theorem (6.3.8).

Corollary 6.3.9

(1) Every monotonic function on an interval I is differentiable almost everywhere on I.

(2) Every bounded variation function on an interval $[a, b]$ is differentiable almost everywhere on $[a, b]$.

(3) Every absolutely continuous function on $[a, b]$ is differentiable almost everywhere on $[a, b]$.

A first attempt to obtain the Leibniz-Newton formula is presented in the following proposition:

Proposition 6.3.10

(1) For every monotonically increasing function $f : [a, b] \to \mathbb{R}$, $f' \in \mathfrak{L}^1_+([a, b])$ and $\int_{[a,b]} f' d\lambda \leq f(b) - f(a)$.

(2) For every bounded variation function (so for any absolutely continuous one) $f : [a, b] \to \mathbb{R}$, $f' \in \mathfrak{L}^1([a, b])$.

Proof

(1) We extend f to $]b, b+1]$ by putting $f(x) = f(b)$, for every $x \in]b, b+1]$; the function f remains monotonically increasing on $[a, b+1]$. According to (1) of the previous corollary, f is differentiable almost everywhere on $[a, b+1]$. Then, f is continuous almost everywhere on $[a, b+1]$ and, according to (2) of Theorem 2.1.13, $f \in \mathfrak{L}([a, b+1])$.

Let $f_n : [a, b] \to \mathbb{R}_+$, defined by $f_n(x) = n \cdot \left[f\left(x + \frac{1}{n}\right) - f(x)\right]$, for any $n \in \mathbb{N}^*$. The sequence $(f_n)_n \subseteq \mathfrak{L}_+([a, b])$ converges a.e. on $[a, b]$ to f'. It follows that $f' \in \mathfrak{L}_+([a, b])$. We apply to $(f_n)_n$ Fatou's lemma (see Corollary 3.1.12), and, since $\liminf_n f_n = f'$ a.e. on $[a, b]$, we obtain

$$\int_{[a,b]} f' d\lambda \leq \liminf_n \int_{[a,b]} f_n d\lambda. \tag{6.1}$$

We calculate the integral of f_n by applying the change variable formula (Theorem 3.5.6) with $g(x) = x + \frac{1}{n}$ (we note that, f being increasing, f_n is the Borel function—see Proposition 2.1.5).

$$\int_{[a,b]} f_n d\lambda = n \cdot \left[\int_{[a,b]} f\left(x + \frac{1}{n}\right) d\lambda(x) - \int_{[a,b]} f(x) d\lambda(x)\right] = \tag{6.2}$$

$$= n \cdot \left[\int_{[a+\frac{1}{n}, b+\frac{1}{n}]} f(x) d\lambda(x) - \int_{[a,b]} f(x) d\lambda(x)\right] =$$

$$= n \cdot \left[\int_{[b,b+\frac{1}{n}]} f(x)d\lambda(x) - \int_{[a,a+\frac{1}{n}]} f(x)d\lambda(x) \right] \leq$$

$$\leq \left[n \cdot f\left(b + \frac{1}{n}\right) \cdot \frac{1}{n} - n \cdot f(a) \cdot \frac{1}{n} \right] = f(b) - f(a).$$

From (6.1) and (6.2), we obtain $\int_{[a,b]} f' d\lambda \leq f(b) - f(a)$. According to the previous inequality, it follows that $f' \in \mathfrak{L}^1([a, b])$.

(2) Let $f \in BV_{[a,b]}$; according to the Jordan theorem (6.3.8), $f = f_1 - f_2$, where f_1, f_2 are monotonically increasing functions on $[a, b]$. The previous point assures us that $f_1', f_2' \in \mathfrak{L}^1_+([a, b])$ and so $f' \in \mathfrak{L}^1([a, b])$.

■

Remarks 6.3.11

(i) The inequality of the previous proposition may be strict. Indeed, we notice that we can modify the function f by giving it in b any value greater than $f(b)$; this change leaves the first member of the inequality unchanged, while the right member can become as large as possible.

(ii) The derivative of a function over an interval is a measurable function over this interval (see Exercise (6) of 2.5) but is not always integrable. An example of a differentiable function for which the derivative, although bounded, is not Riemann integrable is given in ▶ Sect. 7.2 (Pompeiu's function). Here is a simple example of a differentiable function with an unbounded and non-integrable Lebesgue derivative. Let be the function $f : [0, 1] \to \mathbb{R}$, $f(x) = \begin{cases} 0, & x = 0 \\ x^2 \cos \frac{1}{x^2}, & x \in]0, 1] \end{cases}$, $f'(x) = \begin{cases} 0, & x = 0 \\ 2x \cos \frac{1}{x^2} + \frac{2}{x} \sin \frac{1}{x^2}, & x \in]0, 1]. \end{cases}$ Then, $f'(x) = u(x) + v(x)$, where $u(x) = \begin{cases} 0, & x = 0 \\ 2x \cos \frac{1}{x^2}, & x \in]0, 1] \end{cases}$ and $v(x) = \begin{cases} 0, & x = 0 \\ \frac{2}{x} \sin \frac{1}{x^2}, & x \in]0, 1]. \end{cases}$ The function u is continuous on $[0, 1]$; then it is Riemann integrable and so it is Lebesgue integrable on $[0, 1]$. By changing the variable $x = y^{-\frac{1}{2}}$, the generalized Riemann integral of the function v on $]0, 1]$ becomes $2 \int_1^{+\infty} \frac{\sin y}{y} dy$. This integral is simply convergent (see (v) of Remark 3.4.3), and then, according to Theorem 3.4.2, $v \notin \mathfrak{L}^1([0, 1])$. Therefore, $f' \notin \mathfrak{L}^1([0, 1])$.

Definition 6.3.12

Let $f \in \mathfrak{L}^1([a, b])$; for every $x \in [a, b]$, $f \in \mathfrak{L}^1([a, x])$, and so we can define the function $F : [a, b] \to \mathbb{R}$ by $F(x) = \int_{[a,x]} f d\lambda$, for every $x \in [a, b]$. The function F is called an **indefinite integral** of f or the **integral depending on the upper limit**.

Proposition 6.3.13 *Let $f \in \mathfrak{L}^1([a,b])$ and let $F : [a,b] \to \mathbb{R}$, $F(x) = \int_{[a,x]} f d\lambda$ be its indefinite integral. Then, F is absolutely continuous on $[a,b]$.*

Proof

For the function f being integrable on $[a,b]$, we can apply point (2) of Theorem 3.3.1—the absolute continuity property of the integral (not by coincidence so-called!). Then, for every $\varepsilon > 0$, there exists $\delta > 0$ such that, for every $A \in \mathscr{L}([a,b])$ with $\lambda(A) < \delta$, $\int_A |f| d\lambda < \varepsilon$.

Let now $\{[x_k, y_k] : k = 1, \ldots, n\}$ be a finite family of closed nonoverlapping subintervals of $[a,b]$ with $\sum_{k=1}^{n}(y_k - x_k) < \delta$. Then, $A = \cup_{k=1}^{n}[x_k, y_k] \in \mathscr{L}([a,b])$ and $\lambda(A) < \delta$. Using property (4) of Theorem 3.3.1, it follows that

$$\sum_{k=1}^{n} |F(y_k) - F(x_k)| = \sum_{k=1}^{n} \left| \int_{]x_k, y_k]} f d\lambda \right| \le \sum_{k=1}^{n} \int_{]x_k, y_k[} |f| d\lambda = \int_A |f| d\lambda < \varepsilon.$$

■

Remark 6.3.14 According to the previous proposition and to Corollary 6.3.9, it follows that F—the indefinite integral of f—is differentiable a.e. on $[a,b]$. If the function f is positive, then F is monotonically increasing, and so $F' \ge \underline{0}$ a.e.

We will show that the derivative of F is equal to f almost everywhere on $[a,b]$. First we need a lemma.

Lemma 6.3.15 *Let $f \in \mathfrak{L}^1([a,b])$ and let F be its indefinite integral; if $F(x) = 0$, for every $x \in [a,b]$, then $f = \underline{0}$ a.e. on $[a,b]$.*

Proof

For every subinterval $I = |c,d| \subseteq [a,b]$, $\int_I f d\lambda = F(d) - F(c) = 0$. Let $D \in \tau_u$ be an open set. According to Theorem 1.1.3, D admits a unique representation as a union of a countable family of pairwise disjoint open intervals $\{I_n : n \in \mathbb{N}^*\}$. Then,

$$\int_{D \cap [a,b]} f d\lambda = \sum_{n=1}^{\infty} \int_{I_n \cap [a,b]} f d\lambda = 0.$$

Let $A \in \mathscr{L}([a,b])$; according to (2) of Theorem 3.3.1), for every $\varepsilon > 0$, there exists $\delta > 0$ such that, for every $B \in \mathscr{L}([a,b])$ with $\lambda(B) < \delta$, $\int_B |f| d\lambda < \varepsilon$. Using the Lebesgue measurability definition of the subsets of $\mathbb{R}$, there exists $D \in \tau_u$ such that $A \subseteq D$ and $\lambda(D \setminus A) < \delta$; then $\lambda(D \cap [a,b] \setminus A) < \delta$ and so $\int_{D \cap [a,b] \setminus A} |f| d\lambda < \varepsilon$. It follows that

$$\left| \int_A f d\lambda \right| = \left| \int_{D \cap [a,b]} f d\lambda - \int_{D \cap [a,b] \setminus A} f d\lambda \right| \le \int_{D \cap [a,b] \setminus A} |f| d\lambda < \varepsilon.$$

$\varepsilon > 0$ being arbitrary, it follows that $\int_A f d\lambda = 0$, for every $A \in \mathscr{L}([a,b])$. Point (c) of Theorem 3.2.5 assures us that $f = \underline{0}$ a.e. on $[a,b]$. ■

The following theorem answers to the first question at the beginning of this paragraph: if f is Lebesgue integrable over an interval, then its indefinite integral has a derivative equal to f almost everywhere.

Theorem 6.3.16

Let $f \in \mathfrak{L}^1([a,b])$ and let F be its indefinite integral; then $F' = f$ a.e. on $[a,b]$ or

$$\left(\int_{[a,x]} f d\lambda\right)' = f(x), \text{ for almost every } x \in [a,b].$$

Proof

We noted above that F' exists almost everywhere on $[a,b]$.

(1) First, we suppose that f is a bounded function; let $k > 0$ such that $|f(x)| \leq k$, for every $x \in [a,b]$.

Let us define the sequence $(F_n)_n$, letting $F_n(x) = n \cdot \left[F\left(x + \frac{1}{n}\right) - F(x)\right] = n \cdot \int_{]x,x+\frac{1}{n}]} f d\lambda$, for any $n \in \mathbb{N}^*$ and for every $x \in [a,b]$. The function F is continuous on $[a,b]$; therefore, F_n are continuous functions and then they are measurable on $[a,b]$.

The sequence $(F_n)_n$ converges almost everywhere on $[a,b]$ to F'. Moreover, for any $n \in \mathbb{N}^*$ and every $x \in [a,b]$, $|F_n(x)| \leq n \cdot k \cdot \frac{1}{n} = k$. The bounded convergence theorem can be applied (see Corollary 3.3.10), and, repeating the calculation performed in the demonstration of Proposition 6.3.10, it turns out that for all $c \in [a,b]$,

$$\int_{[a,c]} F' d\lambda = \lim_n \int_{[a,c]} F_n d\lambda = \lim_n n \cdot \int_{[a,c]} \left[F\left(x + \frac{1}{n}\right) - F(x)\right] d\lambda =$$

$$= \lim_n \left[n \cdot \int_c^{c+\frac{1}{n}} F(x)dx - n \cdot \int_a^{a+\frac{1}{n}} F(x)dx\right] =$$

$$= \lim_n [F(c_n) - F(a_n)] = F(c) - F(a) = \int_{[a,c]} f d\lambda$$

(since F is continuous, we have applied a mean-value theorem for the Riemann integral; here $c_n \in]c, c + \frac{1}{n}[$ and $a_n \in]a, a + \frac{1}{n}[$). It follows that $\int_{[a,c]}(F' - f)d\lambda = 0$, for every $c \in [a,b]$. The above lemma assures us that $F' = f$, a.e. on $[a,b]$.

(2) Let us now suppose that $f \in \mathfrak{L}^1_+([a,b])$; then its indefinite integral F is an increasing function on $[a,b]$. According to Proposition 6.3.10, $F' \in \mathfrak{L}^1_+([a,b])$ and

$$\int_{[a,b]} F' d\lambda \leq F(b) - F(a). \tag{6.1}$$

For any $n \in \mathbb{N}$, $f_n = \min\{f, n\} : [a,b] \to [0,n]$, defined by $f_n(x) = \begin{cases} f(x), & \text{if } f(x) \leq n, \\ n, & \text{if } f(x) > n. \end{cases}$ Then, $f_n \in \mathfrak{L}^1_+([a,b])$ and f_n is bounded. Let F_n be the indefinite integral of f_n; according to the first step of the demonstration, $F_n' = f_n$ a.e. on $[a,b]$, for any $n \in \mathbb{N}$.

We remark that $F - F_n$ is the indefinite integral of positive function $f - f_n$ ($(F - F_n)(x) = \int_{[a,x]}(f - f_n)d\lambda$). It follows that $F - F_n$ is an increasing function and then $(F - F_n)' \geq \underline{0}$ a.e. But $(F - F_n)' = F' - F_n' = F' - f_n \geq \underline{0}$ a.e. Then, for any $n \in \mathbb{N}$, $F' \geq f_n$ a.e. Since $f_n \uparrow f$, it follows that $F' \geq f$ a.e. on $[a,b]$. F' and f are integrable on $[a,b]$; it follows that

$$\int_{[a,b]} F' d\lambda \geq \int_{[a,b]} f d\lambda = F(b) - F(a). \tag{6.2}$$

From (6.1) and (6.2), we deduce that

$$\int_{[a,b]} F' d\lambda = F(b) - F(a) = \int_{[a,b]} f d\lambda.$$

Therefore, $\int_{[a,b]}(F' - f)d\lambda = 0$, and, since $F' - f \geq \underline{0}$ a.e., it follows that $F' = f$ a.e. (see Proposition 3.1.13).

(3) Let $f \in \mathfrak{L}^1([a,b])$ be an arbitrary function and let $f^+, f^- \in \mathfrak{L}^1_+([a,b])$ be the positive, respectively the negative part of f (see Definition 2.1.21). We denote by F_+ and F_- the indefinite integrals of f^+ and respectively of f^-. From the second step of the demonstration, $F_+' = f^+$ and $F_-' = f^-$, a.e. on $[a,b]$. Then, the derivative of $F = F_+ - F_-$ is $F' = F_+' - F_-' = f^+ - f^- = f$ a.e. on $[a,b]$.

■

We will now try to answer the second question posed at the beginning of this section. It is possible that a function can be differentiable almost everywhere over an interval, and its derivative can be Lebesgue integrable, yet the formula $(L - N)$ is not verified. One such example is offered by the Cantor staircase.

The Cantor staircase 6.3.17 ("Devil's Staircase") We recall some of the notations of Cantor Ternary Set 1.3.16, where we built the triadic set of Cantor. So $I = [0,1]$, $J_1 = \left]\frac{1}{3}, \frac{2}{3}\right[$, $J_2 = \left]\frac{1}{3^2}, \frac{2}{3^2}\right[\cup \left]\frac{7}{3^2}, \frac{8}{3^2}\right[$, $J_3 = \left]\frac{1}{3^3}, \frac{2}{3^3}\right[\cup \left]\frac{7}{3^3}, \frac{8}{3^3}\right[\cup \left]\frac{19}{3^3}, \frac{20}{3^3}\right[\cup \left]\frac{25}{3^3}, \frac{26}{3^3}\right[$, and so on. J_n will be a union of 2^{n-1} pairwise disjoint intervals of length $\frac{l}{3^n}$ each. We have denote $J = \bigcup_{n=1}^{\infty} J_n$ and $D = I \setminus J$ and we have shown that $\lambda(J) = 1$. Using Exercise (16) of 1.5, it follows that J is dense in I (for every $x, y \in I$ with $x < y$, there exists $z \in J$ such that $x < z < y$).

We will build a function $F : [0, 1] \to [0, 1]$ as follows. First we will give the values of F on the set J:

- $F(x) = \frac{1}{2}$, if $x \in \left]\frac{1}{3}, \frac{2}{3}\right[$,
- $F(x) = \begin{cases} \frac{1}{2^2}, & \text{if } x \in \left]\frac{1}{3^2}, \frac{2}{3^2}\right[, \\ \frac{3}{2^2}, & \text{if } x \in \left]\frac{7}{3^2}, \frac{8}{3^2}\right[, \end{cases}$
- $F(x) = \begin{cases} \frac{1}{2^3}, & \text{if } x \in \left]\frac{1}{3^3}, \frac{2}{3^3}\right[, \\ \frac{3}{2^3}, & \text{if } x \in \left]\frac{7}{3^3}, \frac{8}{3^3}\right[, \\ \frac{5}{2^3}, & \text{if } x \in \left]\frac{19}{3^3}, \frac{20}{3^3}\right[, \\ \frac{7}{2^3}, & \text{if } x \in \left]\frac{25}{3^3}, \frac{26}{3^3}\right[, \end{cases}$

and so on.

Generally, $F(x) = \frac{2k-1}{2^n}$, if $x \in J_{k,n}$ where, for any $n \in \mathbb{N}^*$, $J_{k,n}$ is the k-th interval between the 2^{n-1} intervals composing J_n.

The graph of the function F on the set J roughly resembles the following Fig. 6.1:

Note that F is increasing on $J = \bigcup_{n=1}^{\infty} J_n$.

Since J is a dense subset of I, we will define F on $D = I \setminus J$ by: $F(0) = 0$ and

$$F(x) = \sup\{F(t) : t \in J, t < x\}, \text{ for every } x \in]0, 1] \setminus J.$$

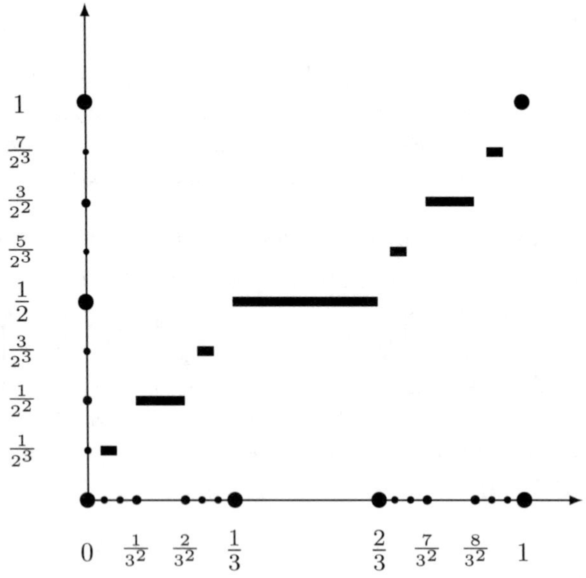

Fig. 6.1 Graph of Cantor staircase

Let us show that F is an increasing function on I. We suppose by absurd that there are two points $x, y \in I$ with $x < y$ and $F(x) > F(y)$. As we mentioned, F is monotonically increasing on J; so there cannot be $x, y \in J$. It remains to study the following cases: (1) $x, y \in]0, 1] \setminus J$, (2) $x \in J$, $y \in]0, 1] \setminus J$, and (3) $x \in]0, 1] \setminus J$ and $y \in J$.

(1) Since $x \in]0, 1] \setminus J$ and $F(y) < F(x)$, there is $t_1 \in J$ such that $t_1 < x$ and $F(y) < F(t_1)$; let $t_2 \in]x, y[\cap J$. Then, $F(t_2) \leq F(y) < F(t_1)$ which is absurd (F is increasing on J).
(2) Since $x \in J$ and $x < y$, $F(x) \leq F(y)$, which is absurd.
(3) Since $x \in]0, 1] \setminus J$ and $F(y) < F(x)$, there is $t \in J, t < x < y$ such that $F(y) < F(t)$ which contradicts $y \in J$.

The hypothesis adopted leads to contradictions in all cases; it turns out that F is increasing over I.

Since the set $\{F(t) : t \in J\} = \{\frac{2k-1}{2^n} : n \in \mathbb{N}^*, k = 1, 2, \cdots, 2^{n-1}\}$ is dense in I, the increasing function F can't have jumps; therefore it is continuous on I.

If $E = \left\{\frac{2}{3}, \frac{2}{3^2}, \frac{8}{3^2}, \frac{2}{3^3}, \frac{8}{3^3}, \frac{20}{3^3}, \frac{26}{3^3}, \cdots\right\}$, then the Cantor set is $C = D \cup E$.

We can immediately notice that the set E is formed by the right ends of the intervals $J_{k,n}$; then, since F is constant over each open interval $J_{k,n} \setminus E$, it follows that $F'(x) = 0$ on $J_{k,n} \setminus E$. The open set $J \setminus E = \bigcup_{n=1}^{\infty} \bigcup_{k=1}^{2^{n-1}} J_{k,n} \setminus E$ has the Lebesgue measure equal to 1 ($\lambda(E) = 0$) and $F'(x) = 0$, for every $x \in J \setminus E$. Therefore, $F' = \underline{0}$, almost everywhere and then $F' \in \mathfrak{L}^1([0, 1])$. However, F does not satisfy the formula $(L - N)$!

Moreover, we can see that this function does not have the Lusin property (see (i) of Remark 6.3.4). Indeed, the Lebesgue measure of the Cantor set (see 1.3.16) is zero, but its image by F has the measure of 1.

From the above, we can see that the Cantor function has a derivative equal to 0 almost everywhere on $[0, 1]$, but it is not constant on $[0, 1]$. We will prove that the absolutely continuous functions whose derivative is 0 almost everywhere are constant. To achieve this result, we need the concept of Vitali cover.

Definition 6.3.18

Let $E \subseteq \mathbb{R}$ and let $\mathcal{V}$ be a family of nondegenerate closed intervals (which are not reduced to a point). $\mathcal{V}$ is said to be a **Vitali cover** for E if, for every $\varepsilon > 0$ and every $x \in E$, there exists $I \in \mathcal{V}$ such that $x \in I$ and $|I| < \varepsilon$.

Lemma 6.3.19 (Vitali) *Let $E \subseteq [a, b]$ be a bounded measurable set and let $\mathcal{V}$ be a Vitali cover for E. Then, for every $\varepsilon > 0$, there exist $I_1, \cdots, I_n \in \mathcal{V}$ pairwise disjoint such that $\lambda(E \setminus \bigcup_{k=1}^{n} I_k) < \varepsilon$.*

Proof

From the definition, for every $x \in E$, there exists $I \in \mathcal{V}$ such that $x \in I$ and $|I| < 1$; then $I \subseteq [a-1, b+1]$. So we can assume, possibly abandoning some of the sets of $\mathcal{V}$, that the set $\bigcup\{I : I \in \mathcal{V}\}$ is bounded.

Let $I_0 \in \mathcal{V}$; if $E \subseteq I_0$, then the demonstration is over.

Let us suppose that there is $x \in E \setminus I_0$; then there exists $\varepsilon_0 > 0$ such that $]x - \varepsilon_0, x + \varepsilon_0[\cap I_0 = \emptyset$, and so there is $I \in \mathcal{V}$ such that $x \in I$ and $|I| < \varepsilon_0$. It follows that $I \subseteq]x - \varepsilon_0, x + \varepsilon_0[$ and then $I \cap I_0 = \emptyset$.

Let $k_1 = \sup\{|I| : I \in \mathcal{V}, I \cap I_0 = \emptyset\}$; since $\bigcup\{I : I \in \mathcal{V}\}$ is bounded, $k_1 < +\infty$. Let then

$$I_1 \in \mathcal{V}, I_0 \cap I_1 = \emptyset \text{ and } |I_1| > \frac{k_1}{2}. \tag{1}$$

If $E \subseteq I_0 \cup I_1$, then the demonstration is over.

Let us suppose that there is $x \in E \setminus (I_0 \cup I_1)$; then there exists $\varepsilon_1 > 0$ such that $]x - \varepsilon_1, x + \varepsilon_1[\cap(I_0 \cup I_1) = \emptyset$ and so there is $I \in \mathcal{V}$ such that $x \in I$ and $|I| < \varepsilon_1$. It follows that $I \subseteq]x - \varepsilon_1, x + \varepsilon_1[$ and then $I \cap (I_0 \cup I_1) = \emptyset$.

Let $k_2 = \sup\{|I| : I \in \mathcal{V}, I \cap (I_0 \cup I_1) = \emptyset\} < +\infty$. Let then

$$I_2 \in \mathcal{V}, I_2 \cap (I_0 \cup I_2) = \emptyset \text{ and } |I_2| > \frac{k_2}{2}. \tag{2}$$

Suppose that we built by induction $I_0, I_1, \cdots, I_n, \cdots \in \mathcal{V}$ pairwise disjoint such that, for any $n \in \mathbb{N}^*$,

$$I_n \cap \left(\bigcup_{j=0}^{n-1} I_j\right) = \emptyset \text{ and } |I_n| > \frac{k_n}{2}, \tag{n}$$

where $k_n = \sup\{|I| : I \in \mathcal{V}, I \cap (I_0 \cup \cdots \cup I_{n-1}) = \emptyset\}$.

If there is $n \in \mathbb{N}^*$ such that $E \subseteq \bigcup_{j=0}^{n} I_j$, then the demonstration is over.

Otherwise, for any $n \in \mathbb{N}$, there exists $x \in E \setminus \bigcup_{j=0}^{n} I_j$.

Since $\bigcup_{n=1}^{\infty} I_n \subseteq \bigcup\{I : I \in \mathcal{V}\}$ is bounded, it follows that $\sum_{n=1}^{\infty} |I_n| < +\infty$.

Let $\varepsilon > 0$ be arbitrary and let $N \in \mathbb{N}$ such that

$$\sum_{n=N+1}^{\infty} |I_n| < \frac{\varepsilon}{5}. \tag{a}$$

We will demonstrate that $\lambda\left(E \setminus \bigcup_{n=0}^{N} I_n\right) < \varepsilon$.

For every $x \in E \setminus \bigcup_{n=0}^{N} I_n$, there exists $I \in \mathcal{V}$ such that $x \in I$ and

$I \cap \left(\bigcup_{n=0}^{N} I_n\right) = \emptyset$. If $I \cap \left(\bigcup_{n=N+1}^{\infty} I_n\right) = \emptyset$, then $I \cap \left(\bigcup_{n=0}^{\infty} I_n\right) = \emptyset$ and so

$$|I| \leq k_n, \text{ for any } n \in \mathbb{N}^*. \tag{b}$$

But $\sum_{n=1}^{\infty} k_n < 2 \cdot \sum_{n=2}^{\infty} |I_n| < +\infty$ and then

$$k_n \to 0. \tag{c}$$

From (b) and (c), it follows that $|I| = 0$ and therefore $I = \{x\}$ is a degenerate interval, which is absurd.

Therefore, $I \cap \left(\bigcup_{n=N+1}^{\infty} I_n\right) \neq \emptyset$. Let

$$n_x = \min\{n > N+1 : I \cap I_n \neq \emptyset\} \geq N+1.$$

Then, $I \cap \left(\bigcup_{n=0}^{n_x - 1} I_n\right) = \emptyset$ and from (n)

$$|I| \leq k_n < 2 \cdot |I_n|, \text{ for any } n = 1, \cdots, n_x. \tag{d}$$

Let a_{n_x} be the middle of the interval I_{n_x}; then, considering (d) and since $x \in I$ and $I \cap I_{n_x} \neq \emptyset$,

$$|x - a_{n_x}| \leq \frac{1}{2} \cdot |I_{n_x}| + |I| < \frac{1}{2} \cdot |I_{n_x}| + 2 \cdot |I_{n_x}| = \frac{5}{2} \cdot |I_{n_x}|. \tag{e}$$

For any $n \in \mathbb{N}$, let J_n be the interval with the same center as I_n and with length $|J_n| = 5 \cdot |I_n|$. It follows from (e) that $x \in J_{n_x}$; therefore, for every $x \in E \setminus \bigcup_{n=0}^{N} I_n$, there is $n_x \geq N+1$ such that $x \in J_{n_x}$. Then, $E \setminus \bigcup_{n=0}^{N} I_n \subseteq \bigcup_{n=N+1}^{\infty} J_n$ and so, according to (a),

$$\lambda\left(E \setminus \bigcup_{n=0}^{N} I_n\right) \leq \sum_{n=N+1}^{\infty} |J_n| = 5 \cdot \sum_{n=N+1}^{\infty} |I_n| < \varepsilon.$$

■

We can now prove the result announced above.

Theorem 6.3.20
Let $F \in AC_{[a,b]}$ such that $F' = \underline{0}$ almost everywhere on $[a, b]$. Then, F is a constant function on $[a, b]$.

Proof
We will show that, for any $c \in]a, b]$, $F(c) = F(a)$.

Let $E = \{x \in]a, c[: F'(x) = 0\}$. Since $F' = \underline{0}$ is almost everywhere on $[a, b]$, the set $N = [a, c] \setminus E$ is a null set. Then,

$$E =]a, c[\setminus N \in \mathscr{L} \text{ and } \lambda(E) = c - a. \tag{6.1}$$

We will highlight a Vitali cover for E after which we will apply the previous lemma.

Since F is absolutely continuous, for every $\varepsilon > 0$, there exists $\delta > 0$ such that, for every finite family of closed nonoverlapping intervals $\{[a_k, b_k] \subseteq [a, b] : k = 1, \cdots, n\}$ such that $\sum_{k=1}^{n}(b_k - a_k) < \delta$,

$$\sum_{k=1}^{n} |F(b_k) - F(a_k)| < \frac{\varepsilon}{2}. \tag{6.2}$$

Since F is differentiable on E, for every $x \in E$, there exists $h_x > 0$ such that

$$\frac{|F(x+h) - F(x)|}{h} < \frac{\varepsilon}{2(c-a)} \equiv \alpha, \text{ for any } h \in]0, h_x[. \tag{6.3}$$

Then $\mathcal{V} = \{[x, x+h] : x \in E, h \in]0, h_x[\}$ is a Vitali cover for E. Since E is bounded, according to Vitali lemma, there exists a finite family of pairwise disjoint intervals $\{I_1, \cdots, I_n\} \subseteq \mathcal{V}$ such that

$$\lambda\left(E \setminus \bigcup_{k=1}^{n} I_k\right) < \delta. \tag{6.4}$$

Suppose that the intervals $\{I_k = [x_k, x_k + h_k] : k = 1, \cdots, n\}$ are numbered so that $a < x_1 < x_1 + h_1 < x_2 < x_2 + h_2 < \cdots < x_n < x_n + h_n < c$. According to (6.1), (6.4), $\lambda\left(]a, c[\setminus \bigcup_{k=1}^{n} I_k\right) = \lambda(]a, x_1[\cup]x_1 + h_1, x_2[\cup \cdots \cup]x_n + h_n, c[) < \delta$.

From (6.2) it follows that

$$|F(x_1) - F(a)| + |F(x_2) - F(x_1 + h_1)| + \cdots + |F(c) - F(x_n + h_n)| < \frac{\varepsilon}{2}. \tag{6.5}$$

On the other hand, from (6.3)

$$\sum_{k=1}^{n} |F(x_k + h_k) - F(x_k)| < \alpha \cdot \sum_{k=1}^{n} h_k < \alpha \cdot (c - a). \tag{6.6}$$

Now from relations (6.5) and (6.6), we get

$$|F(c) - F(a)| \leq |F(x_1) - F(a)| + |F(x_2) - F(x_1 + h_1)| + \cdots + |F(c) - F(x_n + h_n)|$$

$$+ \sum_{k=1}^{n} |F(x_k + h_k) - F(x_k)| < \frac{\varepsilon}{2} + \alpha \cdot (c - a) = \varepsilon.$$

Since ε is arbitrarily positive, it follows that $F(c) = F(a)$. ■

Remark 6.3.21 The absolute continuity condition of the previous theorem cannot be removed. Cantor's staircase is an example of a monotonic continuous function (moreover, uniformly continuous) with the zero derivative almost everywhere which is not constant. Clearly, Cantor's function is not absolutely continuous.

We have noticed that if F is the indefinite integral of the function $f \in \mathfrak{L}^1([a, b])$, then, $F \in AC_{[a,b]}$ and $\int_{[a,b]} F' d\lambda = F(b) - F(a)$. The following result answers to the second question posed at the beginning of this paragraph: any absolutely continuous function is the indefinite integral of its derivative.

Theorem 6.3.22
Let $f : [a, b] \to \mathbb{R}$ be an arbitrary real function; the following conditions are equivalent:
(1) f is absolutely continuous on $[a, b]$.
(2) f is differentiable almost everywhere on $[a, b]$, $f' \in \mathfrak{L}^1([a, b])$ and

$$f(x) = f(a) + \int_{[a,x]} f' d\lambda, \text{ for every } x \in [a, b].$$

(3) There exists a function $g : [a, b] \to \mathbb{R}$, $g \in \mathfrak{L}^1([a, b])$ such that

$$f(x) = f(a) + \int_{[a,x]} g d\lambda, \text{ for every } x \in [a, b].$$

Proof
(1) $\Longrightarrow$ (2): Let $f \in AC_{[a,b]}$; according to Proposition 6.3.10, $f' \in \mathfrak{L}^1([a, b])$. Let g be its indefinite integral; we remember that $g : [a, b] \to \mathbb{R}$ is defined by $g(x) = \int_{[a,x]} f' d\lambda$, for every $x \in [a, b]$. According to Proposition 6.3.13, g is absolutely continuous on $[a, b]$, and, from Theorem 6.3.16, $g' = f'$ is almost everywhere on $[a, b]$. Then, $g - f$ is absolutely

continuous and $(g - f)' = 0$, a.e. on $[a, b]$. From Theorem 6.3.20, it follows that $g - f$ is constant on $[a, b]$. Therefore, for every $x \in [a, b]$, $g(x) - f(x) = g(a) - f(a)$, from where it follows that

$$f(x) = f(a) + \int_{[a,x]} f' d\lambda, \quad \text{for every } x \in [a, b].$$

(2) $\Longrightarrow$ (3) is obvious.

(3) $\Longrightarrow$ (1): According to the property of absolute continuity of the Lebesgue integral (see (2) of Theorem 3.3.1), for every $\varepsilon > 0$, there exists $\delta > 0$ such that

$$\int_B |g| d\lambda < \varepsilon, \quad \text{for every } B \in \mathscr{L}([a, b]) \text{ with } \lambda(B) < \delta. \tag{$*$}$$

Let $\{[x_k, y_k] : k = 1, \cdots, n\}$ be a finite family of nonoverlapping closed subintervals of $[a, b]$ with $\sum_{k=1}^{n}(y_k - x_k) < \delta$, and let $B = \bigcup_{k=1}^{n}[x_k, y_k] \in \mathscr{L}([a, b])$. Then, $\lambda(B) < \delta$ and so, from $(*)$,

$$\sum_{k=1}^{n} |f(y_k) - f(x_k)| = \sum_{k=1}^{n} \left| \int_{]x_k, y_k]} g d\lambda \right| \leq \sum_{k=1}^{n} \int_{[x_k, y_k]} |g| d\lambda = \int_B |g| d\lambda < \varepsilon.$$

■

As an immediate consequence of Theorems 6.3.6 and 6.3.22, we have

Corollary 6.3.23 *Let $f : [a, b] \to \mathbb{R}$ be a function differentiable at every point of $[a, b]$. If $f' \in \mathfrak{L}^1([a, b])$, then, for every $x \in [a, b]$,*

$$f(x) = f(a) + \int_{[a,x]} f' d\lambda.$$

Another consequence of Theorem 6.3.22 is the formula for integration by parts.

Theorem 6.3.24
Let $f, g \in AC_{[a,b]}$; then $fg', f'g \in \mathfrak{L}^1([a, b])$ and

$$\int_{[a,b]} (f \cdot g') d\lambda = f(b) \cdot g(b) - f(a) \cdot g(a) - \int_{[a,b]} (f' \cdot g) d\lambda.$$

Proof
According to Theorem 6.3.22, f and g are differentiable almost everywhere on $[a, b]$ and $f', g' \in \mathfrak{L}^1([a, b])$. Since f, g are continuous on $[a, b]$, they are bounded; let $M > 0$ such that, for every $x \in [a, b]$, $|f(x)| \leq M$ and $|g(x)| \leq M$. Then, $|fg'| \leq M|g'|$ and $|f'g| \leq M|f'|$. By Theorem 3.2.6, $fg', f'g \in \mathfrak{L}^1([a, b])$.

Now $f \cdot g \in AC_{[a,b]}$; indeed, since $f, g \in AC_{[a,b]}$, for every $\varepsilon > 0$, there exists $\delta > 0$ such that for every family of closed nonoverlapping subintervals of $[a, b]$, $\{[x_k, y_k] : k = 1, \cdots, n\}$ with $\sum_{k=1}^{n}(y_k - x_k) < \delta$,

$$\sum_{k=1}^{n} |f(y_k) - f(x_k)| < \tfrac{\varepsilon}{2M},$$
$$\sum_{k=1}^{n} |g(y_k) - g(x_k)| < \tfrac{\varepsilon}{2M}.$$

Then

$$\sum_{k=1}^{n} |f(y_k)g(y_k) - f(x_k)g(x_k)| \leq \sum_{k=1}^{n} |f(y_k) - f(x_k)| \cdot |g(y_k)| +$$

$$+ \sum_{k=1}^{n} |f(x_k)| \cdot |g(y_k) - g(x_k)| \leq M \cdot \frac{\varepsilon}{2M} + M \cdot \frac{\varepsilon}{2M} = \varepsilon.$$

According to (2) of Theorem 6.3.22, for every $x \in [a, b]$,

$$f(x) \cdot g(x) = f(a) \cdot g(a) + \int_{[a,x]} (f \cdot g)' d\lambda.$$

Then

$$f(b) \cdot g(b) - f(a) \cdot g(a) = \int_{[a,b]} (f' \cdot g) d\lambda + \int_{[a,b]} (f \cdot g') d\lambda.$$

■

6.4 Exercises

(1) Let $\nu : \mathscr{L}(\mathbb{R}) \to \mathbb{R}$ be defined by $\nu(B) = \int_B x \cdot e^{-x^2} d\lambda(x)$, for every $B \in \mathscr{L}(\mathbb{R})$. Determine the ν-positive and ν-negative sets. Find a Hahn decomposition of the set $\mathbb{R}$ with respect to ν.

(2) Let $(a_n)_n \subseteq \mathbb{R}$ be an absolute summable sequence ($\sum_{n=0}^{\infty} |a_n| < +\infty$), and let $\nu : \mathcal{P}(\mathbb{N}) \to \mathbb{R}$, defined by $\nu(A) = \sum_{n \in A} a_n$, for every $A \subseteq \mathbb{N}$, $A \neq \emptyset$ and $\nu(\emptyset) = 0$. Show that ν is well defined and that it is a finite signed measure on $\mathbb{N}$ (the measure generated by the sequence $(a_n)_n$). Determine a Hahn decomposition of $\mathbb{N}$ with respect to ν. Find Jordan's decomposition of ν. What is the total variation $|\nu|$ of ν?

Indication: Let γ be the counting measure on $\mathbb{N}$ (see (ii) of Example 1.4.6). It will be noted that $\nu(A) = \int_A f d\gamma$ where $f : \mathbb{N} \to \mathbb{R}$, $f(n) = a_n$ (see also Remark 4.5.5). Then $\mathbb{N}^- = (f < 0)$, $\mathbb{N}^+ = (f \geq 0)$ (see Definition 6.1.8). For the Jordan decomposition, see Remark 6.1.12.

(3) Let γ be the counting measure on $\mathbb{N}$ (see (ii) of Example 1.4.6). Prove that any finite signed measure $\nu : \mathcal{P}(\mathbb{N}) \to \mathbb{R}$ is absolutely continuous with respect to γ. Which is the Radon-Nikodym derivative $f = \frac{d\nu}{d\gamma}$?

Indication: It will be shown that $f(n) = \nu(\{n\})$, for every $n \in \mathbb{N}$.

(4) Let $\gamma : \mathcal{A} \to \mathbb{R}$ be a positive σ-finite measure on X, and let $\nu_1, \nu_2 : \mathcal{A} \to \mathbb{R}$ be the two finite signed measures absolutely continuous with respect to γ; show that $\nu_1 + \nu_2 \ll \gamma$ and $|\nu_1| \ll \gamma$.

Show that $\frac{d(\nu_1+\nu_2)}{d\gamma} = \frac{d\nu_1}{d\gamma} + \frac{d\nu_2}{d\gamma}$ and $\frac{d|\nu_1|}{d\gamma} = \left|\frac{d\nu_1}{d\gamma}\right|$.

(5) Let $\gamma, \nu : \mathcal{A} \to \mathbb{R}_+$ be the two finite positive measures on X such that $\nu \ll \gamma$ and $\gamma \ll \nu$. Show that $\frac{d\nu}{d\gamma} \cdot \frac{d\gamma}{d\nu} = \underline{1}$ a.e. on X.

(6) Let $f : [a, b] \to \mathbb{R}$ be a differentiable function with the derivative bounded on $[a, b]$.

(a) Show that $\lim_{\varepsilon \to 0} \frac{1}{\varepsilon} \int_{[c,c+\varepsilon]} f d\lambda = f(c)$, for every $c \in [a, b]$.

(b) Show that $f' \in \mathfrak{L}^1([a, b])$ and $\int_{[a,b]} f' d\lambda = f(b) - f(a)$ (see Exercise (6) of 2.5).

Appendices

7.1 Riemann Integral

The concepts and the results concerning the Riemann integral necessary to read this book are presented in this appendix. For some demonstrations we used section 8.6 of [9].

A finite set $\Delta = \{x_0, x_1, \cdots, x_n\} \subseteq [a, b]$ is a **partition** of the interval $[a, b]$ if $a = x_0 < x_1 < \cdots < x_n = b$; $\|\Delta\| = \max\{x_k - x_{k-1} : k = 1, \cdots, n\}$ is the **mesh** of partition Δ. We denote with $\mathcal{D}([a, b])$ the set of all partitions of $[a, b]$.

Another partition Δ' of $[a, b]$ is said to be a **refinement** of Δ if $\Delta \subseteq \Delta'$.

Let $f : [a, b] \to \mathbb{R}$ be a bounded function and let $\Delta = \{x_0, x_1, \cdots, x_n\}$ be a partition of $[a, b]$; for any $k = 1, \cdots, n$, let

$$M_k = \sup\{f(x) : x \in [x_{k-1}, x_k]\} \text{ and } m_k = \inf\{f(x) : x \in [x_{k-1}, x_k]\}.$$

The sums $S_\Delta = \sum_{k=1}^{n} M_k(x_k - x_{k-1})$ and $s_\Delta = \sum_{k=1}^{n} m_k(x_k - x_{k-1})$ are called the **upper**, respectively **lower, Darboux sums** of f corresponding to the partition Δ. If Δ' is a refinement of Δ, alors $s_\Delta \leq s_{\Delta'} \leq S_{\Delta'} \leq S_\Delta$.

The **upper Darboux integral** and , respectively, the **lower Darboux integral** of f are

$$\bar{I}_f = \inf\{S_\Delta : \Delta \in \mathcal{D}([a, b])\}, \underline{I}_f = \sup\{s_\Delta : \Delta \in \mathcal{D}([a, b])\}.$$

Definition 7.1.1

The bounded function $f : [a, b] \to \mathbb{R}$ is **Riemann integrable** on $[a, b]$ if, for every $\varepsilon > 0$, there exists $\Delta \in \mathcal{D}([a, b])$ such that $S_\Delta - s_\Delta < \varepsilon$.

We denote with $\mathcal{R}_{[a,b]}$ the set of all Riemann integrable functions on $[a, b]$.

The following theorem presents, without proof, several conditions equivalent to Riemann's integrability.

L. C. Florescu, *Lebesgue Integral*, Compact Textbooks in Mathematics,
https://doi.org/10.1007/978-3-030-60163-8_7

Theorem 7.1.2

Let $f : [a, b] \to \mathbb{R}$ be a bounded function; the following conditions are equivalent:

1. *f is Riemann integrable on $[a, b]$.*
2. *$\bar{I}_f = \underline{I}_f$ (the common value $I = \int_a^b f(x)dx$ is the* **Riemann integral** *of f on $[a, b]$).*
3. *For every $\varepsilon > 0$, there exists $\Delta_\varepsilon \in \mathcal{D}([a, b])$ such that, for every $\Delta \supseteq \Delta_\varepsilon$, $S_\Delta - s_\Delta < \varepsilon$.*
4. *For every $\varepsilon > 0$, there exists $\delta > 0$ such that, for every $\Delta \in \mathcal{D}([a, b])$ with $\|\Delta\| < \delta$, $S_\Delta - s_\Delta < \varepsilon$.*
5. *There is $I \in \mathbb{R}$ such that, for every $\varepsilon > 0$, there exists $\Delta_\varepsilon \in \mathcal{D}([a, b])$ such that, for every $\Delta = \{x_0, \cdots, x_n\} \supseteq \Delta_\varepsilon$ and any $\{c_1, \cdots, c_n\}$ with $x_0 \le c_1 \le x_1 \le c_2 \le x_2 \le \cdots \le x_{n-1} \le c_n \le x_n$,*

$$\left| I - \sum_{k=1}^{n} f(c_k)(x_k - x_{k-1}) \right| < \varepsilon$$

   (the real number I is unique and $I = \int_a^b f(x)dx$).

6. *There is $I \in \mathbb{R}$ such that, for every $\varepsilon > 0$, there exists $\delta > 0$ such that, for every $\Delta = \{x_0, \cdots, x_n\} \in \mathcal{D}([a, b])$ with $\|\Delta\| < \delta$ and any $\{c_1, \cdots, c_n\}$ with $x_0 \le c_1 \le x_1 \le c_2 \le x_2 \le \cdots \le x_{n-1} \le c_n \le x_n$,*

$$\left| I - \sum_{k=1}^{n} f(c_k)(x_k - x_{k-1}) \right| < \varepsilon.$$

In his famous doctoral dissertation ("Intégrale, longueur, aire", 1902),. H. Lebesgue presents a Riemann integrability criterion which shows to what point the integrable Riemann functions can be discontinuous. To demonstrate this criterion, we must introduce the notion of oscillation of a function and present some preliminary results.

Definition 7.1.3

Let I be an interval, $J \subseteq I$ be a subinterval, and $f : I \to \mathbb{R}$ be a bounded function. The **oscillation of f on J** is

$$\omega_f(J) = \sup\{|f(x) - f(y)| : x, y \in J\} = \sup_{x \in J} f(x) - \inf_{y \in J} f(y).$$

Let $x \in I$; the **oscillation of f at x** is

$$\omega_f(x) = \inf_{\delta > 0} \omega_f(]x - \delta, x + \delta[\cap I) = \inf_{\delta > 0} \left[\sup_{u,v \in]x-\delta, x+\delta[\cap I} |f(u) - f(v)| \right].$$

The demonstration of the following proposition is an elementary exercise.

Proposition 7.1.4 *The function f is continuous at $x \in I$ if and only if $\omega_f(x) = 0$.*

Lemma 7.1.5 *Let I be a closed interval and let $f : I \to \mathbb{R}$ be a bounded function. For every $\alpha \geq 0$, $D_\alpha = \{x \in I : \omega_f(x) \geq \alpha\}$ is a closed set.*

Proof

Let us suppose that there is a sequence $(x_n)_n \subseteq D_\alpha$, $x_n \to x$ such that $x \notin D_\alpha$. Since I is a closed interval, it is a closed set (1.1.7) and then $x \in I$. Therefore $\omega_f(x) < \alpha$, and then there is β and $\delta > 0$ such that

$$\sup\{|f(u) - f(v)| : u, v \in]x - \delta, x + \delta[\cap I\} < \beta < \alpha.$$

Let $p \in \mathbb{N}$ such that $|x_p - x| < \frac{\delta}{2}$. Then, for every $u, v \in]x_p - \frac{\delta}{2}, x_p + \frac{\delta}{2}[\cap I$, $|u - x| < \delta$, and $|v - x| < \delta$ and so $|f(u) - f(v)| < \beta$ from where we deduce that $\omega_f(x_p) = \inf_{\delta>0}\left[\sup_{u,v\in]x_p-\frac{\delta}{2},x_p+\frac{\delta}{2}[\cap I} |f(u) - f(v)|\right] \leq \beta < \alpha$ which contradicts $x_p \in D_\alpha$. ■

The following topological lemma is formulated for the compacts of $\mathbb{R}$; however, the demonstration can be reproduced with small modifications for the subsets of $\mathbb{R}^2$, $\mathbb{R}^3$, etc.

Lemma 7.1.6 *A set $K \subseteq \mathbb{R}$ is compact if and only if for every open cover of K, $\{D_\gamma : \gamma \in \Gamma\} \subseteq \tau_u$, $(K \subseteq \bigcup_{\gamma\in\Gamma} D_\gamma)$, there is $\gamma_1, \cdots, \gamma_n \in \Gamma$ such that $K \subseteq \bigcup_{i=1}^n D_{\gamma_i}$ (any open cover of K, has a finite subcover).*

Proof

Let's remember that a set $K \subseteq \mathbb{R}$ is compact if it is closed and bounded or, equivalently, if every sequence of K has a subsequence convergent to a point of K.

Let K be a compact set and let $\mathcal{A} = \{D_\gamma : \gamma \in \Gamma\}$ be an open cover of K. The set of rational numbers, $\mathbb{Q}$, is countable, and then in the family $\mathcal{D} = \{]q, r[: q, r \in \mathbb{Q}$, there is $\gamma \in \Gamma$ such that $]q, r[\subseteq D_\gamma\}$ is also countable. It is easy to see that $\bigcup_{\gamma\in\Gamma} D_\gamma = \bigcup_{D\in\mathcal{D}} D$.

Then $\mathcal{D} = \{D_1, \cdots, D_n, \cdots\}$ is a countable open cover of K: $K \subseteq \bigcup_{n=1}^\infty D_n$.

Let's show that $\mathcal{D}$ has a finite subcover. If not, for any $n \in \mathbb{N}^*$ and every $\{D_1, \cdots, D_n\} \subseteq \mathcal{D}$, there exists $x_n \in K \setminus \bigcup_{i=1}^n D_i$. The sequence $(x_n)_n \subseteq K$ has a subsequence $(x_{k_n})_n$ which converges to $x \in K$. Let $p \in \mathbb{N}^*$ such that $x \in D_p$. Since D_p is open, there is $n_0 \in \mathbb{N}^*$ such that $x_{k_n} \in D_p$, for any $n \geq n_0$. Let $n \geq n_0$ such that $n > p$; then $x_{k_n} \in K \setminus \bigcup_{i=1}^{k_n} D_i$, and, since $p < n < k_n$, it follows that $x_{k_n} \notin D_p$, which is a contradiction.

Therefore there is $n \in \mathbb{N}^*$ such that $K \subseteq \bigcup_{i=1}^n D_i$. The way we built the family $\mathcal{D}$, for any $i = 1, \cdots n$, there is $\gamma_i \in \Gamma$ such that $D_i \subseteq D_{\gamma_i}$. Then $\{D_{\gamma_1}, \cdots, D_{\gamma_n}\}$ is a finite subcover of $\mathcal{A}$.

Conversely, suppose that any open cover of K has a finite subcover. Then, since $K \subseteq \mathbb{R} = \bigcup_{n=1}^\infty]-n, n[$, there is $p \in \mathbb{N}^*$ such that $K \subseteq \bigcup_{n=1}^p]-n, n[=]-p, p[$, and so K is bounded.

Let us suppose that K is not closed; and then there exists a sequence $(x_n)_n \subseteq K$, $x_n \to x$, and $x \notin K$. For every $y \in K$, $r_y = |x - y| > 0$ and $K \subseteq \bigcup_{y\in K} \left]y - \frac{r_y}{3}, y + \frac{r_y}{3}\right[$. This cover has a finite subcover; let $y_1, \cdots y_n \in K$ such that $K \subseteq \bigcup_{i=1}^{n} \left]y_i - \frac{r_{y_i}}{3}, y_i + \frac{r_{y_i}}{3}\right[$. There is $i \in \{1, \cdots, n\}$ and $k_n \uparrow +\infty$ such that the sequence $(x_{k_n})_n \subseteq \left]y_i - \frac{r_{y_i}}{3}, y_i + \frac{r_{y_i}}{3}\right[$. Then $x \in \left[y_i - \frac{r_{y_i}}{3}, y_i + \frac{r_{y_i}}{3}\right]$ and so $r_{y_i} = |x - y_i| < 2 \cdot \frac{r_{y_i}}{3}$, which is absurd. Therefore K is bounded and closed, hence a compact set. ■

Theorem 7.1.7 (Lebesgue criterion for Riemann integrability)
The bounded function $f : [a, b] \to \mathbb{R}$ is Riemann integrable if and only if f is continuous almost everywhere on $[a, b]$.

Proof
Let D be the set of discontinuities of f on $[a, b]$; according to Proposition 7.1.4, $D = \{x \in [a, b] : \omega_f(x) > 0\}$. We remark that $D = \bigcup_{p=1}^{\infty} D_p$ where $D_p = \{x \in [a, b] : \omega_f(x) \geq \frac{1}{p}\}$, for any $p \in \mathbb{N}^*$. The theorem says that $f \in \mathcal{R}_{[a,b]}$ if and only if $\lambda^*(D) = 0$.
Necessity. Let $f \in \mathcal{R}_{[a,b]}$. We will prove that, for any $p \in \mathbb{N}^*$, $\lambda^*(D_p) = 0$ and then it will result that $\lambda^*(D) = 0$.

Let so $p \in \mathbb{N}^*$; for every $\varepsilon > 0$, there exists $\Delta = \{x_0, \cdots, x_n\} \in \mathcal{D}([a, b])$ such that

$$S_\Delta - s_\Delta < \frac{\varepsilon}{p}. \tag{7.1}$$

For any $k = 1, \cdots, n$, let $I_k = [x_{k-1}, x_k]$ and then

$$S_\Delta - s_\Delta = \sum_{k=1}^{n} (M_k - m_k)(x_k - x_{k-1}) = \sum_{k=1}^{n} \omega_f(I_k) \cdot |I_k|. \tag{7.2}$$

We divide the points of D_p into two categories: some can be placed inside the intervals I_k, and others can coincide with some of the points of Δ. Let $N = \left\{k \in \{1, \cdots, n\} :]x_{k-1}, x_k[\cap D_p \neq \emptyset\right\}$. Then

$$D_p \subseteq \bigcup_{k\in N} I_k \cup \{x_0, \cdots, x_n\}. \tag{7.3}$$

For any $k \in N$, there exists $x_0 \in]x_{k-1}, x_k[\cap D_p$; it follows that there exists $\delta > 0$ such that $]x_0 - \delta, x_0 + \delta[\subseteq]x_{k-1}, x_k[$ and that $\omega_f(x_0) \geq \frac{1}{p}$. Then

$$\frac{1}{p} \leq \omega_f(x_0) \leq \sup_{u\in]x_0-\delta, x_0+\delta[\cap I_k} f(u) - \inf_{v\in]x_0-\delta, x_0+\delta[\cap I_k} f(v) \leq M_k - m_k,$$

from where

$$\frac{1}{p} \cdot |I_k| \le (M_k - m_k) \cdot |I_k| \text{ for any } k \in N. \tag{7.4}$$

From (7.1), (7.2), and (7.4)

$$\frac{1}{p} \cdot \sum_{k \in N} |I_k| \le \sum_{k \in N} (M_k - m_k) \cdot |I_k| \le \sum_{k=1}^{n} (M_k - m_k) \cdot |I_k| = S_\Delta - s_\Delta < \frac{\varepsilon}{p},$$

from where

$$\sum_{k \in N} |I_k| < \varepsilon. \tag{7.5}$$

From (7.3) and (7.5),

$$\lambda^*(D_p) \le \sum_{k \in N} |I_k| + \lambda^*(\{x_0, \cdots, x_n\}) < \varepsilon, \text{ for every } \varepsilon > 0$$

and then $\lambda^*(D_p) = 0$.

Sufficiency. Let $M > 0$ such that $|f(x)| \le M$, for every $x \in [a, b]$.

Let us suppose that $\lambda^*(D) = 0$ and let $\varepsilon > 0$; we denote $\alpha = \frac{\varepsilon}{2(b-a)}$. Since $D_\alpha = \{x \in [a, b] : \omega_f(x) \ge \alpha\} \subseteq D$, it follows that $\lambda^*(D_\alpha) = 0$; then there exists a sequence of open intervals $(I_n)_{n \in \mathbb{N}} \subseteq \mathcal{I}$ such that

$$D_\alpha \subseteq \bigcup_{n=1}^{\infty} I_n \text{ and } \sum_{n=1}^{\infty} |I_n| < \frac{\varepsilon}{4M}.$$

According to Lemma 7.1.5, D_α is closed, and since $D_\alpha \subseteq [a, b]$, D_α is a compact. We can use Lemma 7.1.6; therefore there exists $n \in \mathbb{N}^*$ such that

$$D_\alpha \subseteq \bigcup_{i=1}^{n} I_i \text{ and } \sum_{i=1}^{n} |I_i| < \frac{\varepsilon}{4M}. \tag{7.1}$$

Without limiting the generality, we can consider that the intervals I_i are pairwise disjoint.

(a) First we suppose that $a, b \notin D_\alpha$; we can renumber the intervals $\{I_i =]a_i, b_i[: i = 1, \cdots, n\}$ so that

$$a \le a_1 < b_1 \le a_2 < b_2 \le \cdots \le a_n < b_n \le b.$$

We denote $J_1 = [a, a_1]$, $J_2 = [b_1, a_2], \cdots, J_n = [b_{n-1}, a_n]$, $J_{n+1} = [b_n, b]$. Then the set $K = [a, b] \setminus \bigcup_{i=1}^{n} I_i = \bigcup_{j=1}^{n+1} J_j$ is compact and, for any $j = 1, \cdots, n+1$ and every $x \in J_j \subseteq K \subseteq [a, b] \setminus D_\alpha$, $\omega_f(x) < \alpha$. Therefore, for every $x \in J_j$, there exists $\delta_x > 0$ such that

$$\sup_{u,v \in]x-\delta_x, x+\delta_x[\cap J_j} |f(u) - f(v)| < \alpha. \tag{7.2}$$

Then $\{]x - \delta_x, x + \delta_x[: x \in J_j, j = 1, \cdots, n+1\}$ is an open cover of K. According to Lemma 7.1.6, it has a finite subcover $\{]x_k - \delta_{x_k}, x_k + \delta_{x_k}[: k = 1, \cdots, p\}$.

It is easy to verify that each closed interval J_j has a partition $\Delta_j = \{x_0^j, x_1^j, \cdots, x_{p_j}^j\}$, so that any interval $[x_{l-1}^j, x_l^j]$ of this partition Δ_j is contained in an interval $]x_k - \delta_{x_k}, x_k + \delta_{x_k}[$ and, from (7.2),

$$|f(u) - f(v)| < \alpha, \forall u, v \in [x_{l-1}^j, x_l^j], l = 1, \cdots, p_j, j = 1, \cdots, n+1. \tag{7.3}$$

Then $\Delta = \{x_0^1 = a, \cdots, x_{p_1}^1 = a_1, x_0^2 = b_1, \cdots, x_{p_2}^2 = a_2, \cdots, x_0^{n+1} = b_n, \cdots, x_{p_{n+1}}^{n+1} = b\}$ is a partition of the interval $[a, b]$. We note that, for every two consecutive points of Δ, let's say x_{r-1}, x_r,

$$M_r - m_r = \sup_{u,v \in [x_{r-1}, x_r]} |f(u) - f(v)| \leq \begin{cases} \alpha & \text{if } x_{r-1}, x_r \in \Delta_j \\ 2M & \text{if } x_{r-1} = a_i, x_r = b_i \end{cases}. \tag{7.4}$$

From (7.4) and (7.1), we obtain $S_\Delta - s_\Delta = \sum_{x_{r-1}, x_r \in \Delta} (M_r - m_r) \cdot (x_r - x_{r-1}) =$

$$= \sum_{j=1}^{n+1} \sum_{x_{r-1}, x_r \in \Delta_j} (M_r - m_r) \cdot (x_r - x_{r-1}) + \sum_{i=1}^{n} \sup_{u,v \in [a_i, b_i]} |f(u) - f(v)| \cdot (b_i - a_i) <$$

$$< \alpha \cdot (b - a) + 2M \cdot \sum_{i=1}^{n} |I_i| \leq \frac{\varepsilon}{2} + 2M \cdot \frac{\varepsilon}{4M} = \varepsilon.$$

(b) If $a \in D_\alpha$ or $b \in D_\alpha$, then $K = \bigcup_{j=2}^{n+1} J_j$, $K = \bigcup_{j=1}^{n} J_j$ or $K = \bigcup_{j=2}^{n} J_j$ and reasoning as in the case (a), we obtain $S_\Delta - s_\Delta < \varepsilon$.

Therefore, for every $\varepsilon > 0$, there is a partition $\Delta \in \mathcal{D}([a, b])$ such that $S_\Delta - s_\Delta < \varepsilon$, which means that f is Riemann integrable on $[a, b]$. ■

The Riemann function is an interesting example of a function continuous in irrational points and discontinuous in rational points of $[0, 1]$ (see Exercise (3) of 2.5). This function is obviously Riemann integrable.

As immediate consequences of the previous theorem, we underline some important classes of Riemann integrable functions. For the demonstrations, it is necessary to take into account the fact that any monotonic function has a set of points of discontinuity at most countable and that any function with bounded variation is a difference of two monotonic functions (see Theorem 6.3.8).

Corollary 7.1.8

1. *Every continuous function on* $[a, b]$ *is Riemann integrable.*
2. *Every monotonic function on* $[a, b]$ *is Riemann integrable.*
3. *Every bounded variation function on* $[a, b]$ *is Riemann integrable.*

7.2 Pompeiu's Function

For a continuous function f over an interval $[a, b]$, the reconstruction of its primitive is given by the formula

$$F(x) = F(a) + \int_a^x f(t)dt, \text{ for every } x \in [a, b].$$

If f is only Riemann integrable (so continues almost everywhere), then $F'(x) = f(x)$ almost everywhere on $[a, b]$ (from Theorem 6.3.16, the result remains valid in the more general case where f is Lebesgue integrable).

However, there are bounded, non-integrable Riemann functions which admit primitives impossible to obtain with the above formula. V. Volterra published in 1881 an example of such a function (Giornale di Matematiche, 19 (1881), 333–372).

D. Pompeiu gives a more interesting example than Volterra in his doctoral dissertation presented in Paris in 1905. Compared to that of Volterra, the derivative of Pompeiu is not Riemann integrable in any subinterval of the domain of definition; the example was published in [7].

Let $Q_1 = \mathbb{Q} \cap [0, 1] = \{q_1, q_2, \cdots, q_n, \cdots\}$ be a numbering of rational numbers of $[0, 1]$ and let $(a_n)_{n\in\mathbb{N}} \subseteq \mathbb{R}_+^* =]0, +\infty[$ such that

$$\sum_{n=1}^{\infty} a_n^{\frac{1}{2}} < +\infty. \tag{7.1}$$

We notice that there is $n_0 \in \mathbb{N}$ such that

$$a_n < a_n^{\frac{1}{2}} < 1, \text{ for any } n \geq n_0. \tag{7.2}$$

So the series $\sum_{n=1}^{\infty} a_n$ is convergent, and then we can deduce from

$$\left| a_n \cdot (x - q_n)^{\frac{1}{3}} \right| < a_n, \text{ for any } n \in \mathbb{N}^* \text{ and every } x \in [0, 1],$$

that the series of functions $\sum_{n=1}^{\infty} a_n \cdot (x - q_n)^{\frac{1}{3}}$ is uniformly convergent on the interval [0, 1]. Therefore, we can define the function

$$f : [0, 1] \to \mathbb{R},\ f(x) = \sum_{n=1}^{\infty} a_n \cdot (x - q_n)^{\frac{1}{3}} \tag{7.3}$$

and this function is continuous on [0, 1]. On the other hand, we remark that f is strictly increasing on [0, 1] and then

$$f : [0, 1] \to [a, b],\ \text{where } a = f(0), b = f(1). \tag{7.4}$$

Because f is strictly increasing and continuous, it is a bijection of [0, 1] on $[a, b]$. Its inverse

$$g = f^{-1} : [a, b] \to [0, 1]$$

is the **Pompeiu's function**; it is also strictly increasing and continuous (f and g are homeomorphisms). We will prove that g is differentiable and that its derivative is a bounded function which is not Riemann integrable in any subinterval $[c, d] \subseteq [a, b]$.

We will first study the differentiability of the function f. We denote

$$\left.\begin{aligned} A &= \left\{ x \in [0,1] \setminus Q_1 : \textstyle\sum_{n=1}^{\infty} \frac{a_n}{(x-q_n)^{\frac{2}{3}}} < +\infty \right\} \\ B &= \left\{ x \in [0,1] \setminus Q_1 : \textstyle\sum_{n=1}^{\infty} \frac{a_n}{(x-q_n)^{\frac{2}{3}}} = +\infty \right\} \end{aligned}\right. . \tag{7.5}$$

Then $[0, 1] = A \cup B \cup Q_1$.

Q_1 is a null set; we will prove that B is also a null set. From (7.1) it follows that

$$\bigcup_{p=1}^{\infty} \bigcap_{n=p}^{\infty} \left\{ x \in [0,1] \setminus Q_1 : \frac{a_n}{(x - q_n)^{\frac{2}{3}}} \le a_n^{\frac{1}{2}} \right\} \subseteq A. \tag{7.6}$$

From (7.5) and (7.6)

$$B \cup Q_1 = [0,1] \setminus A \subseteq \bigcap_{p=1}^{\infty} \bigcup_{n=p}^{\infty} \left\{ x \in [0,1] \setminus Q_1 : \frac{a_n}{(x - q_n)^{\frac{2}{3}}} > a_n^{\frac{1}{2}} \right\} \cup Q_1 \text{ or}$$

$$B \cup Q_1 \subseteq \bigcap_{p=1}^{\infty} \bigcup_{n=p}^{\infty} \left] q_n - a_n^{\frac{3}{4}}, q_n + a_n^{\frac{3}{4}} \right[.$$

Using (7.2) we obtain

$$\lambda^*(B\cup Q_1)\le\sum_{n=p}^{\infty}\lambda\left(]q_n-a_n^{\frac{3}{4}},q_n+a_n^{\frac{3}{4}}[\right)\le 2\sum_{n=p}^{\infty}a_n^{\frac{1}{2}},\ \text{for any } p\ge n_0. \tag{7.7}$$

From (7.1), $\lim_{p\to\infty}\sum_{n=p}^{\infty}a_n^{\frac{1}{2}}=0$, and then from (7.7), $\lambda^*(B\cup Q_1)=0$, from where we get that B is a null set.

It follows that $\lambda(A)=1$ and therefore the series $\sum_{n=1}^{\infty}\frac{a_n}{(x-q_n)^{\frac{2}{3}}}$ is convergent almost everywhere on $[0,1]$. Moreover A and $B\cup Q_1$ are dense in $[0,1]$ (see Exercise (16) of 1.5).

Proposition 7.2.1 *The function f has a derivative in every point x of the interval* $[0,1]$ *and*

$$f'(x)=\begin{cases}\frac{1}{3}\sum_{n=1}^{\infty}\frac{a_n}{(x-q_n)^{\frac{2}{3}}}, & x\in A,\\ +\infty, & x\in B\cup Q_1.\end{cases}$$

Proof

(1) Let x be an arbitrary point in A; then, for every $y\in\mathbb{R}^*=\mathbb{R}\setminus\{0\}$,

$$\frac{f(x+y)-f(x)}{y}=\sum_{n=1}^{\infty}a_n\frac{(x+y-q_n)^{\frac{1}{3}}-(x-q_n)^{\frac{1}{3}}}{y}=$$

$$=\sum_{n=1}^{\infty}\frac{a_n}{(x+y-q_n)^{\frac{2}{3}}+(x+y-q_n)^{\frac{1}{3}}(x-q_n)^{\frac{1}{3}}+(x-q_n)^{\frac{2}{3}}}=$$

$$=\sum_{n=1}^{\infty}\frac{a_n}{(x-q_n)^{\frac{2}{3}}}\cdot\frac{1}{\left[\left(\frac{x+y-q_n}{x-q_n}\right)^{\frac{1}{3}}\right]^2+\left(\frac{x+y-q_n}{x-q_n}\right)^{\frac{1}{3}}+1}\le$$

$$\le\frac{4}{3}\cdot\sum_{n=1}^{\infty}\frac{a_n}{(x-q_n)^{\frac{2}{3}}}<+\infty.$$

It follows that the series $\sum_{n=1}^{\infty}a_n\frac{(x+y-q_n)^{\frac{1}{3}}-(x-q_n)^{\frac{1}{3}}}{y}$ is uniformly convergent with respect to $y\in\mathbb{R}^*$. Therefore there exists $\lim_{y\to 0}\frac{f(x+y)-f(x)}{y}=$

$$=\sum_{n=1}^{\infty}a_n\lim_{y\to 0}\frac{(x+y-q_n)^{\frac{1}{3}}-(x-q_n)^{\frac{1}{3}}}{y}=\frac{1}{3}\cdot\sum_{n=1}^{\infty}\frac{a_n}{(x-q_n)^{\frac{2}{3}}}.$$

(2) Let $x \in B$; then, for every $M > 0$ there is $n_1 \in \mathbb{N}$ such that

$$\frac{1}{3} \cdot \sum_{n=1}^{n_1} \frac{a_n}{(x - q_n)^{\frac{2}{3}}} > M.$$

Since $\frac{1}{3} \cdot \sum_{n=1}^{n_1} \frac{a_n}{(x - q_n)^{\frac{2}{3}}} = \lim_{y \to 0} \sum_{n=1}^{n_1} a_n \cdot \frac{(x + y - q_n)^{\frac{1}{3}} - (x - q_n)^{\frac{1}{3}}}{y}$, there exists $\delta > 0$ such that $\sum_{n=1}^{n_1} a_n \cdot \frac{(x + y - q_n)^{\frac{1}{3}} - (x - q_n)^{\frac{1}{3}}}{y} > M$, for $0 < |y| < \delta$.

Since the function f is increasing

$$\frac{f(x+y) - f(x)}{y} = \sum_{n=1}^{\infty} a_n \cdot \frac{(x + y - q_n)^{\frac{1}{3}} - (x - q_n)^{\frac{1}{3}}}{y} >$$

$$> \sum_{n=1}^{n_1} a_n \cdot \frac{(x + y - q_n)^{\frac{1}{3}} - (x - q_n)^{\frac{1}{3}}}{y} > M, \text{ for } 0 < |y| < \delta.$$

Therefore $f'(x) = \lim_{y \to 0} \frac{f(x+y) - f(x)}{y} = +\infty$.

(3) If $x = q_k \in Q_1$, then

$$\frac{(x - q_n)^{\frac{1}{3}} - (q_k - q_n)^{\frac{1}{3}}}{x - q_k} > 0, \text{ for any } n \in \mathbb{N}^* \text{ and every } x \neq q_k.$$

Therefore

$$\frac{f(x) - f(q_k)}{x - q_k} = \sum_{n=1}^{\infty} a_n \cdot \frac{(x - q_n)^{\frac{1}{3}} - (q_k - q_n)^{\frac{1}{3}}}{x - q_k} \geq a_k \cdot \frac{(x - q_k)^{\frac{1}{3}}}{x - q_k} = \frac{a_k}{(x - q_k)^{\frac{2}{3}}}$$

and then $\lim_{x \to q_k} \frac{f(x) - f(q_k)}{x - q_k} = +\infty$.

■

Corollary 7.2.2 *The inverse function* $g = f^{-1} : [a, b] \to [0, 1]$ *is differentiable on* $[a, b]$ *and has a bounded derivative given by the formula*

$$g'(y) = \begin{cases} \frac{1}{f'(g(y))}, & y \in f(A), \\ 0, & y \in f(B \cup Q_1). \end{cases}$$

Proof

It suffices to show that the derivative of g is bounded. Because g is monotonically increasing, $g'(y) \geq 0$, for every $y \in [a, b]$.

We remark that, for every $x \in [0, 1]$ and for any $n \in \mathbb{N}^*$, $|x - q_n| \leq 1$. Therefore

$$f'(x) = \frac{1}{3} \cdot \sum_{n=1}^{\infty} \frac{a_n}{(x - q_n)^{\frac{2}{3}}} \geq \frac{1}{3} \cdot \sum_{n=1}^{\infty} a_n, \text{ for every } x \in A.$$

So

$$0 \leq g'(y) \leq \frac{3}{\sum_{n=1}^{\infty} a_n}, \text{ for every } y \in [a, b].$$

■

The continuous function f carries dense sets into dense sets, so $f(A)$, $f(Q_1)$, and $f(B \cup Q_1)$ are dense in $[a, b]$; this is used in the demonstration of the following proposition.

Proposition 7.2.3 $g' \in \mathfrak{L}^1([a, b])$ *but* $g' \notin \mathcal{R}_{[c,d]}$ *for every* $[c, d] \subseteq [a, b]$.

Proof
g' is measurable and bounded on $[a, b]$; therefore, it is Lebesgue integrable (see Exercise (6).b of 6.4) and $\int_{[a,b]} g' d\lambda = g(b) - g(a) = 1$.

Let $a \leq c < d \leq b$; if we suppose that $g' \in \mathcal{R}_{[c,d]}$, then $g(d) - g(c) = \int_c^d g'(y)dy$. Let $\Delta = \{c = y_0, y_1, \cdots, y_n = d\} \in \mathcal{D}([c, d])$. Since $f(Q_1)$ is dense in $[a, b]$, for any $k = 1, \cdots, n$, there is $q_k \in Q_1$ such that $c_k = f(q_k) \in]y_{k-1}, y_k[\cap f(Q_1)$ and then $\sum_{k=1}^{n} g'(c_k)(y_k - y_{k-1}) = 0$. From point (5) of Theorem 7.1.2, it follows that $\int_c^d g'(y)dy = 0$, from where $g(c) = g(d)$; it is absurd because the function g is injective. ■

Remark 7.2.4 According to Theorem 6.3.6, g is absolutely continuous on $[a, b]$. Then we can apply Theorem 6.3.22, and therefore, for every $x \in [a, b]$, $g(x) = \int_{[a,x]} g' d\lambda$.

7.3 Differentiability of Monotonic Functions

In this appendix, we discuss one of the most surprising results of the analysis of the real functions of a real variable. In short, this result indicates that any monotone function over an interval I is differentiable almost everywhere on I. The theorem has been proved for monotonic continuous functions by H. Lebesgue in 1904 (*Leçons sur l'intégration et la recherche de fonctions primitives*, Paris, Gauthier-Villars, 1904); his demonstration calls Vitali's lemma as well as the measurability of Dini derivatives. Subsequently, F. Riesz gave another demonstration using the so-called "rising sun" lemma, or the "flowing water" lemma. (*Sur l'existence de la dérivée des fonctions monotones et sur quelques problèmes qui s'y rattachent*, Acta. Sci. Math., Szeged, 5 (1932), 208–221).

In what follows, we will present an elementary demonstration given by M. W. Botsko (see [2]); instead of the Dini derivatives, the demonstration uses only the upper and lower derivatives and does not require that they be measurable.

We first show two lemmas.

Lemma 7.3.1 *Let* $f : [c, d] \to \mathbb{R}$, *let* $\Delta = \{x_0, x_1, \cdots, x_n\} \in \mathcal{D}([c, d])$, $S \subseteq \{1, 2, \cdots, n\}$ *and let* $a > 0$. *If*

$$(a)\ f(c) \le f(d) \text{ and } \frac{f(x_k) - f(x_{k-1})}{x_k - x_{k-1}} < -a, \text{ for any } k \in S \text{ or}$$
$$(b)\ f(c) \ge f(d) \text{ and } \frac{f(x_k) - f(x_{k-1})}{x_k - x_{k-1}} > a, \text{ for any } k \in S,$$

then

$$V_\Delta(f) = \sum_{k=1}^{n} |f(x_k) - f(x_{k-1})| > |f(d) - f(c)| + a \cdot \sum_{k\in S}(x_k - x_{k-1}).$$

($V_\Delta(f)$ is the variation of f with respect to Δ—see Definition 6.3.1).

Proof
In the case (a), $|f(d) - f(c)| = f(d) - f(c) = \sum_{k=1}^{n}[f(x_k) - f(x_{k-1})] = \sum_{k\in S}[f(x_k) - f(x_{k-1})] + \sum_{k\notin S}[f(x_k) - f(x_{k-1})] < -a\sum_{k\in S}(x_k - x_{k-1}) + \sum_{k=1}^{n} |f(x_k) - f(x_{k-1})|$.

In the case (b) the function $g = -f$ verifies the hypotheses of (a) and then $\sum_{k=1}^{n} |f(x_k)-f(x_{k-1})| = \sum_{k=1}^{n} |g(x_k)-g(x_{k-1})| > |g(d)-g(c)|+a\cdot\sum_{k\in S}(x_k-x_{k-1}) = |f(d) - f(c)| + a \cdot \sum_{k\in S}(x_k - x_{k-1})$. ■

Lemma 7.3.2 *Let* $A \subseteq]a, b[$ *such that* $\lambda^*(A) > 0$ (*A is not a null set*). *Then there exists* $\varepsilon_0 > 0$ *such that, for every cover of A with the open intervals,* $\mathcal{A} = \{I \subseteq [a, b] : I \in \mathcal{I}\}$, $A \subseteq \bigcup_{I\in\mathcal{A}} I$, *there exists a finite subfamily of pairwise disjoint intervals,* $\{I_1, \cdots, I_p\} \subseteq \mathcal{A}$, *such that* $\sum_{k=1}^{p} |I_k| > \frac{\varepsilon_0}{3}$.

Proof
Since $\lambda^*(A) > 0$, there exists $\varepsilon_0 > 0$ such that, for every countable cover of A with open intervals, $\{I_n : \sum_{n=1}^{\infty} |I_n| \ge \varepsilon_0$. The set $D = \bigcup_{I\in\mathcal{A}} I$ is open; according to the structure theorem of open sets in $\mathbb{R}$ open sets in $\mathbb{R}$ (Theorem 1.1.3), $D = \bigcup_{n=1}^{\infty}]a_n, b_n[$ and the intervals $]a_n, b_n[$ are pairwise disjoint. Since $A \subseteq D = \bigcup_{n=1}^{\infty}]a_n, b_n[$, it follows that

$$\sum_{n=1}^{\infty}(b_n - a_n) \ge \varepsilon_0. \tag{7.1}$$

For any $n \in \mathbb{N}^*$, let be the closed interval $[c_n, d_n] \subseteq]a_n, b_n[$ such that $d_n - c_n = \frac{3}{4}\cdot(b_n - a_n)$. Then

$$\sum_{n=1}^{\infty}(d_n - c_n) \ge \frac{3}{4} \cdot \varepsilon_0. \tag{7.2}$$

Let's fix $n \in \mathbb{N}^*$; for every $x \in [c_n, d_n] \subseteq]a_n, b_n[\subseteq \bigcup_{I \in \mathcal{A}} I$, there exists $I_x \in \mathcal{A}$ such that

$$x \in I_x \subseteq]a_n, b_n[. \tag{7.3}$$

(the intervals $]a_n, b_n[$ form a pairwise disjoint cover of D).

The family $\{I_x : x \in [c_n, d_n]\}$ is an open cover of the compact $[c_n, d_n]$. According to Lemma 7.1.6, there exists a finite subcover $\{I_{x_i} =]u_i, v_i[: i = 1, \cdots, q\}$. Therefore

$$[c_n, d_n] \subseteq \bigcup_{i=1}^{q} I_{x_i} = \bigcup_{i=1}^{q}]u_i, v_i[. \tag{7.4}$$

We can suppose that, for any $i \in \{1, \cdots, q\}$, $I_{x_i} \nsubseteq \bigcup_{j \neq i} I_{x_j}$ (otherwise, we remove the interval I_{x_i} from the subcover). Then, for any $i \in \{1, \cdots, q\}$, there exists

$$y_i \in I_{x_i} \setminus \bigcup_{j \neq i} I_{x_j}. \tag{7.5}$$

We can renumber the intervals I_{x_i} so that

$$y_1 < y_2, \cdots < y_q. \tag{7.6}$$

We remark that

$$y_i \leq u_{i+1}, \text{ for any } i = 1, \cdots, q-1. \tag{7.7}$$

Indeed, if we suppose that there is i such that $y_i > u_{i+1}$, then, or $v_{i+1} \leq y_i$ from where $y_{i+1} < v_{i+1} \leq y_i$ which contradicts (7.6), or $y_i < v_{i+1}$ from where $y_i \in]u_{i+1}, v_{i+1}[= I_{x_{i+1}}$ which contradicts (7.5).

From (7.7) it follows that

$$v_i \leq u_{i+2}, \text{ for any } i = 1, \cdots, q-2. \tag{7.8}$$

Indeed, if we suppose that there is i such that $u_{i+2} < v_i$, then, using (7.7), $u_i < y_i \leq u_{i+1} < y_{i+1} \leq u_{i+2} < v_i$ and so $y_{i+1} \in I_{x_i}$ which contradicts (7.5).

The relation (7.8) says that

$$I_{x_i} \cap I_{x_{i+2}} = \emptyset, \text{ for any } i = 1, \cdots, q-2. \tag{7.9}$$

From (7.9) it follows that the intervals $I_{x_1}, I_{x_3}, \cdots$ are pairwise disjoint just like the intervals $I_{x_2}, I_{x_4}, \cdots$.

It is obvious that one of the following inequalities must be satisfied:

$$\begin{cases} \sum_i |I_{x_{2i-1}}| \geq \frac{1}{2} \cdot \sum_{i=1}^{q} |I_{x_i}| \geq \frac{1}{2} \cdot (d_n - c_n) \text{ or} \\ \sum_i |I_{x_{2i}}| \geq \frac{1}{2} \cdot \sum_{i=1}^{q} |I_{x_i}| \geq \frac{1}{2} \cdot (d_n - c_n). \end{cases}$$

It follows that, for any $n \in \mathbb{N}^*$, we get a finite family of pairwise disjoint intervals $\mathcal{A}_n$, so that

$$\sum_{I \in \mathcal{A}_n} |I| \geq \frac{1}{2} \cdot (d_n - c_n). \tag{7.10}$$

($\mathcal{A}_n = \{I_{x_{2i-1}} : 1 \leq i \leq \frac{q+1}{2}\}$ or $\mathcal{A}_n = \{I_{x_{2i}} : 1 \leq i \leq \frac{q}{2}\}$.

By summing the two sides of the inequality (7.10) after $n \in \mathbb{N}^*$ and taking into account (7.2), we obtain

$$\sum_{n=1}^{\infty} \sum_{I \in \mathcal{A}_n} |I| \geq \frac{3}{8} \cdot \varepsilon_0 > \frac{\varepsilon_0}{3}, \tag{7.11}$$

from where there is $n_0 \in \mathbb{N}^*$ such that

$$\sum_{n=1}^{n_0} \sum_{I \in \mathcal{A}_n} |I| > \frac{\varepsilon_0}{3}. \tag{7.12}$$

Note that after (7.3), for all $I \in \mathcal{A}_n$, $J \in \mathcal{A}_m$, $I \subseteq]a_n, b_n[$, $J \subseteq]a_m, b_m[$ and so $I \cap J = \emptyset$. It follows from (7.12) that $\bigcup_{n=1}^{n_0} \mathcal{A}_n = \{I_1, \cdots, I_p\} \subseteq \mathcal{A}$ is a finite family of pairwise disjoint open intervals for which $\sum_{k=1}^{p} |I_k| > \frac{\varepsilon_0}{3}$. ■

Corollary 7.3.3 *Let $A \subseteq]a, b[$ with $\lambda^*(A) > 0$. There exists $\varepsilon_0 > 0$ such that, for every partition $\Delta = \{x_0, \cdots, x_p\} \in \mathcal{D}([a, b])$ and every cover $\mathcal{B} = \{J \subseteq [a, b] : J \in \mathcal{I}\}$ of $A \setminus \Delta$ with open intervals, there exists $\{J_1, \cdots, J_q\} \subseteq \mathcal{B}$, pairwise disjoint, such that $\sum_{l=1}^{q} |J_l| > \frac{\varepsilon_0}{4}$.*

Proof

Let $\varepsilon_0 > 0$ with the meaning of the previous lemma and let $\Delta = \{x_0, \cdots, x_p\} \in \mathcal{D}([a, b])$ be an arbitrary partition of $[a, b]$; we choose $0 < \delta < \min\left\{\frac{\varepsilon_0}{24p}, \|\Delta\|\right\}$. Let $\mathcal{B} = \{J \subseteq [a, b] : J \in \mathcal{I}\}$ be an arbitrary cover of $A \setminus \Delta$ with open intervals and let $\mathcal{C} = \{J_i =]x_i - \delta, x_i + \delta[: i = 1, \cdots, p-1\}$ be a family of open subintervals of $[a, b]$. Then $\mathcal{B} \bigcup \mathcal{C}$ is a cover of A with open subintervals of $[a, b]$. According to previous lemma, there exists a finite subfamily of pairwise disjoint intervals $\{J_1, \cdots, J_q\} \subseteq \mathcal{B} \bigcup \mathcal{C}$ such that $\sum_{l=1}^{q} |J_l| > \frac{\varepsilon_0}{3}$. Only $p - 1$ of the intervals $J_1, \cdots, J_q$ can belong to $\mathcal{C}$; therefore

$$\sum_{\substack{l \in \{1, \cdots, q\} \\ J_l \in \mathcal{B}}} |J_l| \geq \sum_{l=1}^{q} |J_l| - \sum_{\substack{l \in \{1, \cdots, q\} \\ J_l \in \mathcal{C}}} |J_l| > \frac{\varepsilon_0}{3} - 2p \cdot \delta \geq$$

$$\geq \frac{\varepsilon_0}{3} - 2p \cdot \frac{\varepsilon_0}{24p} = \frac{\varepsilon_0}{3} - \frac{\varepsilon_0}{12} = \frac{\varepsilon_0}{4}.$$

■

Remark 7.3.4 We note that ε_0, whose existence is proved in the previous corollary, depends only on the set A and not on the partition Δ.

Let $f : [a, b] \to \mathbb{R}$; we define the upper and lower derivatives of f at $x \in [a, b]$ by

$$\bar{D}f(x) = \limsup_{y \to x} \frac{f(y) - f(x)}{y - x} = \inf_{\delta > 0} \sup_{\substack{0<|y-x|<\delta \\ y \in [a,b]}} \left[\frac{f(y) - f(x)}{y - x} \right],$$

$$\underline{D}f(x) = \liminf_{y \to x} \frac{f(y) - f(x)}{y - x} = \sup_{\delta > 0} \inf_{\substack{0<|y-x|<\delta \\ y \in [a,b]}} \left[\frac{f(y) - f(x)}{y - x} \right].$$

We can immediately see that $-\infty \le \underline{D}f(x) \le \bar{D}f(x) \le +\infty$, for every $x \in [a, b]$.

The function f is **differentiable at** x if $\underline{D}f(x) = \bar{D}f(x) \in \mathbb{R}$; the common value of the two extreme derivatives is called the derivative of the function f at x, and it is denoted by $f'(x)$.

Theorem 7.3.5 (Lebesgue's differentiability theorem)
An increasing function $f : [a, b] \to \mathbb{R}$ is differentiable almost everywhere on $[a, b]$.

Proof
Since f is increasing,

$$0 \le \underline{D}f(x) \le \bar{D}f(x) \le +\infty, \text{ for every } x \in [a, b].$$

We denote

$$A = \{x \in]a, b[\setminus C : \underline{D}f(x) < \bar{D}f(x)\}, \quad B = \{x \in]a, b[\setminus C : \underline{D}f(x) = +\infty\},$$

where C is the set of discontinuity points of f. We remark that

$$\{x \in [a, b] : f \text{ is not differentiable at } x\} \subseteq A \cup B \cup C \cup \{a, b\}. \tag{7.1}$$

Since the set C is at most countable, $\lambda^*(C) = 0$ and $\lambda^*(\{a, b\}) = 0$.

From (7.1), it follows that, to show that f is differentiable almost everywhere, it suffices to show that A and B are a null sets.

(1) **A is a null set**. We remark that

$$A = \bigcup_{\substack{r,s \in \mathbb{Q} \\ r > s}} A_{r,s}, \text{ where } A_{r,s} = \{x \in]a, b[\setminus C : \bar{D}f(x) > r > s > \underline{D}f(x)\}.$$

To show that A is a null set, it suffices to show that, for all $r, s \in \mathbb{Q}$, $r > s$, $\lambda^*(A_{r,s}) = 0$.

We suppose by absurd that there exist $r, s \in \mathbb{Q}, r > s$ such that $\lambda^*(A_{r,s}) > 0$; let $\varepsilon_0 > 0$ such that, for every countable cover of $A_{r,s}$ with the open intervals, $\{I_n : n \in \mathbb{N}^*\}$, $\sum_{n=1}^{\infty} |I_n| \geq \varepsilon_0$.

Let $g : [a, b] \to \mathbb{R}$, $g(x) = f(x) - \frac{r+s}{2} \cdot x$ and let $t = \frac{r-s}{2} > 0$. It is obvious that

$$A_{r,s} = \{x \in]a, b[: g \text{ is continuous at } x \text{ and } \overline{D}g(x) > t, \underline{D}g(x) < -t\}.$$

For every partition $\Delta = \{x_0, \cdots, x_n\} \in \mathcal{D}([a, b])$, the variation of g with respect to Δ is

$$V_\Delta(g) = \sum_{k=1}^{n} |g(x_k) - g(x_{k-1})| = \sum_{k=1}^{n} |f(x_k) - f(x_{k-1}) - \frac{r+s}{2} \cdot (x_k - x_{k-1})| \leq$$

$$\leq \sum_{k=1}^{n} [f(x_k) - f(x_{k-1})] + \frac{r+s}{2} \cdot (b-a) = f(b) - f(a) + \frac{r+s}{2} \cdot (b-a).$$

Therefore g is a bounded variation function. We denote

$$T = V_a^b(g) = \sup_{\Delta \in \mathcal{D}([a,b])} V_\Delta(g).$$

Let $\Delta = \{x_0, \cdots, x_n\} \in \mathcal{D}([a, b])$ such that

$$V_\Delta(g) = \sum_{k=1}^{n} |g(x_k) - g(x_{k-1})| > T - \frac{t \cdot \varepsilon_0}{4}. \tag{7.2}$$

We fix a point $x \in A_{r,s} \setminus \Delta$. There is $k \in \{1, \cdots, n\}$ such that $x \in]x_{k-1}, x_k[$, g is continuous at x and $\underline{D}g(x) < -t$, $\overline{D}g(x) > t$. Let $\delta = \min\{x - x_{k-1}, x_k - x\}$; then

$$\inf_{\substack{0<|y-x|<\delta \\ y\in[a,b]}} \left[\frac{g(y) - g(x)}{y - x}\right] < -t \text{ and } \sup_{\substack{0<|y-x|<\delta \\ y\in[a,b]}} \left[\frac{g(y) - g(x)}{y - x}\right] > t. \tag{$*$}$$

(a) Let us suppose that $g(x_{k-1}) \leq g(x_k)$. From the first inequality of $(*)$, there exists y such that $0 < |y - x| < \delta$ and $\frac{g(y) - g(x)}{y - x} < -t$; it is obvious that $y \in]x_{k-1}, x_k[$.

If $y < x$, let $y = a_x$; since g is continuous at x, $\lim_{z \downarrow x} \frac{g(z) - g(a_x)}{z - a_x} = \frac{g(x) - g(a_x)}{x - a_x} < -t$. Therefore there is $b_x \in]x, x_k[$ such that $\frac{g(b_x) - g(a_x)}{b_x - a_x} < -t$.

If $y > x$, let $y = b_x$; since g is continuous at x, $\lim_{z \uparrow x} \frac{g(b_x) - g(z)}{b_x - z} = \frac{g(b_x) - g(x)}{b_x - x} < -t$. Therefore there is $a_x \in]x_{k-1}, x[$ such that $\frac{g(b_x) - g(a_x)}{b_x - a_x} < -t$.

Hence, if $g(x_{k-1}) \leq g(x_k)$, then there exists a_x, b_x such that

$$x \in]a_x, b_x[\subseteq]x_{k-1}, x_k[\text{ and } \frac{g(b_x) - g(a_x)}{b_x - a_x} < -t. \tag{7.3}$$

(b) If $g(x_k) \leq g(x_{k-1})$, then, from the second inequality of $(*)$, there is y such that $0 < |y - x| < \delta$ and $\frac{g(y) - g(x)}{y - x} > t$. Based on reasoning similar to that of a) are found a_x, b_x such that

$$x \in]a_x, b_x[\subseteq]x_{k-1}, x_k[\text{ and } \frac{g(b_x) - g(a_x)}{b_x - a_x} > t. \tag{7.4}$$

From (a) and (b), for every $x \in A_{r,s} \setminus \Delta$, there is $k \in \{1, \cdots, n\}$, and there exist a_x, b_x such that $x \in]a_x, b_x[\subseteq]x_{k-1}, x_k[$ and

$$\begin{cases} \frac{g(b_x)-g(a_x)}{b_x-a_x} < -t, & \text{if } g(x_{k-1}) \leq g(x_k), \\ \frac{g(b_x)-g(a_x)}{b_x-a_x} > t, & \text{if } g(x_{k-1}) \geq g(x_k). \end{cases} \tag{7.5}$$

The family $\mathcal{B} = \{]a_x, b_x[: x \in A_{r,s} \setminus \Delta\}$ is a cover of $A_{r,s} \setminus \Delta$ with open intervals. According to Corollary 7.3.3, there exists $x_1, \cdots, x_q \in A_{r,s} \setminus \Delta$ such that $]a_{x_l}, b_{x_l}[$ are pairwise disjoint and

$$\sum_{l=1}^{q} (b_{x_l} - a_{x_l}) > \frac{\varepsilon_0}{4}. \tag{7.6}$$

We consider the partition $\Delta' = \Delta \cup \{a_{x_l}, b_{x_l} : l = 1, \cdots, q\}$; reordering the terms, $\Delta' = \{y_0, y_1, \cdots, y_m\} \in \mathcal{D}([a, b])$. For any $k \in \{1, \cdots, n\}$ let $N_k = \{l \in \{1, \cdots, q\} :]a_{x_l}, b_{x_l}[\subseteq [x_{k-1}, x_k]\}$ (for some values of k, it may be that $N_k = \emptyset$). Then $\bigcup_{k=1}^{n} N_k = \{1, \cdots, q\}$.

For any $k = 1, \cdots, n$, we apply Lemma 7.3.1 on the interval $[x_{k-1}, x_k]$. $\Delta_k = \{a_{x_l}, b_{x_l} : l \in N_k\} \cup \{x_{k-1}, x_k\} - \{z_0^k, z_1^k, \cdots, z_{m_k}^k\} \in \mathcal{D}([x_{k-1}, x_k])$. Then, from (7.5),

$$V_{\Delta_k}(g) = \sum_{j=1}^{m_k} |g(z_j^k) - g(z_{j-1}^k)| > |g(x_k) - g(x_{k-1}| + \sum_{l \in N_k} (b_{x_l} - a_{x_l}). \tag{7.7}$$

By summing after k in (7.7), we get

$$V_{\Delta'} = \sum_{j=1}^{m} |g(y_j) - g(y_{j-1})| = \sum_{k=1}^{n} \sum_{j=1}^{m_k} |g(z_j^k) - g(z_{j-1}^k)| > \sum_{k=1}^{n} |g(x_k) - g(x_{k-1})| +$$

$$+t \cdot \sum_{l=1}^{q} (b_{x_l} - a_{x_l}) = V_\Delta(g) + t \cdot \sum_{l=1}^{q} (b_{x_l} - a_{x_l}).$$

Using (7.2) and (7.6) in the above relation, we obtain

$$V_{\Delta'}(g) > T - \frac{t\varepsilon_0}{4} + t \cdot \frac{\varepsilon_0}{4} = T,$$

which contradicts $T = \sup_{\Delta \in \mathcal{D}([a,b])} V_\Delta(g)$.

(2) **B is a null set**. Let us suppose, by reduction to the absurd, that $\lambda^*(B) > 0$.

According to Lemma 7.3.2, there exists $\varepsilon_0 > 0$ such that, for every cover of B with open subintervals of $[a,b]$, $\mathcal{A} = \{I \subseteq [a,b] : I \in \mathcal{I}\}$, there exists $\{I_1, \cdots, I_p\} \subseteq \mathcal{A}$ pairwise disjoint, such that $\sum_{k=1}^{p} |I_k| > \frac{\varepsilon_0}{3}$.

For every $x \in B$ and every $M > 0$, $\underline{D}f(x) > M$. By reasoning as in case (a) above, there is a_x, b_x such that

$$x \in]a_x, b_x[\subseteq]a, b[\text{ and } \frac{f(b_x) - f(a_x)}{b_x - a_x} > M. \tag{7.8}$$

The family $\{]a_x, b_x[: x \in B\}$ is a cover of B with open intervals. According to Lemma 7.3.2, there exist $x_1, \cdots, x_p \in B$ such that the intervals $]a_{x_i}, b_{x_i}[$ are pairwise disjoint and $\sum_{i=1}^{p}(b_{x_i} - a_{x_i}) > \frac{\varepsilon_0}{3}$. Then, since f is monotonically increasing, $f(b) - f(a) \geq \sum_{i=1}^{p}[f(b_{x_i}) - f(a_{x_i})] > m \cdot \sum_{i=1}^{p}(b_{x_i} - a_{x_i}) > M \cdot \frac{\varepsilon_0}{3}$, for every $M > 0$, which is absurd.

■

Bibliography

[1] Bogachev, V.I.: Measure Theory, vol. 1, Springer, Berlin/Heidelberg (2007)

[2] Botsko, M.W.: An elementary proof of Lebesgue's differentiation theorem. Am. Math. Mon. **110**, 834–838 (2003)

[3] Brezis, H.: Analyse fonctionnelle: Théorie et applications. Masson, Paris (1983)

[4] Dugundji, J.: Topology. Allyn and Bacon, Boston (1966)

[5] Folland, G.B.: Real Analysis. Modern Techniques and Their Applications. Wiley Inc., New York/Chichester/Weinheim/Brisbane/Singapore/Toronto (1999)

[6] Halmos, P.R.: Measure Theory. Springer, New York/Heidelberg/Berlin (1974)

[7] Pompeiu, D.: Sur les fonctions dérivées. Math. Ann. **63**(3), 326–332 (1907)

[8] Srivastava, S.M.: A Course on Borel Sets. Springer, New York/Heidelberg/Berlin (1998)

[9] Thomson, B.S., Bruckner, J.B., Bruckner, A.M.: Elementary Real Analysis. Prentice-Hall, New York (2001)

[10] Yeh, J.: Real Analysis. Theory of Measure and Integration. World Scientific, London/Singapore/Beijing (2006)

L. C. Florescu, *Lebesgue Integral*, Compact Textbooks in Mathematics,
https://doi.org/10.1007/978-3-030-60163-8

Index

Symbols

L. C. Florescu, *Lebesgue Integral*, Compact Textbooks in Mathematics,
https://doi.org/10.1007/978-3-030-60163-8

Printed in the United States
By Bookmasters